Matrix Perturbation Theory

This is a volume in
COMPUTER SCIENCE AND SCIENTIFIC COMPUTING

Werner Reinboldt and Daniel Siewiorek, editors

Matrix Perturbation Theory

G. W. Stewart
Computer Science Department
Institute for Advanced Computer Studies
University of Maryland
College Park, Maryland

Ji-guang Sun
Computing Center of the Chinese Academy of Sciences
Beijing, China

ACADEMIC PRESS, INC.
Harcourt Brace Jovanovich, Publishers
Boston San Diego New York
London Sydney Tokyo Toronto

This book is printed on acid-free paper. ∞

Copyright © 1990 by Academic Press, Inc.
All rights reserved.
No part of this publication may be reproduced or
transmitted in any form or by any means, electronic
or mechanical, including photocopy, recording, or
any information storage and retrieval system, without
permission in writing from the publisher.

ACADEMIC PRESS, INC.
1250 Sixth Avenue, San Diego, CA 92101

United Kingdom Edition published by
ACADEMIC PRESS LIMITED
24–28 Oval Road, London NW1 7DX

Library of Congress Cataloging-in-Publication Data

Stewart, G. W. (Gilbert W.)
  Matrix perturbation theory.

  (Computer science and scientific computing),
  Includes bibliographical references.
  1. Perturbation (Mathematics). 2. Matrices.
I. Sun, Ji-guang. II. Title.
QA871.S775 1990   512.9'434   90-33378
ISBN 0-12-670230-6 (alk. paper)

Printed in the United States of America
90 91 92 93   9 8 7 6 5 4 3 2 1

*This book is dedicated
by Pete Stewart to his father
Pete Stewart
and
by Ji-guang Sun to his wife
Jing-han Guo*

Contents

Preface		**xiii**
I	**Preliminaries**	**1**
1	Notation	1
	Notes and References	4
	Exercises	5
2	The QR Decomposition — Projections	6
	2.1 The QR Decomposition	6
	2.2 Hadamard's Inequality	8
	2.3 Projections	9
	Notes and References	10
	Exercises	11
3	Eigenvalues and Eigenvectors	14
	3.1 Definitions and Elementary Properties	14
	3.2 The Schur Decomposition	17
	3.3 The Jordan Canonical Form	20
	3.4 Invariant Subspaces	21
	3.5 The Field of Values	23
	3.6 Sums of Hermitian Matrices	25
	Notes and References	26
	Exercises	27
4	The Singular Value Decomposition	30
	4.1 The Singular Value Decomposition	30
	4.2 Two Inequalities	33

		Notes and References	34
		Exercises	36
	5	Pairs of Subspaces	37
		5.1 The CS Decomposition	37
		5.2 Pairs of Subspaces	40
		5.3 Pairs of Projections	43
		Notes and References	45
		Exercises	45

II	**Norms and Metrics**		**49**
	1	Vector Norms	50
		1.1 Definition	50
		1.2 Examples	51
		1.3 Equivalence and Limits	53
		1.4 Linear Functionals and Dual Norms	56
		Notes and References	59
		Exercises	60
	2	Matrix Norms	64
		2.1 Basic Concepts	64
		2.2 Operator Norms	67
		Notes and References	71
		Exercises	71
	3	Unitarily Invariant Norms	74
		3.1 Von Neumann's Theory	74
		3.2 Properties of Unitarily Invariant Norms	79
		3.3 Doubly Stochastic Matrices and Fan's Theorem	81
		Notes and References	87
		Exercises	88
	4	Metrics on Subspaces of $\mathbf{C}^n$	89
		4.1 The Gap	90
		4.2 Unitarily Invariant Metrics	94
		Notes and References	98
		Exercises	99

III	**Linear Systems and Least Squares Problems**		**101**
	1	The Pseudo-Inverse and Least Squares	102
		1.1 Generalized Inverses and the Pseudo-Inverse	102
		1.2 Projections and Least Squares	106

	Notes and References 108
	Exercises 109
2	Inverses and Linear Systems 114
	2.1 Absolute and Relative Errors 115
	2.2 The Inverse Matrix 117
	2.3 Linear Systems 124
	2.4 Asymptotic Forms and Derivatives 130
	Notes and References 132
	Exercises 134
3	The Pseudo-Inverse 136
	3.1 Projections and Acute Perturbations 137
	3.2 General Results 140
	3.3 Acute Perturbations 146
	3.4 Asymptotic Forms and Derivatives 150
	Notes and References 151
	Exercises 152
4	Projections 153
	Notes and References 155
5	The Linear Least Squares Problem 155
	5.1 Perturbation of the Coefficients 156
	5.2 The Residual 160
	5.3 Backward Perturbations 160
	5.4 Asymptotic Forms and Derivatives 162
	Notes and References 163
	Exercises 163

IV The Perturbation of Eigenvalues   **165**

1	General Perturbation Theorems 166
	1.1 Continuity: Ostrowski–Elsner Theorems 166
	1.2 The Bauer–Fike and Henrici Theorems 170
	1.3 Residual Bounds 174
	Notes and References 176
	Exercises 178
2	Gerschgorin Theory: Differentiability 180
	2.1 Gerschgorin's Theorem 181
	2.2 Diagonal Similarities 182
	Notes and References 186

		Exercises 187
	3	Normal and Diagonalizable Matrices 189
		3.1 The Hoffman–Wielandt Theorem 189
		3.2 Diagonalizable Matrices 192
		Notes and References 193
		Exercises 194
	4	Hermitian Matrices 196
		4.1 Inertia and Interlacing 196
		4.2 Wielandt's Theorem and Its Consequences 198
		4.3 Mirsky's Theorem 203
		4.4 Residual Bounds 205
		4.5 Approximation by a Low-Rank Matrix 208
		Notes and References 209
		Exercises 210
	5	Some Further Results 211
		5.1 Non-Hermitian Perturbations 212
		5.2 Similarity Bounds 215
		Notes and References 217
		Exercises 217
V	**Invariant Subspaces**		**219**
	1	The Theory of Simple Invariant Subspaces 220
		1.1 Definition 220
		1.2 The Operator $\mathbf{T} = X \mapsto AX - XB$ 222
		1.3 The Spectral Resolution 223
		Notes and References 227
		Exercises 227
	2	Perturbation of Invariant Subspaces 229
		2.1 The Approximation Problem 230
		2.2 Perturbation Theorems 236
		2.3 Eigenvectors 240
		2.4 Solution of a Nonlinear Equation 242
		Notes and References 244
		Exercises 245
	3	Hermitian Matrices 246
		3.1 The Approximation Theorem 246
		3.2 Generalized Rayleigh Quotients 248

	3.3 Direct Bounds 249
	3.4 Residual Bounds for Eigenvalues 254
	Notes and References 258
	Exercises 258
4	The Singular Value Decomposition 259
	4.1 Two $\sin\Theta$ Theorems 260
	4.2 A Perturbation Expansion 263
	Notes and References 266
	Exercises 267

VI Generalized Eigenvalue Problems   **271**

	1 Background 273
	1.1 Matrix Pairs 273
	1.2 Triangular and Weierstrass Forms 276
	1.3 Definite Pairs 281
	1.4 Metrics and Their Limitations 283
	Notes and References 289
	Exercises 290
2	Regular Matrix Pairs 291
	2.1 Continuity, First Order Theory 291
	2.2 Gerschgorin Theory 294
	2.3 Diagonalizable Pairs 300
	2.4 Eigenspaces 303
	Notes and References 311
	Exercises 312
3	Definite Matrix Pairs 312
	3.1 Eigenvalues of Definite Pairs 313
	3.2 Eigenspaces 317
	3.3 Direct Bounds 322
	Notes and References 324
	Exercises 324

References   **325**

Notation   **347**

Index   **351**

Preface

The central question of perturbation theory is: How does a function change when its argument is subject to a perturbation? The function may be almost anything—the modes of a vibrating system, the solution of an ordinary or partial differential equation, the states of an electron. This book is concerned with the perturbation of matrix functions, such as the solution of a linear system or the singular values of a matrix.

The result of a perturbation analysis may be a perturbation expansion or a perturbation bound. A perturbation expansion approximates the perturbation in the function in terms of a known perturbation in the argument. Perturbation expansions are widely used in the physical sciences.

A perturbation bound starts with a bound on a perturbation in the argument—here the perturbations are often called errors—and uses it to bound the resulting error in the function. Matrix perturbation theory has traditionally emphasized perturbation bounds. The reasons are varied, but two are paramount.

The first reason is the widespread use of backward rounding-error analysis, a technique in which errors made in executing an algorithm are thrown back on the original data. To complete the analysis, perturbation theory is used to assess the effects of these backward errors on the accuracy of the computed solution. Since only bounds on the error are known, only bounds on the error in the solution can be expected.

The second reason is that perturbation bounds often give insight into the behavior of a function of a matrix under perturbation. For example, it may be possible to determine a multiplier, called a CON-DITION NUMBER, that converts a bound on the error in the argument into a bound on the error induced in the function. The knowledge of a condition number enables one to say whether a function is sensitive or insensitive to perturbations of its argument.

Although we shall give some perturbation expansions, we will be chiefly concerned with perturbation bounds. Deriving perturbation bounds is like cutting a diamond. Tap a problem in just the right way and it decomposes into one or two informative expressions. Smash it with a hammer and it shatters into ugly, uninterpretable pieces. One of the purposes of this book is to introduce the reader to the art of deriving perturbation bounds.

Our book began life as a translation of a book in Chinese of Sun [229, 1987] on perturbation theory. As we progressed, however, it became clear that an expanded treatment was needed. The result is an entirely new book, in which the spirit of the old still lives.

Matrix perturbation theory is too large a field to fit between the covers of a single book, and we have had to be selective in our choice of topics. We have chosen to treat the solution of linear systems and least squares problems (Chapter III), the eigenvalue problem (Chapters IV and V), and the generalized eigenvalue problem (Chapter VI). Aesthetics has played a part in our choice of what results to present; we have generally preferred simple, informative bounds to more complicated, though technically sharper bounds. In particular, we have not been greatly concerned to seek out optimal bounds when nearly optimal ones are available for less effort.

The book is divided into chapters, sections, and subsections, with bibliographical notes and exercises at the end of each section. We have tried to keep the notation uniform. Our hero is the intrepid, yet sensitive matrix A. Our villain is E, who keeps perturbing A. When A is perturbed he puts on a crumpled hat: $\tilde{A} = A + E$. There are many parts for A to play—he is variously square, rectangular, Hermitian, normal, and unitary. To avoid confusion, A's current guise is posted at the beginning of each chapter or, if necessary, section.

The book is largely self contained. We have made a deliberate effort to keep important material outside the proofs, which in most cases can be skipped without loss of continuity. When a proof illustrates a general technique, we point out the fact explicitly.

The notes and references emphasize original sources and historical development. They also point the reader to topics not treated, many of which are developed in the exercises. However, the bibliography, like the book, is not comprehensive, and we apologize in advance to all those whose work we may have slighted.

The assigning of names to theorems is complicated by the fact that some theorems have been discovered twice — as theorems in linear algebra and as theorems in functional analysis. We have adopted the practice of naming the first claimant, whatever his field and adding names of people who made substantial generalizations. Thus, the theorem on low-rank approximation, known to many as the Eckart–Young theorem, is called the Schmidt–Mirsky theorem after Schmidt, who proved it for integral equations, and Mirsky who showed it held for all unitarily invariant norms. Difficult cases are treated in the notes.

The exercises are intended to amplify the material in the text and to introduce new results. They range from the trivial to the very difficult. We have not graded them, but the reader will do well to assume that any exercise with a reference attached is a major undertaking, especially in the later chapters.

We debated whether to include numerical examples. Our decision not to was directed by the increasing availability of interactive systems that manipulate matrices. With these systems it is a small matter to play with a perturbation bound, watching it perform under a variety of circumstances. Beside such lively exercises, a printed example is a thing of lead and stone.

We wish to thank Nick Higham, Xiaobai Sun, and Guodong Zhang for reading and commenting on parts of the text. Sven Hammarling and Vince Fernando furnished us with historical material on the singular value decomposition. We are particularly indebted to the staff of the library of the National Institute of Standards and Technology, who contributed much to the completeness of the bibliography. We are also indebted to the National Science Foundation of the People's Republic of China for its support.

College Park
Beijing
1990

Chapter I
Preliminaries

This chapter prepares the ground for the chapters that follow. To keep it reasonably short it contains only material that is used in two or more chapters. Background for the individual chapters is developed in initial introductory sections.

The first section introduces some notation. In Section 2 we introduce the QR decomposition. In Section 3 we review the elementary theory of eigenvalues, eigenvectors, and invariant subspaces. In Section 4 we introduce the singular value decomposition. The chapter concludes with a section on the geometry of pairs of subspaces.

1. Notation

In this section we will review our basic notation. Other notation will be introduced as needed. A summary of all notation is given in an appendix at the end of the book.

The real numbers will be denoted by $\mathbf{R}$. The space of all n-dimensional column vectors with components in $\mathbf{R}$ will be denoted by $\mathbf{R}^n$. The set of all $m \times n$ matrices with elements in $\mathbf{R}$ will be denoted by $\mathbf{R}^{m \times n}$. The complex numbers, their vectors, and matrices will be denoted by $\mathbf{C}$, $\mathbf{C}^n$, and $\mathbf{C}^{m \times n}$. To avoid clutter, we will use this notation sparingly, giving dimensions only when they are required and cannot be inferred from context.

Generally we will use lower-case Greek letters for scalars, lower-case Latin letters for vectors, and upper-case Latin and Greek letters for matrices. However, we will not make a fetish of this convention, especially for scalar valued functions like the determinant and rank. Sets of all kinds will be denoted by calligraphic letters (but note **R** and **C** in the last paragraph.)

We will maintain a loose association between the letter denoting a vector or matrix and the lower-case Greek letter denoting its elements. Thus α_{ij} will usually denote the (i,j)-element of a matrix A, and β_i the ith component of b. The reader should keep in mind the association of ξ with x and η with y.

The zero matrix, vector, or scalar will all be written 0. The identity matrix will be written I — or I_n when it is necessary to specify the order. The vector of all ones of any dimension will be written **1** (a boldface one). The ith unit vector, whose ith component is one and whose other components are zero, is $\mathbf{1}_i$ (this unusual notation is due to the fact that in matrix perturbation theory the more conventional symbols e and e_i are needed to represent errors).

The transpose of the matrix A will be written A^T, and its conjugate transpose A^H; i.e., $A^\mathrm{H} = \bar{A}^\mathrm{T}$. As usual A^{-1} will denote the inverse of A. The inverse of the transpose of A will be written $A^{-\mathrm{T}}$, and the inverse conjugate transpose $A^{-\mathrm{H}}$.

Operations on or between sets will run through all combinations of the members of the sets. For example, if $\mathcal{X}, \mathcal{Y} \subset \mathbf{C}^n$, then

$$\mathcal{X} + \mathcal{Y} = \{x + y : x \in \mathcal{X}, y \in \mathcal{Y}\}$$

and

$$A\mathcal{X} = \{Ax : x \in \mathcal{X}\}.$$

The COLUMN SPACE of $A \in \mathbf{C}^{m \times n}$ is

$$\mathcal{R}(A) = \{Ax : x \in \mathbf{C}^n\}.$$

Its NULL SPACE is

$$\mathcal{N}(A) = \{x : Ax = 0\}.$$

The RANK of A is $\mathrm{rank}(A) = \dim[\mathcal{R}(A)]$, where $\dim(\mathcal{X})$ denotes the dimension of the subspace $\mathcal{X}$. The DETERMINANT of a square matrix A will be written $\det(A)$, and its TRACE written $\mathrm{trace}(A)$.

1. NOTATION

We will write $\|x\|_2$ for the 2-norm of a vector, which is defined as the positive square root of $\sum_i |\xi_i|^2 = x^H x$. The 2-norm is also called Euclidean norm, since in real two or three dimensional space it is the Euclidean length of its argument. The properties of this and other vector norms are treated in detail in Section II.1.

The matrix $|A|$ is the matrix whose elements are $|\alpha_{ij}|$. We write $A \geq B$ to mean that $\alpha_{ij} \geq \beta_{ij}$ for all i, j, with similar definitions for the relations $>$, $\leq$, $<$. Note that this notation is inconsistent with the convention by which $A > B$ means that $A - B$ is positive definite.

The symbol "$\stackrel{\text{def}}{=}$" is used in definitions to introduce new terminology. The relation "$\equiv$" is used for implicit definitions; for example, $(\alpha, \beta + \gamma) \equiv (\alpha, \delta)$ defines δ to be $\beta + \gamma$.

Some special types of matrices are collected in the following definition.

Definition 1.1. *A matrix A is*

1. SYMMETRIC (HERMITIAN) *if $A^T = A$ ($A^H = A$);*

2. POSITIVE DEFINITE (POSITIVE SEMI-DEFINITE, NEGATIVE DEFINITE, NEGATIVE SEMI-DEFINITE) *if it is Hermitian and $x^H A x > (\geq, <, \leq) \, 0$ for all $x \neq 0$;*

3. UNITARY, *or* ORTHOGONAL *in the real case, if $A^H A = A A^H = I$;*

4. NORMAL *if $A^H A = A A^H$;*

5. UPPER TRIANGULAR *if it is square and $i > j \Rightarrow \alpha_{ij} = 0$; i.e., if it is zero below its diagonal;*

6. LOWER TRIANGULAR *if it is square and $i < j \Rightarrow \alpha_{ij} = 0$; i.e., if it is zero above its diagonal;*

7. DIAGONAL *if it is upper and lower triangular; i.e., its nonzero elements are on its diagonal;*

8. *a* PERMUTATION MATRIX *if it is obtained by permuting rows and columns of the identity matrix.*

The notation $\operatorname{diag}(\delta_1, \delta_2, \ldots, \delta_n)$ will mean a DIAGONAL MATRIX whose diagonal elements are $\delta_1, \delta_2, \ldots, \delta_n$. The scalars δ_i may be replaced by square matrices, in which case the matrix will be said to be BLOCK DIAGONAL. BLOCK TRIANGULAR matrices are defined similarly.

Notes and References

This book presupposes a knowledge of basic matrix theory, for which there are any number of good introductory texts. For more advanced treatments see [98, 140]. The books by Gantmacher [81, 1959] and Bellman [21, 1970] and the little pamphlet by Marcus [151, 1960] are classics. The perturbation theorist will find the survey of matrix inequalities by Marcus and Minc [152, 1964] particularly useful. Other useful references on inequalities are [20, 160]. Mac Duffee [150, 1946] and Wedderburn [257, 1934] contain extensive references to the older literature. We have drawn heavily on the former in preparing the notes for this book.

Much of matrix perturbation theory comes from numerical linear algebra. For an entry into this area see the books by Varga [252, 1962], Householder [121, 1964], Wilkinson [269, 1965], Stewart [203, 1974], and Golub and Van Loan [93, 1983]. Particular mention should be made of Parlett's excellent book on the symmetric eigenvalue problem [175, 1980], which contains a host of perturbation results. Another source of matrix perturbation theory is the specialization of perturbation theorems for operators in infinite dimensional spaces — a vast area with an extensive literature. The definitive reference is by Kato [135, 1966]. A third, hybrid source is the approximation of linear operators by finite dimensional operators, for which see the book by Chatelin [43, 1983].

There is no standard notation for matrix theory. The one adopted here will not be unfamiliar to numerical analysts.

When $A \in \mathbf{C}^{m \times n}$ is regarded as an operator on $\mathbf{C}^n$, its range is the same as the space spanned by its columns — hence the use of $\mathcal{R}(A)$ to denote the column space of A.

Our definition of positive definite carries the implication that the matrix is symmetric. Some authors (e.g., see [276]) drop this requirement.

1. NOTATION

Exercises

1. Let A be nonsingular. Show that if $T = I + V^H A^{-1} U$ is nonsingular, then
$$(A + UV^H)^{-1} = A^{-1} - A^{-1} U T^{-1} V^H A^{-1}.$$
For a history of this useful formula, which is sometimes called the Sherman–Morrison–Woodbury formula, see [108].

2. Show that $\mathcal{R}(A) \cap \mathcal{N}(A^H) = \{0\}$.

3. Show that $\det(I - uv^H) = 1 - v^H u$.

4. (Cauchy inequality). Show that $\|x^H y\|_2 \leq \|x\|_2 \|y\|_2$, with equality if and only if x and y are linearly dependent.

5. (Triangle inequality). Show that $\|x + y\|_2 \leq \|x\|_2 + \|y\|_2$.

6. Show that $\operatorname{trace}(A^H A) = \sum_{i,j} |\alpha_{ij}|^2$.

7. Show that the diagonal elements of a Hermitian matrix are real.

8. A matrix A is SKEW HERMITIAN if $A^H = -A$. Show that the diagonal elements of a skew Hermitian matrix are imaginary.

9. Show that any square matrix A can be written uniquely in the form
$$A = \Re A + i \Im A$$
where $\Re A$ and $\Im A$ are Hermitian.

10. Show that if A is positive definite, then $Ax = 0 \Rightarrow x = 0$.

11. Show that for any matrix A the cross-product matrix $A^T A$ is positive semi-definite. Show that $A^T A$ is positive definite if and only if the columns of A are linearly independent.

12. Let $\|u\|_2 = 1$. Show that $H = I - 2uu^H$ is Hermitian and unitary. The matrix H is called a HOUSEHOLDER TRANSFORMATION.

13. Show that triangular matrix is normal if and only if it is diagonal.

14. A matrix is STRICTLY UPPER TRIANGULAR if it is upper triangular with zero diagonal elements. Show that if A is a strictly upper triangular matrix of order n then $A^n = 0$.

2. The QR Decomposition — Projections

Throughout this book we will be confronted with the following problem: Given a matrix A, find an orthonormal basis for $\mathcal{R}(A)$ and an orthonormal basis for the orthogonal complement of $\mathcal{R}(A)$. In this section we will take a constructive approach to the problem via the QR decomposition. We will then use the resulting bases to construct orthogonal projections onto a subspace and its complement.

2.1. The QR Decomposition

To establish the existence of the QR decomposition, we will use a lemma, which is useful in its own right. Its proof is purely computational and is left as an exercise.

Lemma 2.1 (Householder). *Let $\|x\|_2 = 1$ and suppose that the first component of x is real and nonnegative. Let*

$$u = \frac{x + \mathbf{1}_1}{\|x + \mathbf{1}_1\|_2}.$$

Then the matrix

$$H = I - 2uu^{\mathrm{H}}$$

is Hermitian, and unitary. Moreover,

$$Hx = -\mathbf{1}_1. \tag{2.1}$$

Equation (2.1) can be written in the form

$$H\mathbf{1}_1 = -x.$$

In other words, there is a unitary matrix whose first column is $-x$. By scaling the first column, we may change it to any multiple of x. Hence,

> given any vector x of 2-norm one, there is a unitary matrix whose first column is x.

We may use this result to reduce a matrix to upper triangular form by a unitary transformation.

2. The QR Decomposition

Theorem 2.2. *Let A be an $m \times n$ matrix with $m \geq n$. Then there is a unitary matrix Q such that*

$$Q^H A = \begin{pmatrix} R \\ 0 \end{pmatrix},$$

where R is upper triangular with nonnegative diagonal elements.

Proof. The proof is by induction on the number of columns of A. Let $n = 1$, so that A is a vector — call it a. Let Q be a unitary matrix whose first column is $a/\|a\|_2$ (if $a = 0$ let $Q = I$). Then

$$Q^H a = \begin{pmatrix} \|a\|_2 \\ 0 \end{pmatrix},$$

which is in the required form.

Now let A have $n > 1$ columns, and partition A in the form $A = (a \; A_*)$. Let H be a unitary matrix whose first column is $a/\|a\|_2$ (if $a = 0$ let $H = I$). Then $H^H A$ can be written

$$H^H A = \begin{pmatrix} \|a\|_2 & b^H \\ 0 & C \end{pmatrix}.$$

By hypothesis there is a unitary transformation V such that

$$V^H C = \begin{pmatrix} S \\ 0 \end{pmatrix},$$

where S is upper triangular with nonnegative diagonal elements. If we set $Q = H\mathrm{diag}(1, V)$, then

$$Q^H A = \begin{pmatrix} \|a\|_2 & b^H \\ 0 & S \\ 0 & 0 \end{pmatrix},$$

which is the required upper triangular form. ∎

Let us now assume that A has rank n. Since R is of order n, it must be nonsingular with positive diagonal elements. Partition $Q = (Q_A \; Q_A^\perp)$, where Q_A has n columns. Then

$$A = (Q_A \; Q_A^\perp) \begin{pmatrix} R \\ 0 \end{pmatrix} = Q_A R.$$

(This decomposition is sometimes called the QR FACTORIZATION of A; it is essentially unique [Exercise 2.6].) It follows that $\mathcal{R}(Q_A) = \mathcal{R}(A)$; i.e., the columns of Q_A form an orthonormal basis for $\mathcal{R}(A)$. Moreover, the columns of $Q_A^\perp$ form an orthonormal basis for the orthogonal complement $\mathcal{R}(A)^\perp$. Thus we have the following corollary.

Corollary 2.3. *Let $\mathcal{X}$ be a subspace of $\mathbf{C}^n$. Then there is a unitary matrix $Q = (Q_\mathcal{X} \; Q_\mathcal{X}^\perp)$ such that $\mathcal{R}(Q_\mathcal{X}) = \mathcal{X}$.*

Proof. Let the columns of A form a basis for $\mathcal{X}$. ∎

2.2. Hadamard's Inequality

The main use we will put the QR decomposition to in this section is to establish the properties of orthogonal projectors. But before we do, we shall digress to establish an important determinantal inequality due to Hadamard.

Theorem 2.4 (Hadamard). *Let $A = (a_1 \; a_2 \; \ldots \; a_n)$ be of order n. Then*

$$|\det(A)| \leq \prod_{j=1}^{n} \|a_j\|_2. \tag{2.2}$$

Moreover that equality holds if and only if A has a zero column or A is unitary.

Proof. Let $Q^H A = R \equiv (r_1, \ldots, r_n)$ be the QR factorization of A. Since premultiplication by an orthogonal matrix does not change the magnitude of the determinant or the norms of the columns of A, we have

$$|\det(A)| = |\det(R)| \leq \prod_i |\rho_{ii}| \leq \prod_i \|r_i\|_2 = \prod_i \|a_i\|_2.$$

If equality holds then either $\det(A) = 0$ and one of the a_i is zero or $\det(A) \neq 0$ and $\rho_{ii} = \|a_i\|_2$ ($i = 1, \ldots, n$), which can only happen if A is unitary. ∎

2. THE QR DECOMPOSITION

2.3. Projections

Let $\mathcal{X}$ be a subspace of $\mathbf{C}^n$ and let the columns of $Q_\mathcal{X}$ form an orthonormal basis for $\mathcal{X}$. The matrix

$$P_\mathcal{X} = Q_\mathcal{X} Q_\mathcal{X}^{\mathrm{H}} \tag{2.3}$$

is called the ORTHOGONAL PROJECTION ONTO $\mathcal{X}$. Informally, $P_\mathcal{X}$ acts like the sun at high noon, projecting a vector z into its shadow $P_\mathcal{X} z$ on the ground $\mathcal{X}$. In this subsection we will treat the elementary properties of projections.

For (2.3) to be a proper definition, it must be independent of the choice of $Q_\mathcal{X}$. In fact, if the columns of $\hat{Q}_\mathcal{X}$ also form an orthonormal basis for $\mathcal{X}$, then $\hat{Q}_\mathcal{X} = Q_\mathcal{X} U$ for some nonsingular U. Since

$$I = \hat{Q}_\mathcal{X}^{\mathrm{H}} \hat{Q}_\mathcal{X} = U^{\mathrm{H}} Q_\mathcal{X}^{\mathrm{H}} Q_\mathcal{X} U = U^{\mathrm{H}} U,$$

it follows that U is unitary. Hence

$$\hat{Q}_\mathcal{X} \hat{Q}_\mathcal{X}^{\mathrm{H}} = Q_\mathcal{X} U U^{\mathrm{H}} Q_\mathcal{X}^{\mathrm{H}} = Q_\mathcal{X} Q_\mathcal{X}^{\mathrm{H}} = P_\mathcal{X}.$$

The matrix $P_\mathcal{X}$ is Hermitian ($P_\mathcal{X}^{\mathrm{H}} = P_\mathcal{X}$) and idempotent ($P_\mathcal{X}^2 = P_\mathcal{X}$). The fact that $P_\mathcal{X}$ is idempotent implies that for any $x \in \mathcal{X}$ we have $P_\mathcal{X} x = x$. In fact, since $\mathcal{R}(P_\mathcal{X}) = \mathcal{X}$, we must have $x = P_\mathcal{X} w$ for some vector w. Hence $P_\mathcal{X} x = P_\mathcal{X}^2 w = P_\mathcal{X} w = x$. On the other hand, the fact that $P_\mathcal{X}$ is Hermitian implies that $P_\mathcal{X} y = 0$ for any $y \in \mathcal{X}_\perp$. In fact, y must be orthogonal to the columns of $P_\mathcal{X}$; that is, $0 = (y^{\mathrm{H}} P_\mathcal{X})^{\mathrm{H}} = P_\mathcal{X}^{\mathrm{H}} y = P_\mathcal{X} y$.

It follows that if we decompose any vector z in the form

$$z = x + y, \quad x \in \mathcal{X}, y \in \mathcal{X}_\perp, \tag{2.4}$$

then $x = P_\mathcal{X} z$ and $y = (I - P_\mathcal{X}) z$. Thus $P_\mathcal{X}$ projects z orthogonally onto its component in $\mathcal{X}$, and $I - P_\mathcal{X} = P_\mathcal{X}^\perp$ projects z onto its component in $\mathcal{X}_\perp$.

Moreover, if P is any Hermitian, idempotent matrix with column space $\mathcal{X}$, then by the above argument $Pz = x$, where x is defined by (2.4). It follows that $P_\mathcal{X} z = Pz$ for all z, and hence $P_\mathcal{X} = P$. In other words, any Hermitian, idempotent matrix is the orthogonal projection onto its column space.

By way of notation, we will write P_A for the orthogonal projection onto $\mathcal{R}(A)$. We will write $P_{\mathcal{X}}^{\perp}$ and $P_A^{\perp}$ for the projections complementary to $P_{\mathcal{X}}$ or P_A. When $\mathcal{X}$ or A can be inferred from context, we will simply write P and $P_{\perp}$ for the complementary projections.

Since the vectors in the decomposition

$$z = P_{\mathcal{X}}z + P_{\perp}z$$

are orthogonal, we have

$$\|z\|_2^2 = (P_{\mathcal{X}}z + P_{\perp}z)^{\mathrm{H}}(P_{\mathcal{X}}z + P_{\perp}z)$$
$$= (P_{\mathcal{X}}z)^{\mathrm{H}}(P_{\mathcal{X}}z) + (P_{\perp}z)^{\mathrm{H}}(P_{\perp}z)$$
$$= \|P_{\mathcal{X}}z\|_2^2 + \|P_{\perp}z\|_2^2,$$

a relation we shall call the PYTHAGOREAN EQUALITY. This equality is the basis of the following important theorem.

Theorem 2.5. *If $\mathcal{X}$ is a subspace of $\mathbf{C}^n$, then $P_{\mathcal{X}}z$ is the unique vector satisfying*

$$\|z - P_{\mathcal{X}}z\|_2 = \min_{x \in \mathcal{X}} \|z - x\|_2.$$

Proof. For any $x \in \mathcal{X}$

$$\|z - x\|_2^2 = \|(P_{\mathcal{X}}z - x) + P_{\perp}z\|_2^2 = \|P_{\mathcal{X}}z - x\|_2^2 + \|P_{\perp}z\|_2^2.$$

The right-hand side of this relation is clearly minimized when its first term is zero; that is, it is minimized if and only if $x = P_{\mathcal{X}}z$. ∎

Thus $P_{\mathcal{X}}z$ is the vector in $\mathcal{X}$ that is nearest z in the 2-norm. It is called the LEAST SQUARES APPROXIMATION to z, since the sum of squares of the absolute values of the components of $z - P_{\mathcal{X}}z$ is minimal.

Notes and References

Although Householder transformations are mentioned in passing by Feller and Forsythe [73, 1951], Householder [120, 1958] was the first to use them systematically to reduce a matrix to a simpler form, in this case upper

2. THE QR DECOMPOSITION

triangular. The requirement that the first component of x be nonnegative is more than a trick to avoid the degenerate case $x = -\mathbf{1}_1$; it is necessary for numerical stability.

In one sense the QR decomposition goes back to Gram [97, 1883], who introduced the idea of orthogonalizing a sequence of functions and gave a determinantal expression for the resulting sequence.* Schmidt [192, 1907] described the orthogonalization technique now known as the Gram-Schmidt algorithm and pointed out that the results are the same as Gram's. If Schmidt's formulas are applied to the columns of A, they compute the columns of Q_A with the elements of R appearing as the coefficients in the expansions (Exercise 2.4). A drawback of the Gram–Schmidt approach to the QR decomposition is that it does not provide an explicit basis for $\mathcal{R}(A)^\perp$.

The name of the decomposition derives from Francis's QR algorithm for finding the eigenvalues of a matrix [75, 1961–2] and its precursor the LR algorithm [191, 1955]. The letter R comes from the German word *recht* — the equivalent of English *upper* in reference to triangular matrices. The letter Q was chosen "somewhat arbitrarily" by Francis.

Projections do not have to be orthogonal. In fact, any idempotent matrix P, Hermitian or not, can be regarded as an oblique projection onto $\mathcal{R}(P)$ along $\mathcal{N}(P^\mathrm{H})$ (see Exercise 2.10). We will use such oblique projections in the perturbation theory for invariant subspaces (Chapter V).

For more on least squares approximations, see the notes and references to Section III.1.

Exercises

1. (Householder [120]). Show that if $x \neq y$, $\|x\|_2 = \|y\|_2$, and $y^\mathrm{H} x$ is real, then there is a Householder transformation H such that $Hx = y$.

2. Let $m \geq n$ and $A \in \mathbf{C}^{m \times n}$ have rank k. Show that there is a permutation matrix P and an orthogonal matrix Q such that

$$Q^\mathrm{H} A P = \begin{pmatrix} R_{11} & R_{12} \\ 0 & 0 \end{pmatrix},$$

where R_{11} is a nonsingular, upper triangular matrix of order k. Moreover,

---

*Precursors may be found in the works of Laplace [141, 1820] and Chebyshev [44, 1859]. But the former is not concerned with the decomposition *qua* decomposition (the formulas are used to determine the variance of a least-squares estimate), and the latter restricts himself to polynomials.

the columns of the matrix

$$P\begin{pmatrix} R_{11}^{-1}R_{12} \\ -I \end{pmatrix}$$

form a basis for $\mathcal{N}(A)$.

3. Use Exercise 2.2 to show that any matrix A can be written in the form $A = FG^H$, where F and G have full column rank. This factorization, which is not unique, is called a FULL RANK FACTORIZATION of A.

THE NEXT TWO EXERCISES TREAT THE GRAM–SCHMIDT ALGORITHM, AN IMPORTANT ALTERNATIVE FOR COMPUTING THE QR DECOMPOSITION.

4. (The Gram–Schmidt algorithm). Let $A = (a_1 \; a_2 \; \ldots \; a_n)$ have rank n, and consider the following algorithm.

```
for j := 1 to n
    q_j := a_j
    for i := 1 to j − 1
        ρ_ij := q_i^T q_j
    end for
    for i := 1 to j − 1
        q_j := q_j − ρ_ij q_i
    end for
    ρ_jj := √(q_j^T q_j)
    q_j := ρ_jj^{-1} q_j
end for
```

Show that the algorithm goes to completion. Moreover, if we set $Q_A = (q_1 \; q_2 \; \ldots \; q_n)$ and

$$R = \begin{pmatrix} \rho_{11} & \rho_{12} & \cdots & \rho_{1n} \\ 0 & \rho_{22} & \cdots & \rho_{2n} \\ \vdots & \vdots & & \vdots \\ 0 & 0 & \cdots & \rho_{nn} \end{pmatrix}$$

then $Q_A^H Q_A = I$ and $A = QR$; i.e., the Gram–Schmidt algorithm computes the QR factorization of A.

5. (The modified Gram–Schmidt algorithm). Show that if the two inner loops in the Gram=-Schmidt algorithm are replaced by the single loop

2. THE QR DECOMPOSITION

```
for i := 1 to j - 1
   ρ_ij := q_i^T q_j
   q_j  := q_j - ρ_ij q_i
end for
```

Then the same decomposition is computed. This modified procedure has superior numerical properties [35].

———◇———

6. Let $A = QR$ be a QR factorization of A. Show that if A has full column rank, then any QR factorizations of A has the form $A = (QD)(DR)$, where $|D| = I$.

7. Let R be the R-factor in the QR factorization of A. Show that $A^H A = R^H R$. The matrix R is called the CHOLESKY FACTOR of $A^H A$.

8. (The partitioned QR factorization). Let $A \in \mathbf{C}^{m \times n}$, where $m \geq n$. Let A be partitioned in the form $A = (A_1 \; A_2)$, where A_1 has full column rank. Let

$$(A_1 \; A_2)^H (A_1 \; A_2) = \begin{pmatrix} C_{11} & C_{12} \\ C_{12}^H & C_{22} \end{pmatrix}$$

be a conformal partitioning of the cross-product matrix $C = A^H A$, and let

$$(A_1 \; A_2) = (Q_1 \; Q_2) \begin{pmatrix} R_{11} & R_{12} \\ 0 & R_{22} \end{pmatrix}$$

be a conformal partition of a QR factorization of A. Show that

1. $A_1 = Q_1 R_{11}$ 1'. $C_{11} = R_{11}^H R_{11}$

2. $P_{A_1} A_2 = Q_1 R_{12}$ 2'. $C_{11}^{-1} C_{12} = R_{11}^{-1} R_{12}$

3. $P_{A_1}^\perp A_2 = Q_2 R_{22}$ 3'. $C_{22} - C_{12}^H C_{11}^{-1} C_{12} = R_{22}^H R_{22}$.

The matrix $C_{22} - C_{12}^H C_{11}^{-1} C_{12}$ is called the SCHUR COMPLEMENT of C_{11} in C [193, 1909] and appears in many applications. For more see [108, 170].

9. Let $A = (a_1 \; \ldots \; a_n)$ be of order n and let $P_i^\perp$ be the projection onto the orthogonal complement of the space spanned by $a_1, \ldots, a_{i-1}$ (take $P_1^\perp = I$). Show that

$$|\det(A)| = \prod_{i=1}^{n} \|P_i^\perp a_i\|_2.$$

Hence deduce Hadamard's inequality.

10. Show that if P is idempotent then $\mathcal{R}(P) \oplus \mathcal{N}(P) = \mathbf{C}^n$. Conclude that any idempotent matrix P is the projection onto $\mathcal{R}(P)$ along $\mathcal{N}(P)$. Such projections are called OBLIQUE PROJECTIONS.

11. Show that if P is any projection then rank(P) = trace(P).

3. Eigenvalues and Eigenvectors

This section is devoted to the eigenvalue problem $Ax = \lambda x$.

Throughout the section, A will denote a matrix of order n.

3.1. Definitions and Elementary Properties

We begin with the usual definition of an eigenvalue and eigenvector of a matrix. Loosely speaking, an eigenvector is a vector that does not change direction when it is multiplied by A, and its eigenvalue is the amount by which it shrinks or expands in the process.

Definition 3.1. *The pair (x, λ) is called an* EIGENPAIR *of the matrix A if $x \neq 0$ and*
$$Ax = \lambda x.$$
The vector x is called an EIGENVECTOR *of x, and λ is its associated* EIGENVALUE. *The set of all eigenvalues of A is written $\mathcal{L}(A)$.*

A word on nomenclature is appropriate here. The prefix *eigen* is German, and in this connection it means something like "characteristic." Purists originally objected to the hybrid translation "eigenvalue" for the German *eigenwert*, preferring one or another of a host of names (characteristic value, proper value, latent root, etc.). By now, however, "eigen" has become a living English prefix that means "pertaining to eigenvalues and eigenvectors," and we will use it with complete freedom — as we did in defining the term eigenpair above.

The equation $Ax = \lambda x$ may be written in the form $(\lambda I - A)x = 0$, from which it is seen that λ is an eigenvalue of A if and only if $\lambda I - A$ is singular or, equivalently, if and only if
$$\phi_A(\lambda) \stackrel{\text{def}}{=} \det(\lambda I - A) = 0.$$

3. Eigenvalues and Eigenvectors

The function $\phi_A(\lambda)$ is a polynomial of degree n in λ and is called the CHARACTERISTIC POLYNOMIAL of A. Consequently, a matrix has exactly n eigenvalues, each distinct eigenvalue being counted according to its multiplicity as a root of the CHARACTERISTIC EQUATION $\phi_A(\lambda) = 0$. An eigenvalue whose multiplicity is one is called a SIMPLE EIGENVALUE.

The characterization of eigenvalues in terms of the characteristic polynomial has some important consequences. First, since $\phi_A(\lambda) = 0$ if and only if $\phi_{A^H}(\bar{\lambda}) = 0$, each eigenvalue λ of A corresponds to an eigenvalue $\bar{\lambda}$ of A^H. Hence there is a vector y such that $A^H y = \bar{\lambda} y$, or equivalently $y^H A = \lambda y^H$. The vector y is called a LEFT EIGENVECTOR of A (and the original eigenvector is called a RIGHT EIGENVECTOR of A when it must be distinguished from y).

Second, if A is real, then its characteristic polynomial is real, and its complex eigenvalues must occur in complex conjugate pairs.

Third, if A is block triangular, say

$$A = \begin{pmatrix} A_{11} & A_{12} & \ldots & A_{1k} \\ 0 & A_{22} & \ldots & A_{2k} \\ \vdots & \vdots & & \vdots \\ 0 & 0 & \ldots & A_{kk} \end{pmatrix},$$

then $\phi_A(\lambda) = \phi_{A_{11}}(\lambda)\phi_{A_{22}}(\lambda)\cdots\phi_{A_{kk}}(\lambda)$. Hence

the eigenvalues of a block triangular matrix are the eigenvalues of its diagonal blocks.

In particular, the eigenvalues of a triangular matrix are its diagonal elements.

If (x, λ) and (y, λ) are eigenpairs of A, then $(\alpha x + \beta y, \lambda)$ is also an eigenpair, provided $\alpha x + \beta y \neq 0$. Thus the set of all eigenvectors corresponding to an eigenvalue λ together with the zero vector form a subspace, which is equal to $\mathcal{N}(\lambda I - A)$. The dimension of this subspace, $\dim[\mathcal{N}(\lambda I - A)] = n - \text{rank}(\lambda I - A)$, is called the GEOMETRIC MULTIPLICITY of λ. It is easy to see that the geometric multiplicity of an eigenvalue is not greater than its ALGEBRAIC MULTIPLICITY as a root of the characteristic equation. It can, however, be smaller as the following example shows.

Example 3.2. Let

$$A = \begin{pmatrix} 1 & 1 \\ 0 & 1 \end{pmatrix}.$$

Then $\phi_A(\lambda) = (\lambda - 1)^2$, so that 1 is an eigenvalue of algebraic multiplicity two. On the other hand, the only eigenvector of A is a multiple of $\mathbf{1}_1$, so that the geometric multiplicity of the eigenvalue 1 is one.

An eigenvalue whose geometric multiplicity is less than its algebraic multiplicity is said to be DEFECTIVE A DEFECTIVE MATRIX is one with at least one defective eigenvalue. Unfortunately defective eigenvalues and their matrices are the bane of matrix perturbation theory, as the following continuation of the above example shows.

Example 3.2 (continued). Let

$$\tilde{A} = \begin{pmatrix} 1 & 1 \\ \epsilon & 1 \end{pmatrix}$$

be a perturbation of A. Then $\phi_{\tilde{A}}(\lambda) = (\lambda - 1)^2 - \epsilon$, so that the eigenvalues of $\tilde{A}$ are $1 \pm \sqrt{\epsilon}$. Thus the eigenvalues of $\tilde{A}$ are not differentiable at $\epsilon = 0$. Moreover, if $\epsilon = 10^{-10}$, then the eigenvalues of A and $\tilde{A}$ differ by 10^{-5}. Thus a perturbation of 10^{-10} in A can cause a perturbation in its eigenvalues that is five orders of magnitude larger.

The use of a tilde (viz. $\tilde{A}$) to denote a perturbed quantity is the first occurrence of a convention that will be used throughout this book. See the introduction to Section III.2 for more details.

If (x, λ) is an eigenpair of A and U is nonsingular, then

$$(U^{-1}AU)U^{-1}x = \lambda U^{-1}x,$$

which shows that $(U^{-1}x, \lambda)$ is an eigenpair of $U^{-1}AU$. Thus the SIMILARITY TRANSFORMATION $A \to U^{-1}AU$ preserves the eigenvalues of A and transforms its eigenvectors by U^{-1}. Since $\operatorname{rank}(\lambda I - A) = \operatorname{rank}(\lambda I - U^{-1}AU)$, the geometric multiplicities of the eigenvalues are invariant under similarity transformations. Since $\phi_A(\lambda) = \phi_{U^{-1}AU}(\lambda)$, the algebraic multiplicities are also preserved. In the next two subsections we will consider how far a matrix may be simplified by similarity transformations.

3. Eigenvalues and Eigenvectors

3.2. The Schur Decomposition

A major theme of matrix theory is the reduction of matrices to a simple form by similarity transformations. For example, it will follow from the results of the next subsection that if a matrix A is not defective then there is a nonsingular matrix X such that $X^{-1}AX = \Lambda \equiv \mathrm{diag}(\lambda_1, \lambda_2, \ldots, \lambda_n)$, where the λ_i are the eigenvalues of A. The advantages of being able to work with a diagonal matrix instead of a full one are obvious.

Unfortunately, similarity transformations can introduce problems of their own. Suppose, for example, that the matrix of the last paragraph is perturbed by an error E. Then the diagonalized problem is $\Lambda + X^{-1}EX$. If X and its inverse are large (or if X is ILL CONDITIONED, as we shall learn to say in Chapter III), then the effect of the similarity transformation is to magnify the error.

In this connection, unitary similarity transformations of the form $A \to U^H A U$ are particularly desirable, since neither U nor its inverse U^H can be large. Specifically, from the fact that $U^H U = I$, it follows that the jth column of U satisfies $\|u_j\|_2 = 1$. Hence no element of U or its inverse can be greater than one in absolute value.

In general we cannot reduce a matrix to diagonal form by unitary similarities (informally, the relation $U^H U = I$ implies that U has roughly $n^2/2$ degrees of freedom, too few to satisfy the roughly n^2 conditions that the off-diagonal elements of $U^H A U$ be zero). However, the following theorem shows that we can reduce an arbitrary matrix to triangular form by a unitary similarity.

Theorem 3.3 (Schur). *There is a unitary matrix U such that $T = U^H A U$ is upper triangular. The matrix U may be chosen so that the eigenvalues of A appear on the diagonal of T in any order.*

Proof. The proof is by induction. The theorem is trivial for a matrix of order one. Assume it is true for all matrices of order less than $n > 1$. Let an ordering of the eigenvalues of A be given, and let λ be the first eigenvalue of the ordering. Let $Ax = \lambda x$, where $\|x\|_2 = 1$, and let $H = (x \ X)$ be a unitary matrix.

The matrix $H^{\mathrm{H}}AH$ has the form

$$H^{\mathrm{H}}AH = \begin{pmatrix} x^{\mathrm{H}}Ax & x^{\mathrm{H}}AX \\ X^{\mathrm{H}}Ax & X^{\mathrm{H}}AX \end{pmatrix}.$$

Since $Ax = \lambda x$ and $x^{\mathrm{H}}x = 1$, we have $x^{\mathrm{H}}Ax = \lambda x^{\mathrm{H}}x = \lambda$. Since $X^{\mathrm{H}}x = 0$, we have $X^{\mathrm{H}}Ax = \lambda X^{\mathrm{H}}x = 0$. Thus

$$H^{\mathrm{H}}AH = \begin{pmatrix} \lambda & x^{\mathrm{H}}AX \\ 0 & X^{\mathrm{H}}AX \end{pmatrix} \equiv \begin{pmatrix} \lambda & b^{\mathrm{H}} \\ 0 & M \end{pmatrix}.$$

Since $H^{\mathrm{H}}AH$ is block triangular, the eigenvalues of the matrix M are the eigenvalues of A other than λ. By hypothesis, there is a unitary matrix V such that $V^{\mathrm{H}}MV$ is upper triangular with its eigenvalues in the correct order. If we set $U = (x\ XV)$, then

$$T = U^{\mathrm{H}}AU = \begin{pmatrix} \lambda & b^{\mathrm{H}}V \\ 0 & V^{\mathrm{H}}MV \end{pmatrix}$$

is the required decomposition. ∎

Although the proof of Schur's theorem is not difficult, the resulting decomposition is one of the most important tools in theoretical and computational linear algebra. Computationally, it is the target of the QR algorithm, the single most successful algorithm for computing eigensystems of general matrices. We will encounter its theoretical uses throughout this book, beginning with some elementary consequences in this chapter.

Recall (Definition 1.1) that a matrix A is normal if $A^{\mathrm{H}}A = AA^{\mathrm{H}}$. This seemingly innocuous equation has important consequences.

Theorem 3.4. *If A is normal then any Schur decomposition of A is diagonal.*

Proof. The theorem is trivial for a matrix of order one. Assume it is true for all matrices of order less than $n > 1$. Let $T = U^{\mathrm{H}}AU$ be a Schur decomposition of A and partition T in the form

$$T = \begin{pmatrix} \tau & t^{\mathrm{T}} \\ 0 & T_* \end{pmatrix}.$$

3. Eigenvalues and Eigenvectors

Now if A is normal, any matrix unitarily similar to A is normal. Hence $T^\mathrm{H} T = T T^\mathrm{H}$, from which it follows that

$$|\tau|^2 = |\tau|^2 + t^\mathrm{H} t,$$

and

$$T_*^\mathrm{H} T_* = T_* T_*^\mathrm{H}.$$

The first of the equations implies that $t = 0$. The second says that T_* is normal, and hence by hypothesis it is diagonal (since it is triangular). Thus T is diagonal. ∎

If we write $T = \Lambda = \mathrm{diag}(\lambda_1, \lambda_2, \ldots, \lambda_n)$ and $U = (u_1 \; u_2 \; \ldots \; u_n)$, then $AU = U\Lambda$ or

$$A u_j = \lambda_j u_j, \qquad j = 1, \ldots, n.$$

In other words the u_j are eigenvectors of A. Since the eigenvectors are pairwise orthogonal, we have the following corollary.

Corollary 3.5. *A normal matrix of order n has a system of orthonormal eigenvectors that span $\mathbf{C}^n$.*

A matrix chosen at random is unlikely to be normal. However, there are two frequently occurring classes of normal matrices: unitary matrices and Hermitian matrices. They are distinguished by the situation of their eigenvalues.

Corollary 3.6. *A unitary matrix is a normal matrix with eigenvalues on the unit circle. A Hermitian matrix is a normal matrix with real eigenvalues.*

The proof of this corollary is left as an exercise. It immediately implies that any Hermitian matrix can be written in the form

$$A = U \Lambda U^\mathrm{H},$$

where U is unitary and Λ is diagonal and real. We will call this the SPECTRAL DECOMPOSITION of A. We will sometimes write it in the form

$$A = \sum_i \lambda_i u_i u_i^\mathrm{H}, \tag{3.1}$$

where u_i is the ith column of U. This form allows us to extend scalar valued functions to Hermitian matrices: namely,

$$\varphi(A) \stackrel{\text{def}}{=} \sum_i \varphi(\lambda_i) u_i u_i^{\text{H}}.$$

In particular if A is positive semi-definite, its eigenvalues are nonnegative and its SQUARE ROOT

$$A^{\frac{1}{2}} \stackrel{\text{def}}{=} \sum_i \lambda_i^{\frac{1}{2}} u_i u_i^{\text{H}}.$$

is well defined.

3.3. The Jordan Canonical Form

Although the Schur decomposition is useful in many applications, in others it does not go far enough in reducing its matrix. The question thus arises of how much we can simplify a matrix using similarity transformations. Example 3.2 shows that in general we cannot hope to reduce a matrix to diagonal form. However, we can reduce it to a simple block diagonal form, as the following classical theorem shows.

Theorem 3.7 (Jordan). *For any A there is a nonsingular X such that*

$$X^{-1}AX = \text{diag}(J_{k_1}(\lambda_1), J_{k_2}(\lambda_2), \ldots, J_{k_l}(\lambda_l)), \tag{3.2}$$

where $J_k(\lambda) \in \mathbf{C}^{k \times k}$ is a JORDAN BLOCK *of the form*

$$J_k(\lambda) \stackrel{\text{def}}{=} \begin{pmatrix} \lambda & 1 & 0 & \cdots & 0 & 0 \\ 0 & \lambda & 1 & \cdots & 0 & 0 \\ \vdots & \vdots & \vdots & & \vdots & \vdots \\ 0 & 0 & 0 & \cdots & \lambda & 1 \\ 0 & 0 & 0 & \cdots & 0 & \lambda \end{pmatrix}.$$

The right-hand side of the decomposition is unique up to the ordering of the blocks.

For a proof of this theorem (which is not easy) see any good linear algebra text.

3. Eigenvalues and Eigenvectors

The eigenvalues of a matrix corresponding to Jordan blocks of order greater than one are defective (n.b., the same eigenvalue can occur in different blocks).. Thus a nondefective matrix has only 1×1 Jordan blocks, or equivalently every nondefective matrix can be diagonalized by a similarity transformation. For this reason a nondefective matrix is also called a DIAGONALIZABLE matrix

If we partition $X = (x_1, \ldots, x_n)$ by columns, then the first k_1 vectors associated with the Jordan block $J_{k_1}(\lambda_1)$ satisfy the equations

$$x_1 = \lambda_1 x_1,$$
$$x_j = \lambda_1 x_j + x_{j-1}, \qquad j = 2, \ldots, k_1.$$

Such a sequence is called a chain of PRINCIPAL VECTORS of A. The columns of X corresponding to the other Jordan blocks also form chains of principal vectors.

In spite of its solid appearance, the Jordan form is a fragile thing. The continuation of Example 3.2 shows that the slightest perturbation can shatter it to bits. Moreover, the ones on the superdiagonal of a Jordan block represent an arbitrary normalization. For example the similarity

$$\begin{pmatrix} \delta & 0 & 0 \\ 0 & \delta^2 & 0 \\ 0 & 0 & \delta^3 \end{pmatrix}^{-1} \begin{pmatrix} \lambda & 1 & 0 \\ 0 & \lambda & 1 \\ 0 & 0 & \lambda \end{pmatrix} \begin{pmatrix} \delta & 0 & 0 \\ 0 & \delta^2 & 0 \\ 0 & 0 & \delta^3 \end{pmatrix} = \begin{pmatrix} \lambda & \delta & 0 \\ 0 & \lambda & \delta \\ 0 & 0 & \lambda \end{pmatrix} \qquad (3.3)$$

puts δ's on the superdiagonal of $J_3(\lambda)$. (Note that as δ approaches zero, the transformation becomes increasingly ill conditioned). For these reasons, there has been a tendency to shy away from the Jordan form, at least in applications.

3.4. Invariant Subspaces

One of the ways of handling instabilities in the eigensystem of a matrix is to decompose the matrix into smaller matrices acting on subspaces. Although the matrices may be ill-behaved within their subspaces, the subspaces themselves often turn out to be insensitive to perturbations.

For example, if in (3.2) we partition $X = (X_1 \ldots X_l)$ conformally with the Jordan blocks and similarly partition $X^{-H} = (Y_1 \ldots Y_l)$, then

$$A = X_1 J_{k_1}(\lambda_1) Y_1^H + X_2 J_{k_2}(\lambda_2) Y_2^H + \cdots + X_l J_{k_l}(\lambda_l) Y_l^H. \qquad (3.4)$$

Since $Y^{\mathrm{H}} = X^{-1}$, it follows that

$$AX_i = X_i J_{k_i}(\lambda_i), \qquad i = 1, \ldots, l. \tag{3.5}$$

In other words, A can be regarded as an operator mapping $\mathcal{R}(X_i)$ into itself. If $u = X_i v$ is a representation of u in the basis formed by the columns of X, then by (3.5)

$$AX_i v = X_i J_{k_i}(\lambda_i) v;$$

i.e., v maps into $J_{k_i}(\lambda_i)v$. This means that $J_{k_i}(\lambda_i)$ is the representation of A on $\mathcal{R}(X_i)$ with respect to the basis formed by the columns of X_i. These considerations suggest the following definition.

Definition 3.8. *The subspace $\mathcal{X}$ is an* INVARIANT SUBSPACE *of A if*

$$A\mathcal{X} \subset \mathcal{X}.$$

Some of the facts about invariant subspaces are contained in the following theorem.

Theorem 3.9. *Let $\mathcal{X}$ be an invariant subspace of A, and let the columns of X form a basis for $\mathcal{X}$. Then there is a unique matrix L such that*

$$AX = XL. \tag{3.6}$$

The matrix L is the representation of A on $\mathcal{X}$ with respect to the basis X. In particular (v, λ) is an eigenpair of L if and only if (Xv, λ) is an eigenpair of A.

Proof. Let $X = (x_1 \ \ldots \ x_k)$. Since $Ax_i \in \mathcal{X}$, it can be expressed as a unique linear combination of the columns of X; that is, $Ax_i = Xl_i$ for some unique vector l_i. The matrix $L = (l_1 \ \ldots \ l_k)$ is the required matrix. The fact that L is the representation of A on X follows as above for the Jordan canonical form. In view of (3.6), the statement about eigenpairs, which amounts to saying that $Lv = \lambda v$ if and only if $AXv = \lambda Xv$, is a triviality. ∎

This is not the whole story — only enough to get us through Chapter IV. We will return to invariant subspaces in Chapter V.

3. Eigenvalues and Eigenvectors

3.5. The Field of Values

The quadratic form $x^H Ax$ plays an important role in many applications. This subsection is devoted to the values that such a form can attain for a given matrix. We begin with a definition.

Definition 3.10. *Let $A \in \mathbf{C}^{n \times n}$. The set*

$$\mathcal{F}(A) \stackrel{\text{def}}{=} \{x^H Ax : \|x\|_2 = 1\}$$

is called the FIELD OF VALUES *of A.*

The set $\mathcal{F}(A)$ is bounded and closed in $\mathbf{C}$. From Definition 3.10, it is easy to verify the following properties of $\mathcal{F}(A)$:

1. $\mathcal{F}(\alpha A + \beta I) = \alpha \mathcal{F}(A) + \beta, \quad \alpha, \beta \in \mathbf{C}$;
2. $\mathcal{L}(A) \subseteq \mathcal{F}(A)$;
3. If U is unitary, $\mathcal{F}(U^H AU) = \mathcal{F}(A)$;
4. $\mathcal{F}(A + B) \subseteq \mathcal{F}(A) + \mathcal{F}(B)$.

An important and far-reaching result is that the field of values of a matrix is convex; that is, it contains any line whose endpoints lie in it. Nothing in the proof of the following theorem is used later, and it may be skipped without loss.

Theorem 3.11 (Toeplitz–Hausdorff). *The field of values $\mathcal{F}(A)$ is a convex set.*

Proof. Let $\rho, \sigma \in \mathcal{F}(A)$. Since $\mathcal{F}(\alpha A + \beta I) = \alpha \mathcal{F}(A) + \beta$ and $\alpha \mathcal{F}(A) + \beta$ is convex if and only if $\mathcal{F}(A)$ is convex, we may assume without loss of generality that $\rho = 0$ and $\sigma = 1$. Thus there are vectors x_0 and x_1 of 2-norm one such that

$$x_0^H A x_0 = 0 \text{ and } x_1^H A x_1 = 1,$$

which implies that x_0 and x_1 are linearly independent. By multiplying x_0 by a scalar of absolute value one we may further assume that

$$\Re\, x_0^H x_1 = 0.$$

We must show that any $\tau \in [0,1]$ is in the field of values of A. By the linear independence of x_0 and x_1, the vector $(1-\lambda)x_0 + \lambda x_1$ is nonzero. Consequently the function
$$\varphi(\lambda) = \frac{[(1-\lambda)x_0 + \lambda x_1]^H A[(1-\lambda)x_0 + \lambda x_1]}{\|(1-\lambda)x_0 + \lambda x_1\|_2^2} = \frac{|\lambda|^2}{\|(1-\lambda)x_0 + \lambda x_1\|_2^2}$$
is continuous and real when λ is real. Moreover, $\varphi(\lambda) \in \mathcal{F}(A)$ for all λ. Since $\varphi(0) = 0$ and $\varphi(1) = 1$, by the intermediate value theorem there is a $\lambda \in [0,1]$ such that $\tau = \varphi(\lambda) \in \mathcal{F}(A)$. ∎

The field of values of a matrix A is closely related to the eigenvalues of A. In particular if $Ax = \lambda x$ with $\|x\|_2 = 1$, then $\lambda = x^H A x \in \mathcal{F}(A)$. Since the field of values is convex, $\mathcal{F}(A)$ must contain the smallest convex set containing all the eigenvalues of A, that is, the set

$$\mathcal{H}[\mathcal{L}(A)] = \left\{ \sum_{\lambda_i \in \mathcal{L}(A)} \theta_i \lambda_i : \theta_i \geq 0, \sum \theta_i = 1 \right\} \tag{3.7}$$

($\mathcal{H}[\mathcal{L}(A)]$ is called the CONVEX HULL of $\mathcal{L}(A)$). Unfortunately, the field of values can be bigger than the convex hull of $\mathcal{L}(A)$, as the following example shows.

Example 3.12. Let
$$A = \begin{pmatrix} 0 & 1 \\ 0 & 0 \end{pmatrix}.$$
Then $\mathcal{H}[\mathcal{L}(A)] = \{0\}$. But $\mathcal{F}(A) = \{z \in \mathbf{C} : |z| \leq \frac{1}{2}\}$.

There is one important class of matrices for which the field of values coincides with the convex hull of its eigenvalues.

Theorem 3.13. *If A is normal, then*
$$\mathcal{F}(A) = \mathcal{H}[\mathcal{L}(A)]. \tag{3.8}$$

Proof. Since the normal matrix A has a decomposition $A = U^H \Lambda U$, where U is unitary and $\Lambda = \text{diag}(\lambda_1, \ldots, \lambda_n)$, we have
$$\mathcal{F}(A) = \{(Ux)^H \Lambda(Ux) : x \in \mathbf{C}^n, \|x\|_2 = 1\}$$
$$= \{y^H \Lambda y : y \in \mathbf{C}^n, \|y\|_2 = 1\}$$
$$= \left\{ \sum_{i=1}^n |y_i|^2 \lambda_i : \sum_i |y_i|^2 = 1 \right\}.$$

Since $\sum |y_i|^2 = 1$, this last equation is clearly equivalent to (3.7). ∎

3.6. Sums of Hermitian Matrices

The main result of this subsection will be found in more general form in Section IV.4, where it is a consequence of a very powerful characterization of the eigenvalues of a Hermitian matrix. Because the result is needed in Chapters II and III, we will give an elementary proof here.

Theorem 3.14. *Let A and B be Hermitian and $\tilde{A} = A + B$. Let the eigenvalues of A be $\lambda_1 \geq \lambda_2 \geq \cdots \geq \lambda_n$, and let the eigenvalues of $\tilde{A}$ be $\tilde{\lambda}_1 \geq \tilde{\lambda}_2 \geq \cdots \geq \tilde{\lambda}_n$. If μ_n is the smallest eigenvalue of B, then*

$$\tilde{\lambda}_i \geq \lambda_i + \mu_n, \qquad i = 1, \ldots, n.$$

Proof. We begin by simplifying the problem. First, since the order of the eigenvalues is not affected by a shift of the origin, we may replace B by $B - \mu_n I$, so that the theorem says that the eigenvalues of $\tilde{A}$ are pairwise greater than the eigenvalues of A.

Second, if $X^H A X = \Lambda$ is the spectral decomposition of A, then we may replace $\tilde{A}$ with $X^H \tilde{A} X$, A with Λ, and B with $X^H B X$. Hence we may assume that $A = \Lambda$ is diagonal.

Third, if $B = \sum_i \mu_i y_i y_i^H$ is the spectral decomposition of B, then it is sufficient to prove the theorem in stages: first for $\Lambda + \mu_1 y_1 y_1^H$, then for $(\Lambda + \mu_1 y_1 y_1^H) + \mu_2 y_2 y_2^H$, and so on. Thus it is sufficient to establish the theorem for the case $B = yy^H$.

Fourth, if the ith component of y is zero, then λ_i is an eigenvalue of $\tilde{A}$. Moreover, the remaining eigenvalues of $\tilde{A}$ are those of the matrix obtained by deleting the ith row and column of $\tilde{A}$. Thus we may assume that $|y| > 0$.

Fifth, the characteristic polynomial of $\tilde{A}$ is

$$\phi_{\tilde{A}}(\lambda) = \det[(\lambda I - \Lambda) - yy^H] = \det(\lambda I - \Lambda)\det[I - (\lambda I - \Lambda)^{-1} yy^H].$$

From the identity $\det(I - uv^H) = 1 - v^H u$, we have

$$\phi_{\tilde{A}}(\lambda) = \phi_\Lambda(\lambda) - \sum_i |\eta_i|^2 \phi_\Lambda^{(i)}(\lambda),$$

where $\phi_\Lambda^{(i)}(\lambda) = \phi_\Lambda(\lambda)/(\lambda - \lambda_i)$. Now if λ_i has multiplicity m then $(\lambda - \lambda_i)^{m-1}$ factors out of $\phi_{\tilde{A}}(\lambda)$. This implies that $m - 1$ copies of λ

stay fixed and we need only be concerned with the change in one of them. Hence we may assume that the λ_i's are distinct.

The result now follows easily from the intermediate value theorem. Since $\phi_\Lambda^{(i)}(\lambda_i) = \phi'_\Lambda(\lambda_i)$, we have $\phi_{\tilde{A}}(\lambda_1) < 0$, $\phi_{\tilde{A}}(\lambda_2) > 0$, $\phi_{\tilde{A}}(\lambda_3) < 0$, and so on. It follows that $\tilde{A}$ has $n-1$ eigenvalues interlaced with the λ_i. Since $\phi_{\tilde{A}}(\lambda_1) < 0$ and $\lim_{\lambda\to\infty} \phi_{\tilde{A}}(\lambda_1) = +\infty$, the remaining eigenvalue of $\tilde{A}$ must be greater than λ_1. ∎

Theorem 3.14 is our first perturbation theorem, in that it restricts the location of the eigenvalues of the perturbed matrix $\tilde{A}$. It is noteworthy that there is no restriction on the size of the perturbation B.

Notes and References

The prefix "eigen" has triumphed because of its brevity and utility. Other coinages are EIGENSYSTEM for the system of eigenvalues and associated vectors and EIGENPROBLEM for the eigenvalue problem. The term "eigenproblem" is actually more precise, since one is seldom concerned with eigenvalues alone; however, the sense does not survive translation back into German — further evidence of the thorough Anglicization of the prefix. Nontheless, our free use of the prefix "eigen" does not extend to revising established nomenclature: there will be no eigenpolynomials in this book ("A foolish consistency is the hobgoblin of little minds ...").

The notation for the set of eigenvalues of a matrix or operator varies. Numerical analysts and some matrix theorists write $\lambda(A)$; functional analysts, who call eigenvalues and their generalizations the spectrum of an operator, write $\sigma(A)$. Unfortunately, the former group uses $\sigma(A)$ to denote the set of singular values of A. We have punted by writing $\mathcal{L}(A)$ for the set of eigenvalues of A.

The proof of the existence of the Schur decomposition is the same as Schur's [193, 1909]. The decomposition is essentially unique, up to the ordering of the eigenvalues; that is, once an ordering has been fixed, the copies of each multiple eigenvalue being placed together, the columns of U corresponding to simple eigenvalues are unique. The columns corresponding to multiple eigenvalues are not unique, but the space spanned by the columns is. This kind of "essential uniqueness" is typical of most decompositions involving eigenvectors and the like.

Since a real matrix can have complex eigenvalues, the Schur decomposition of a real matrix can be complex, something that is undesirable in numerical applications. Reality can be restored by allowing the final form to be block triangular, with 2×2 blocks representing the complex eigenvalues. The details will be found in Exercises 3.22, 3.23, and 3.24.

The Jordan canonical form [124, 1870] represents an extreme reduction of its matrix, which is achieved at the expense of stability. For more on the computation of the

3. Eigenvalues and Eigenvectors

Jordan form and intermediaries see [273, 18, 128].

The term "field of values" is a reasonable translation of *Wertvorrat*, which seems to be due to Toeplitz [241, 1918]. It is also called the numerical range. Toeplitz proved that the boundary of the field of values is a convex curve. Hausdorff [105, 1919] showed that the set itself is convex. The proof given here, which is really a souped-up version of Hausdorff's, was adapted from [122].

The result on sums of Hermitian matrices is an easy consequence of Fischer's characterization of the eigenvalues of a Hermitian matrix (see Corollary IV.4.9). The proof given here owes much to Wilkinson [269, 1965; pp. 94–98]

Exercises

1. (Cayley–Hamilton). Prove that $\phi_A(A) = 0$.

2. Prove that if A is real and $\lambda \in \mathcal{L}(A)$ then $\bar{\lambda} \in \mathcal{L}(A)$.

3. Let A be $m \times n$ and B be $n \times m$. Show that the matrices

$$\begin{pmatrix} AB & 0 \\ B & 0 \end{pmatrix}$$

and

$$\begin{pmatrix} 0 & 0 \\ B & BA \end{pmatrix}$$

are similar. Conclude that the nonzero eigenvalues of AB are the same as those of BA. [Note: This elegant proof is due to Kahan [130].]

4. Show that the geometric multiplicity of an eigenvalue is not greater than its algebraic multiplicity.

5. Show that if A is SKEW-HERMITIAN (i.e., $A^H = -A$), then all its eigenvalues lie on the imaginary axis.

6. (Loewy [148]). Let K be skew Hermitian. Show that the matrix

$$U = (I + K)^{-1}(I - K) \qquad (3.9)$$

is unitary. Moreover, if U is unitary and $1 \notin \mathcal{L}(U)$, then U can be represented in the form (3.9) for some skew-Hermitian matrix K.

7. Suppose that H_1 and H_2 are Hermitian with H_1 positive definite. Show that $H_1 + H_2$ is positive definite if and only if all the eigenvalues of $H_1^{-1}H_2$ are greater than -1.

8. Let A be idempotent (i.e., $A^2 = A$). Show that A is nondefective with eigenvalues zero and one.

9. Let A be nondefective. Show (without appealing to the Jordan canonical form) that if the columns of X form a set of linearly independent eigenvectors of A, then $\Lambda = X^{-1}AX$ is diagonal. What are the diagonal elements of Λ?

10. Let A be diagonalizable and let $X^{-1}AX = \Lambda$, where Λ is diagonal. Let $X = QR$ be the QR factorization of X. Show that $Q^H AQ$ is upper triangular; i.e., it is a Schur decomposition of A.

11. Let
$$\phi(\xi) = \alpha_0 + \alpha_1 \xi + \cdots + \alpha_{n-1}\xi^{n-1} + \xi^n,$$
and let
$$C_\phi = \begin{pmatrix} 0 & 1 & 0 & \cdots & 0 & 0 \\ 0 & 0 & 1 & \cdots & 0 & 0 \\ \vdots & \vdots & \vdots & & \vdots & \vdots \\ 0 & 0 & 0 & \cdots & 0 & 1 \\ -\alpha_0 & -\alpha_1 & -\alpha_2 & \cdots & -\alpha_{n-2} & -\alpha_{n-1} \end{pmatrix}.$$
Show that the characteristic polynomial of C_ϕ is ϕ. The matrix C_ϕ is called the COMPANION MATRIX OF ϕ.

12. Let $\phi(\xi) = (\xi - \lambda)^n$. Show that the companion matrix C_ϕ is similar to the Jordan block $J_n(\lambda)$.

13. Show that $J_k(0)^k = 0$. What do the matrices $J_k(0)^i$ ($i = 1, \ldots, k-1$) look like?

14. Show that
$$J_k(\lambda)^n = \lambda^n I + \lambda^{n-1}\binom{n}{1}J + \lambda^{n-2}\binom{n}{2}J^2 + \cdots + \lambda^{n-k+1}\binom{n}{k-1}J^{k-1},$$
where $J = J_k(0)$.

15. Prove that the field of values of a matrix A is real if and only if A is Hermitian. What if the field of values is imaginary?

16. Prove that the field of values of the matrix in Example 3.12 is $\{z \in \mathbf{C} : |z| \leq \frac{1}{2}\}$.

17. Let A and B be diagonalizable matrices. Show that the following statements are equivalent.

3. Eigenvalues and Eigenvectors

1. $AB = BA$.
2. There is a nonsingular matrix X such that $X^{-1}AX$ and $X^{-1}BX$ are diagonal.
3. There are polynomials p and q and a diagonalizable matrix C such that $A = p(C)$ and $B = q(C)$.

18. Let λ be a simple eigenvalue of A and let x and y be its right and left eigenvectors. Show that $y^H x \neq 0$.

19. Let ϕ be a polynomial and $J_k(\lambda)$ be a Jordan block of order k. Show that

$$\phi[J_k(\lambda)] = \begin{pmatrix} \phi(\lambda) & \phi'(\lambda) & \phi''(\lambda)/2 & \phi^{(3)}(\lambda)/3! & \cdots \\ 0 & \phi(\lambda) & \phi'(\lambda) & \phi''(\lambda)/2 & \phi^{(3)}(\lambda)/3! & \cdots \\ 0 & 0 & \phi(\lambda) & \phi'(\lambda) & \phi''(\lambda)/2 & \phi^{(3)}(\lambda)/3! & \cdots \\ \vdots & \vdots & \vdots & \vdots & \vdots & \vdots \end{pmatrix}.$$

20. Let $r(\lambda) = p(\lambda)/q(\lambda)$ be a rational function, and let A be such that no zero of q is in $\mathcal{L}(A)$. Show that $q(A)$ is nonsingular and hence $r(A) = p(A)/q(A)$ is well defined. Prove that

$$\mathcal{L}[r(A)] = \{r(\lambda) : \lambda \in \mathcal{L}(A)\}.$$

What are the corresponding eigenvectors?

21. Let B be nilpotent (i.e., let there be an integer $k > 0$ such that $B^k = 0$). Show that if $AB = BA$ then $\det(A + B) = \det(A)$.

The following three exercises develop an analogue of the Schur form for real matrices.

22. Let the columns of X form an orthonormal basis for an invariant subspace of A. Let $AX = XL$, and let $(X\ Y)$ be unitary. Show that

$$\begin{pmatrix} X^H \\ Y^H \end{pmatrix} A (X\ Y) = \begin{pmatrix} L & H \\ 0 & M \end{pmatrix}.$$

23. Let A be real, and let λ be a complex eigenvalue of A with eigenvector $x + iy$. Show that the space spanned by x and y is an invariant subspace of A.

24. (Real Schur decomposition). Show that if A is real there is an orthogonal matrix U such that $U^{\mathrm{T}}AU$ is block triangular with 1×1 and 2×2 blocks on its diagonal. The 1×1 blocks contain the real eigenvalues of A, and the eigenvalues of the 2×2 blocks are the complex eigenvalues of A.

———◇———

25. (Bendixson–Hirsch–Toeplitz theorem). Let $A = B + iC$, where $B = (A + A^{\mathrm{H}})/2$ and $C = (A - A^{\mathrm{H}})/2i$ are Hermitian. Let the eigenvalues of B be $\beta_1 \geq \cdots \geq \beta_n$ and those of C be $\gamma_1 \geq \cdots \geq \gamma_n$ Show that $\mathcal{L}(A)$ lies in the rectangle $[\beta_n, \beta_1] \times [\gamma_n, \gamma_1]$. [Note: Bendixson [25, 1902] proved the result for real matrices, giving a weaker bound than the above for the imaginary part. Hirsh [115, 1902] pointed out that the result holds for complex matrices and gave a sharper bound for the imaginary part. In the form stated here, the theorem is due to Toeplitz [241, 1918].]

26. The KRONECKER PRODUCT or TENSOR PRODUCT of two matrices A and B is

$$A \otimes B = \begin{pmatrix} \alpha_{11}B & \alpha_{12}B & \alpha_{13}B & \cdots \\ \alpha_{21}B & \alpha_{22}B & \alpha_{23}B & \cdots \\ \alpha_{31}B & \alpha_{32}B & \alpha_{33}B & \cdots \\ \vdots & \vdots & \vdots & \end{pmatrix}.$$

Show that if A and B are square, then

$$\mathcal{L}(A \otimes B) = \mathcal{L}(A)\mathcal{L}(B).$$

4. The Singular Value Decomposition

4.1. The Singular Value Decomposition

The QR decomposition has the advantage that it does not mix up the columns of a matrix, since it involves only premultiplication by a unitary matrix. The price to be paid is that the resulting matrix is merely upper triangular. If we are also willing to postmultiply by a unitary matrix, we can reduce an arbitrary matrix to a diagonal form. The result is the SINGULAR VALUE DECOMPOSITION.

Theorem 4.1. Let $A \in \mathbf{C}^{m \times n}$ have rank r. Then there are unitary matrices U and V such that

$$U^{\mathrm{H}}AV = \begin{pmatrix} \Sigma_+ & 0 \\ 0 & 0 \end{pmatrix}, \tag{4.1}$$

4. THE SINGULAR VALUE DECOMPOSITION

where $\Sigma_+ = \text{diag}(\sigma_1, \ldots, \sigma_r)$ with $\sigma_1 \geq \cdots \geq \sigma_r > 0$.

Proof. Let the eigenvalues of $A^{\text{H}}A$ be $\sigma_1^2 \geq \cdots \geq \sigma_r^2 > 0 = \sigma_{r+1}^2 = \cdots = \sigma_n^2$. If $V = (V_1\ V_2)$ $(V_1 \in \mathbf{C}^{n \times r})$ is a unitary matrix formed from the corresponding eigenvectors of $A^{\text{H}}A$, then

$$V^{\text{H}}A^{\text{H}}AV = \begin{pmatrix} \Sigma_+^2 & 0 \\ 0 & 0 \end{pmatrix},$$

where Σ_+ is defined as above. Thus we have

$$V_1^{\text{H}}A^{\text{H}}AV_1 = \Sigma_+^2, \qquad V_2^{\text{H}}A^{\text{H}}AV_2 = 0, \qquad (4.2)$$

and from the second of these relations we conclude that

$$AV_2 = 0. \qquad (4.3)$$

Now let

$$U_1 = AV_1\Sigma_+^{-1}. \qquad (4.4)$$

Then from (4.2) we have $U_1^{\text{H}}U_1 = I$. Choose U_2 so that $(U_1\ U_2)$ is unitary. Then from (4.2)–(4.4) we get

$$U^{\text{H}}AV = \begin{pmatrix} U_1^{\text{H}}AV_1 & U_1^{\text{H}}AV_2 \\ U_2^{\text{H}}AV_1 & U_2^{\text{H}}AV_2 \end{pmatrix} = \begin{pmatrix} \Sigma_+ & 0 \\ 0 & 0 \end{pmatrix}. \blacksquare$$

The diagonal elements of Σ_+ are called SINGULAR VALUES of A. Conventions differ on how to handle zero singular values. The choices are to say that A has $\min\{m, n\} - \text{rank}(A)$ zero singular values or that A has $\max\{m, n\} - \text{rank}(A)$ zero singular values. Whenever the choice makes a difference, it will be clear from context what is meant, and we will therefore use either convention at our convenience.

We will denote the set of singular values of A by $\mathcal{S}(A)$. It is easy to see that $\mathcal{S}(A)$ consists of the nonnegative square roots of the eigenvalues of $A^{\text{H}}A$ or of AA^{H}, depending on how we count zero singular values.

The columns of U and V are called LEFT and RIGHT SINGULAR VECTORS. The columns of U are eigenvectors of AA^{H}, while those of V are eigenvectors of $A^{\text{H}}A$. The singular vectors are not unique, but they are by no means arbitrary. The columns of the matrix U_1 must form an orthonormal basis for the column space of A, while the columns of V_1 must form an orthonormal basis for the column space of A^{H}; and these bases are related by (4.4). Moreover, if $U_1' = U_1 W_U$ and $V_1' = V_1 W_V$ also consist of singular vectors of A, then W_U and W_V are unitary and $W_U \Sigma_+ W_V^{\mathrm{H}} = \Sigma_+$. The columns of the matrices U_2 and V_2 may be arbitrary orthonormal bases for the orthogonal complements of $\mathcal{R}(A)$ and $\mathcal{R}(A^{\mathrm{H}})$. Thus the singular value decomposition, like the QR decomposition, can be used to compute the projection onto the column space of A. However, it can also be used to compute the projection onto the row space of A.

The singular value decomposition (4.1) can be written in the form

$$A = U_1 \Sigma_+ V_1^{\mathrm{H}}, \tag{4.5}$$

which is called the SINGULAR VALUE FACTORIZATION of A. If we write $F = U_1$ and $G = V_1 \Sigma_+$, then (4.5) is an example of a FULL RANK FACTORIZATION of A into the product FG^{H} of matrices whose rank is the same as A.

The relation between the singular value decomposition of A and the spectral decomposition of $A^{\mathrm{H}} A$ allows one to obtain results on the singular value decomposition from results for Hermitian matrices. There is another way.

Theorem 4.2 (Jordan–Wielandt). *Let $A \in \mathbf{C}^{m \times n}$ ($m \geq n$) have the singular value decomposition*

$$U^{\mathrm{H}} A V = \mathrm{diag}(\Sigma, 0),$$

where Σ is of order n. Then the matrix

$$C = \begin{pmatrix} 0 & A \\ A^{\mathrm{H}} & 0 \end{pmatrix}$$

has eigenvalues $\pm \sigma_1, \ldots, \pm \sigma_n$, corresponding to the eigenvectors

$$\begin{pmatrix} u_i \\ \pm v_i \end{pmatrix}, \qquad i = 1, \ldots, n,$$

where u_i is the ith column of U and v_i is the ith column of V. In addition C has $m - n$ zero eigenvalues whose eigenvectors are $(u_i^{\mathrm{T}} \ 0)^{\mathrm{T}}$ ($i = n+1, \ldots, n$).

4. The Singular Value Decomposition

The proof is purely computational and is left as an exercise.

An important consequence of Theorem 3.13 is the following characterization of the largest and smallest singular values of a matrix. In stating it we take the opportunity to introduce some useful notation.

Theorem 4.3. *For any matrix* $A \in \mathbf{C}^{m \times n}$

$$\|A\|_2 \stackrel{\text{def}}{=} \max_{\|x\|_2=1} \|Ax\|_2 = \max \mathcal{S}(A).$$

If $m \geq n$, *then*

$$\inf{}_2(A) \stackrel{\text{def}}{=} \min_{\|x\|_2=1} \|Ax\|_2 = \min \mathcal{S}(A).$$

Proof. We have

$$\|A\|_2^2 = \max_{\|x\|_2=1} \|Ax\|_2^2 = \max_{\|x\|_2=1} x^{\text{H}}(A^{\text{H}}A)x = \max \mathcal{F}(A^{\text{H}}A).$$

Since $A^{\text{H}}A$ is Hermitian, the largest member of its field of values is its largest eigenvalue, which is the square of the largest singular value of A. A similar argument shows that $\inf_2(A)$ is the smallest singular value of A. ∎

As the notation suggests, the function $\|\cdot\|_2$ is actually a matrix norm. It will be treated in detail in the next chapter.

4.2. Two Inequalities

In this subsection we will prove two useful inequalities. They are consequences of Theorem 3.14.

Theorem 4.4. *Let* $A \in \mathbf{C}^{m \times n}$ *be partitioned in the form*

$$A = \begin{pmatrix} A_1 \\ A_2 \end{pmatrix}.$$

Let the singular values of A *be* $\sigma_1 \geq \sigma_2 \geq \cdots \geq \sigma_n$ *and those of* A_1 *be* $\tau_1 \geq \tau_2 \cdots \geq \tau_n$. *Then*

$$\sigma_i \geq \tau_i, \quad i = 1, \ldots, n.$$

Proof. The squares of the singular values of A are the eigenvalues of $A^H A$, and the squares of the singular values of A_1 are the eigenvalues of $A_1^H A_1$. But $A^H A = A_1^H A_1 + A_2^H A_2$. Since $A_2^H A_2$ is positive semi-definite, the eigenvalues of $A^H A$ taken in descending order are greater than or equal to the corresponding eigenvalues of $A_1^H A_1$. ∎

The second inequality concerns the product of two matrices.

Theorem 4.5. *Let $B \in \mathbb{C}^{m \times n}$ have singular values $\sigma_1 \geq \sigma_2 \geq \cdots \geq \sigma_n$ and let $C = AB$ have singular values $\tau_1 \geq \tau_2 \geq \cdots \geq \tau_n$. Then*

$$\tau_i \leq \sigma_i \|A\|_2, \qquad i = 1, \ldots, n.$$

Proof. We will establish the inequality by comparing the eigenvalues of $C^H C$ with those of $\|A\|_2^2 B^H B$. Let $A^H A = Q \Lambda^2 Q^H$ be the spectral decomposition of $A^H A$. Let $D = Q(\|A\|_2^2 I - \Lambda^2) Q^H$. Then D is positive semi-definite and $A^H A + D = \|A\|_2^2 I$. Now

$$\|A\|_2^2 B^H B = B^H (A^H A + D) B = C^H C + B^H D B.$$

Since $B^H D B$ is positive semi-definite, the eigenvalues of $\|A\|_2^2 B^H B$ are not less than the corresponding eigenvalues of $C^H C$. ∎

Both of the above theorems have trivial variants and corollaries, which we will use freely in the sequel.

Notes and References

The singular value decomposition was discovered independently by Beltrami [23, 1873] and Jordan [125, 1874]. Both cast their derivations in terms of simplifying a real bilinear form $\phi(x, y) = y^H A x$ by orthogonal transformations of the variables x and y. Beltrami's derivation is close to the one in the text. Jordan's is something else again. He begins by asking for the largest value of ϕ when $\|x\|_2 = \|y_2\| = 1$ and shows that the vector $(x^H \ y^H)^H$ is an eigenvector of the matrix

$$\begin{pmatrix} 0 & A^T \\ A & 0 \end{pmatrix}$$

whose eigenvalue is our σ_1. He then transforms A to the form

$$\begin{pmatrix} \sigma_1 & 0 \\ 0 & A_* \end{pmatrix}$$

4. THE SINGULAR VALUE DECOMPOSITION

and proceeds by induction à la Schur. Thus, Jordan can claim precedence for both variational characterizations in matrix analysis and the recursive definition of matrix decompositions.

The decomposition has been frequently rediscovered, first by Sylvester [235, 236, 1889, 1990]. Autonne [8, 1913] generalized it to complex matrices, and Eckart and Young [63, 1936] to rectangular matrices, where they used it to approximate a matrix by another of lower rank (see Theorem III.4.18).

The use of the word "singular" in this connection apparently comes from the literature of integral equations. The story begins with Schmidt [192, 1907], who expanded the kernal of an integral operator in the form

$$\kappa(x,y) = \sum_{i=1}^{\infty} \frac{1}{\lambda_i} \mu_i(x)\nu_i(y),$$

where the μ_i and ν_i are eigenfunctions of the iterated kernals

$$\int \kappa(x,t)\kappa(y,t)\,dt \quad \text{and} \quad \int \kappa(t,x)\kappa(t,y)\,dt.$$

This is equivalent to the matrix representation

$$K = U\Sigma V^{\mathrm{T}} = \sum_{i=1}^{n} \sigma_i u_i v_i^{\mathrm{T}}$$

A little later Bateman [12, 1908] refers to numbers that are essentially the reciprocals of the eigenvalues of κ as as singular values, but does not relate them to the numbers introduced by Schmidt (he will continue this usage through 1922). Picard [181, 1910] notes that for symmetric kernals Schmidt's λ_i are real, and in this case (but not the general case) he calls them singular values. By 1937, Smithes [198] was calling Schmidt's numbers singular values. Exactly when and how the usage changed remains to be determined.

Theorem 4.2 is implicit in Jordan's derivation of the singular value decomposition; however, Wielandt seems to be responsible for its widespread use today (e.g., see [71, p.113]).

The singular value decomposition is closely related to the analysis, due to Hotelling [119, 1933], of a multivariate random variable into principal components. Specifically if the rows of A form a centered sample of a normally distributed random vector a, then V estimates an orthogonal transformation of a into a vector whose components are uncorrelated.

Exercises

Throughout these exercises A will be an $m \times n$ ($m \geq n$) matrix with the singular value decomposition (4.1).

1. Show that if A is square, $|\det(A)| = \prod_{i=1}^{n} \sigma_i$.

2. (Autonne [7, 1902]). Prove that any square matrix A has a POLAR DECOMPOSITION $A = HQ$, where Q is unitary and H is positive semi-definite. Moreover, if A is nonsingular, then H is positive definite and the polar decomposition of A is unique.

3. What is the singular value decomposition of a Hermitian matrix? Of a unitary matrix?

4. Let A be nonsingular. Show that $\inf_2(A) = \|A^{-1}\|_2^{-1}$.

5. Show that
$$\|A\|_2 = \max_{\|y\|_2=1} \max_{\|x\|_2=1} |y^H A x|.$$

6. Let $B \in \mathbf{C}^{m \times n}$ have singular values $\sigma_1 \geq \sigma_2 \geq \cdots \geq \sigma_n$ and let $C = AB$ have singular values $\tau_1 \geq \tau_2 \geq \cdots \geq \tau_n$. Then
$$\tau_i \geq \sigma_i \inf_2(A), \quad i = 1, \ldots, n.$$

[Hint: Assume without loss of generality that A is square and nonsingular and apply Theorem 4.5 to $A^{-1}C$.]

7. Let the eigenvalues of a square matrix A be ordered so that $|\lambda_1| \geq \cdots \geq |\lambda_n|$. Show that $|\lambda_1| \leq \sigma_1$.

8. The matrix W_n is illustrated below for $n = 5$:
$$W_5 = \begin{pmatrix} 1 & -1 & -1 & -1 & -1 \\ 0 & 1 & -1 & -1 & -1 \\ 0 & 0 & 1 & -1 & -1 \\ 0 & 0 & 0 & 1 & -1 \\ 0 & 0 & 0 & 0 & 1 \end{pmatrix}.$$

Show that $\inf_2(W_n) = O(2^{-n})$. What are the eigenvalues of W_n? What is its determinant?

5. Pairs of Subspaces

The problem to be treated in this section is that of comparing two subspaces of $\mathbf{C}^n$. Our tool will be the CS DECOMPOSITION (cosine–sine decomposition) of a partitioned unitary matrix. This decomposition allows us to define canonical angles for pairs of subspaces in such a way that as the largest canonical angle approaches zero the subspaces approach one another. This in turn leads to some useful results on the singular values of products and differences of projections.

5.1. The CS Decomposition

We begin by establishing the existence of the CS decomposition.

Theorem 5.1. *Let the unitary matrix* $W \in \mathbf{C}^{n \times n}$ *be partitioned in the form*

$$W = \begin{matrix} l \\ n-l \end{matrix} \begin{pmatrix} W_{11} & W_{12} \\ W_{21} & W_{22} \end{pmatrix},$$

where $2l \leq n$. *Then there are unitary matrices* $U = \mathrm{diag}(U_{11}, U_{22})$ *and* $V = \mathrm{diag}(V_{11}, V_{22})$ *with* $U_{11}, V_{11} \in \mathbf{C}^{l \times l}$ *such that*

$$U^{\mathrm{H}} W V = \begin{matrix} l \\ l \\ n-2l \end{matrix} \begin{pmatrix} \Gamma & -\Sigma & 0 \\ \Sigma & \Gamma & 0 \\ 0 & 0 & I \end{pmatrix}, \tag{5.1}$$

where

$$\Gamma = \mathrm{diag}(\gamma_1, \ldots, \gamma_l) \geq 0,$$
$$\Sigma = \mathrm{diag}(\sigma_1, \ldots, \sigma_l) \geq 0,$$

and

$$\Gamma^2 + \Sigma^2 = I.$$

Proof. The proof, which is long and tedious, is included mainly for completeness. Let

$$U_{11}^{\mathrm{H}} W_{11} V_{11} = \Gamma$$

be the singular value decomposition of W_{11}, and suppose that

$$\Gamma = \mathrm{diag}(\Gamma_1, I_{l-k}),$$

where the diagonal elements of Γ_1 satisfy

$$0 \le \gamma_1 \le \cdots \le \gamma_k < 1$$

(n.b., since W is orthogonal the singular values of W_{11} cannot be greater than one). Clearly the matrix

$$\begin{pmatrix} W_{11} \\ W_{21} \end{pmatrix} V_{11}$$

has orthonormal columns. Hence

$$I = \left[\begin{pmatrix} W_{11} \\ W_{21} \end{pmatrix} V_{11}\right]^{\mathrm{H}} \left[\begin{pmatrix} W_{11} \\ W_{21} \end{pmatrix} V_{11}\right] = \Gamma^2 + (W_{21}V_{11})^{\mathrm{H}}(W_{21}V_{11}),$$

that is,

$$(W_{21}V_{11})^{\mathrm{H}}(W_{21}V_{11}) = \mathrm{diag}(I - \Gamma_1^2, 0_{l-k}).$$

This means that the columns of $W_{21}V_{11}$ are orthogonal with its last $l-k$ columns being zero. Thus there is a unitary matrix $\hat{U}_{22} \in \mathbf{C}^{(n-l)\times(n-l)}$ such that

$$\hat{U}_{22} W_{21} V_{11} = \begin{matrix} l \\ n-2l \end{matrix} \begin{pmatrix} \Sigma \\ 0 \end{pmatrix},$$

where

$$\Sigma = \mathrm{diag} \begin{pmatrix} \overset{k}{\Sigma_1} & \overset{l-k}{0} \end{pmatrix}, \qquad (5.2)$$

and $\Sigma_1 = \mathrm{diag}(\sigma_1, \ldots, \sigma_k)$ has positive diagonal elements. Since

$$\mathrm{diag}(U_{11}, \hat{U}_{22})^{\mathrm{H}} \begin{pmatrix} W_{11} \\ W_{21} \end{pmatrix} V_{11} = \begin{pmatrix} \Gamma \\ \Sigma \\ 0 \end{pmatrix}$$

has orthogonal columns, we must have

$$\gamma_i^2 + \sigma_i^2 = 1, \qquad i = 1, \ldots, l. \qquad (5.3)$$

5. Pairs of Subspaces

In a similar manner we may determine a unitary matrix $V_{22} \in \mathbf{C}^{(n-l)\times(n-l)}$ such that

$$U_{11}^{\mathrm{H}} W_{12} V_{22} = (T\ 0),$$

where $T = \mathrm{diag}(\tau_1,\ldots,\tau_l)$ with $\tau_i \leq 0 (i = 1,\ldots,l)$. Since

$$U_{11}^{\mathrm{H}}(W_{11}\ W_{12})\mathrm{diag}(V_{11}, V_{22}) = (\Gamma\ T\ 0)$$

has orthogonal rows, we must have $\gamma_i^2 + \tau_i^2 = 1$ ($i = 1,\ldots,l$), and it follows from (5.2) and (5.3) that $T = -\Sigma$.

Set $\hat{U} = \mathrm{diag}(U_{11}, \hat{U}_{22})$ and $V = \mathrm{diag}(V_{11}, V_{22})$. Then the foregoing shows that the matrix $X = \hat{U}^{\mathrm{H}} W V$ can be partitioned in the form

$$X = \begin{array}{c} k \\ l-k \\ k \\ l-k \\ n-2l \end{array} \begin{pmatrix} \overset{k}{\Gamma_1} & \overset{l-k}{0} & \overset{k}{-\Sigma_1} & \overset{l-k}{0} & \overset{n-2l}{0} \\ 0 & I & 0 & 0 & 0 \\ \Sigma_1 & 0 & X_{33} & X_{34} & X_{35} \\ 0 & 0 & X_{43} & X_{44} & X_{45} \\ 0 & 0 & X_{53} & X_{54} & X_{55} \end{pmatrix}.$$

Since X is unitary and Σ_1 has positive diagonal elements, we have $X_{33} = \Gamma_1$. Moreover, X_{34}, X_{35}, X_{43}, and X_{53} are zero. Therefore

$$X = \begin{pmatrix} \Gamma_1 & 0 & -\Sigma_1 & 0 & 0 \\ 0 & I & 0 & 0 & 0 \\ \Sigma_1 & 0 & \Gamma_1 & 0 & 0 \\ 0 & 0 & 0 & X_{44} & X_{45} \\ 0 & 0 & 0 & X_{54} & X_{55} \end{pmatrix},$$

where

$$\begin{pmatrix} X_{44} & X_{45} \\ X_{54} & X_{55} \end{pmatrix}$$

is unitary.

Set

$$U_{33} = \begin{pmatrix} X_{44} & X_{45} \\ X_{54} & X_{55} \end{pmatrix} \in \mathbf{C}^{(n-l-k)\times(n-l-k)},$$

then we have

$$\mathrm{diag}(I^{(l+k)}, U_{33}^{\mathrm{H}})X = \begin{pmatrix} \Gamma_1 & 0 & -\Sigma_1 & 0 & 0 \\ 0 & I & 0 & 0 & 0 \\ \Sigma_1 & 0 & \Gamma_1 & 0 & 0 \\ 0 & 0 & 0 & I & 0 \\ 0 & 0 & 0 & 0 & I \end{pmatrix} = \begin{matrix} l \\ l \\ n-2l \end{matrix} \begin{pmatrix} \Gamma & -\Sigma & 0 \\ \Sigma & \Gamma & 0 \\ 0 & 0 & I \end{pmatrix}.$$

Observe that

$$\mathrm{diag}(I^{(l+k)}, U_{33}^{\mathrm{H}})X = \mathrm{diag}(I^{(l+k)}, U_{33}^{\mathrm{H}})\hat{U}^{\mathrm{H}}WV.$$

Hence, if we set

$$\begin{aligned} U &= \hat{U}\mathrm{diag}(I^{(l+k)}, U_{33}) \\ &= \mathrm{diag}(U_{11}, \hat{U}_{22})\mathrm{diag}(I^{(l)}, \mathrm{diag}(I^{(k)}, U_{33})) \\ &= \mathrm{diag}(U_{11}, \hat{U}_{22}\mathrm{diag}(I^{(k)}, U_{33})) \\ &= \mathrm{diag}(U_{11}, U_{22}), \end{aligned}$$

then $U^{\mathrm{H}}WV$ has the form (5.1), where U and V are diagonal block unitary matrices. ∎

5.2. Pairs of Subspaces

Armed with the CS decomposition, we may now attack the problem of determining how two subspaces are situated with respect to one another. The following theorem shows that there are natural bases for two subspaces that exhibit their relation.

Theorem 5.2. Let $X_1, Y_1 \in \mathbf{C}^{n \times l}$ with $X_1^{\mathrm{H}} X_1 = I$ and $Y_1^{\mathrm{H}} Y_1 = I$. If $2l \leq n$, there are unitary matrices Q, U_{11} and V_{11} such that

$$QX_1 U_{11} = \begin{matrix} l \\ l \\ n-2l \end{matrix} \begin{pmatrix} I \\ 0 \\ 0 \end{pmatrix}, \qquad (5.4)$$

$$QY_1 V_{11} = \begin{matrix} l \\ l \\ n-2l \end{matrix} \begin{pmatrix} \Gamma \\ \Sigma \\ 0 \end{pmatrix}, \qquad (5.5)$$

5. Pairs of Subspaces

where
$$\Gamma = \operatorname{diag}(\gamma_1, \ldots, \gamma_l) \text{ and } \Sigma = \operatorname{diag}(\sigma_1, \ldots, \sigma_l)$$
satisfy
$$0 \leq \gamma_1 \leq \ldots \leq \gamma_l,$$
$$\sigma_1 \geq \ldots \geq \sigma_l \geq 0, \qquad (5.6)$$
$$\gamma_i^2 + \sigma_i^2 = 1, \quad i = 1, \ldots, l.$$

On the other hand, if $2l > n$, then Q, U_{11}, and V_{11} may be chosen so that

$$QX_1 U_{11} = \begin{array}{c} n-l \\ 2l-n \\ n-l \end{array} \overset{\begin{array}{cc} n-l & 2l-n \end{array}}{\left(\begin{array}{cc} I & 0 \\ 0 & I \\ 0 & 0 \end{array} \right)}, \qquad (5.7)$$

$$QY_1 V_{11} = \begin{array}{c} n-l \\ 2l-n \\ n-l \end{array} \overset{\begin{array}{cc} n-l & 2l-n \end{array}}{\left(\begin{array}{cc} \Gamma & 0 \\ 0 & I \\ \Sigma & 0 \end{array} \right)}, \qquad (5.8)$$

where
$$\Gamma = \operatorname{diag}(\gamma_1, \ldots, \gamma_{n-l}) \text{ and } \Sigma = \operatorname{diag}(\sigma_1, \ldots, \sigma_{n-l})$$
satisfy
$$0 \leq \gamma_1 \leq \ldots \leq \gamma_{n-l},$$
$$\sigma_1 \geq \ldots \geq \sigma_{n-l} \geq 0,$$
$$\gamma_i^2 + \sigma_i^2 = 1, \quad i = 1, \ldots, n-l.$$

Proof. We will prove the theorem for $2l \leq n$, leaving the other case as an exercise. Let X_2 and Y_2 be chosen so that $X = (X_1 \ X_2)$ and $Y = (Y_1 \ Y_2)$ are unitary. Let

$$W = X^H Y = \begin{array}{c} l \\ n-l \end{array} \overset{\begin{array}{cc} l & n-l \end{array}}{\left(\begin{array}{cc} W_{11} & W_{12} \\ W_{21} & W_{22} \end{array} \right)}.$$

By Theorem 5.1, there are unitary matrices U_{11}, U_{22}, V_{11} and V_{22} such that the relation (5.1) holds. Therefore, if we set

$$\hat{X}_i = X_i U_{ii}, \quad \hat{Y}_i = Y_i V_{ii}, \qquad i = 1, 2,$$

and
$$\hat{X} = (\hat{X}_1\ \hat{X}_2), \quad \hat{Y} = (\hat{Y}_1\ \hat{Y}_2),$$
then
$$\hat{X}^H\hat{Y} = \begin{matrix} \\ l \\ l \\ n-2l \end{matrix} \begin{pmatrix} \overset{l}{\Gamma} & \overset{l}{-\Sigma} & \overset{n-2l}{0} \\ \Sigma & \Gamma & 0 \\ 0 & 0 & I \end{pmatrix}.$$

Moreover, by permuting the columns of U_{ii} and V_{ii} we can insure that (5.6) holds. Equations (5.4) and (5.5) now follow on setting $Q = \hat{X}^H$. ∎

Theorem 5.2 has the following geometric significance. Let $\mathcal{X}_1$ and $\mathcal{Y}_1$ be l-dimensional subspaces of $\mathbf{C}^n$. If $2l \leq n$ we can transform $\mathbf{C}^n$ by unitary transformation Q so that the columns of the matrices

$$\begin{pmatrix} I \\ 0 \\ 0 \end{pmatrix} \text{ and } \begin{pmatrix} \Gamma \\ \Sigma \\ 0 \end{pmatrix}$$

form orthogonal bases of $Q\mathcal{X}_1$ and $Q\mathcal{Y}_1$. The space spanned by the columns for which $\gamma_i = 0$ is the subspace of $Q\mathcal{Y}_1$ that is orthogonal to $Q\mathcal{X}_1$.

When $2l > n$, the columns of the matrices

$$\begin{pmatrix} I & 0 \\ 0 & I \\ 0 & 0 \end{pmatrix} \text{ and } \begin{pmatrix} \Gamma & 0 \\ 0 & I \\ \Sigma & 0 \end{pmatrix}$$

form orthogonal bases for $Q\mathcal{X}_1$ and $Q\mathcal{Y}_1$. As above, the space spanned by the columns for which $\gamma_i = 0$ is the subspace of $Q\mathcal{Y}_1$ that is orthogonal to $Q\mathcal{X}_1$. The last $2l - n$ columns represent the smallest possible intersection of $Q\mathcal{X}_1$ and $Q\mathcal{Y}_1$.

Since the numbers σ_i and γ_i satisfy $\sigma_i^2 + \gamma_i^2 = 1$, they can be regarded as sines and cosines of angles between the bases. Moreover, $\mathcal{X} = \mathcal{Y}$ if and only if $\Sigma = 0$. This means that the size of Σ is a measure of how $\mathcal{X}$ and $\mathcal{Y}$ differ. Since Σ depends only on the subspaces $\mathcal{X}$ and $\mathcal{Y}$, we may make the following definition.

5. PAIRS OF SUBSPACES

Definition 5.3. *Let $\mathcal{X}$ and $\mathcal{Y}$ be subspaces of the same dimension. The* CANONICAL ANGLES *between $\mathcal{X}$ and $\mathcal{Y}$ are the diagonals of the matrix*

$$\Theta(\mathcal{X}, \mathcal{Y}) \stackrel{\text{def}}{=} \sin^{-1} \Sigma,$$

where Σ is the matrix of Theorem 5.2.

The following corollary of Theorem 5.2 shows how to compute the canonical angles. Its proof is left as an exercise.

Corollary 5.4. *Let $\mathcal{X}$ and $\mathcal{Y}$ be subspaces of $\mathbf{C}^n$ of the same dimension. Let the columns of $X_\perp$ form an orthonormal basis for $\mathcal{X}_\perp$ and the columns of Y form an orthonormal basis for $\mathcal{Y}$. Then the nonzero singular values of $X_\perp^\mathrm{H} Y$ are the sines of the nonzero canonical angles between $\mathcal{X}$ and $\mathcal{Y}$.*

5.3. Pairs of Projections

A difficulty with the results of the last subsection is that they are cast in terms of explicit bases for the subspaces involved. In many instances it is more convenient to represent subspaces by their orthogonal projections. As one would expect, pairs of projections are closely related to the canonical angles between their subspaces.

Theorem 5.5. *Let $\mathcal{X}$ and $\mathcal{Y}$ be l-dimensional subspaces of $\mathbf{C}^n$. Let*

$$k = \begin{cases} l, & \text{if } 2l \leq n \\ n - l, & \text{if } 2l > n \end{cases}$$

Let $\sigma_1 \geq \sigma_2 \geq \cdots \geq \sigma_k$ be the sines of the canonical angles between $\mathcal{X}$ and $\mathcal{Y}$. Then:

1. *The singular values of $P_\mathcal{X}(I - P_\mathcal{Y})$ are*

$$\sigma_1, \sigma_2, \ldots, \sigma_k, 0, \ldots, 0.$$

2. *The singular values of $P_\mathcal{X} - P_\mathcal{Y}$ are*

$$\sigma_1, \sigma_1, \sigma_2, \sigma_2, \ldots, \sigma_k, \sigma_k, 0, \ldots, 0.$$

Proof. We will prove the theorem for the case $2l \leq n$, leaving the other case as an exercise. If in Theorem 5.2 we let X_1 and Y_1 be bases for $\mathcal{X}$ and $\mathcal{Y}$, then $P_{\mathcal{X}} = X_1 X_1^{\mathrm{H}}$ and $P_{\mathcal{Y}} = Y_1 Y_1^{\mathrm{H}}$. It follows from (5.4) and (5.5) that

$$QP_{\mathcal{X}}(I - P_{\mathcal{Y}})Q^{\mathrm{H}} = \begin{pmatrix} I & 0 & 0 \\ 0 & 0 & 0 \\ 0 & 0 & 0 \end{pmatrix} \begin{pmatrix} I - \Gamma^2 & -\Gamma\Sigma & 0 \\ -\Sigma\Gamma & I - \Sigma^2 & 0 \\ 0 & 0 & I \end{pmatrix}$$

$$= \begin{pmatrix} I - \Gamma^2 & -\Gamma\Sigma & 0 \\ 0 & 0 & 0 \\ 0 & 0 & 0 \end{pmatrix}$$

$$= \begin{pmatrix} \Sigma^2 & -\Gamma\Sigma & 0 \\ 0 & 0 & 0 \\ 0 & 0 & 0 \end{pmatrix}$$

$$= \begin{pmatrix} \Sigma \\ 0 \\ 0 \end{pmatrix} (\Sigma \quad -\Gamma \quad 0).$$

Since the rows of $(\Sigma \quad -\Gamma \quad 0)$ are orthonormal, the singular values of $QP_{\mathcal{X}}(I - P_{\mathcal{Y}})Q^{\mathrm{H}}$ are $\sigma_1, \sigma_2, \ldots, \sigma_l, 0, \ldots, 0$. This proves the first assertion.

In a similar manner we find that

$$Q(P_{\mathcal{X}} - P_{\mathcal{Y}})Q^{\mathrm{H}} = \begin{pmatrix} \Sigma^2 & -\Gamma\Sigma & 0 \\ -\Sigma\Gamma & -\Sigma^2 & 0 \\ 0 & 0 & 0 \end{pmatrix}.$$

Since Σ and Γ are diagonal, the nonzero singular values of $P_{\mathcal{X}} - P_{\mathcal{Y}}$ are just the singular values of the 2×2 matrices

$$S_i = \begin{pmatrix} \sigma_i^2 & -\gamma_i \sigma_i \\ -\sigma_i \gamma_i & -\sigma_i^2 \end{pmatrix}, \qquad i = 1, \ldots, l.$$

But the singular values of S_i are just σ_i twice over. ∎

5. Pairs of Subspaces

Notes and References

The notion of canonical angles between subspaces goes back to Jordan [126, 1875], and has been frequently rediscovered; e.g., in the very readable papers by Afriat [2, 3]. Davis and Kahan [53, 1970] give a unified treatment of the subject in Hilbert space and a bibliographical note with further references. Wedin [261] gives a lucid survey with an emphasis on geometry.

The CS decomposition was introduced by Stewart [205, 1977], although it is implicit in the paper of Davis and Kahan just cited. Paige and Saunders [174, 1981] consider the case where the diagonal blocks are not square.

It is not a trivial matter to compute canonical angles or the CS decomposition. Algorithms are given in [37, 209].

Exercises

IN THE FOLLOWING EXERCISES $\mathcal{X}$ AND $\mathcal{Y}$ ARE SUBSPACES OF $\mathbf{C}^n$ OF DIMENSION l, WHERE $2l \le n$. THE CANONICAL ANGLES OF $\mathcal{X}$ AND $\mathcal{Y}$ ARE $\theta_1 \ge \cdots \ge \theta_l$. BY THE ANGLE BETWEEN TWO NONZERO VECTORS x AND y WE MEAN

$$\angle(x,y) \stackrel{\text{def}}{=} \cos^{-1} \frac{|y^{\text{H}}x|}{\|x\|_2 \|y\|_2}.$$

1. Let $\|x\|_2 = \|y\|_2 = 1$ and $y^{\text{H}}x \ge 0$. Show that

$$\|y - x\|_2 = 2\sin\frac{\angle(x,y)}{2}.$$

2. Let the columns of X and Y form orthonormal bases for $\mathcal{X}$ and $\mathcal{Y}$. Show that the singular values of $Y^{\text{H}}X$ are the cosines of the canonical angles between $\mathcal{X}$ and $\mathcal{Y}$.

3. Show that

$$\cos(\theta_1) = \min_{\substack{x \in \mathcal{X} \\ x \ne 0}} \max_{\substack{y \in \mathcal{Y} \\ y \ne 0}} \cos \angle(x,y).$$

Equivalently,

$$\theta_1 = \max_{\substack{x \in \mathcal{X} \\ x \ne 0}} \min_{\substack{y \in \mathcal{Y} \\ y \ne 0}} \angle(x,y).$$

4. Let $v^\mathrm{H} u = 1$. Show that there are matrices U and V such that $U^\mathrm{H} U = \mathrm{diag}(\|u\|_2 \|v\|_2, I)$, $V^\mathrm{H} V = I$, and

$$(u\ U)^{-1} = \begin{pmatrix} v^\mathrm{H} \\ V^\mathrm{H} \end{pmatrix}.$$

5. (Halmos [101]). A matrix A of order n is a contraction in the 2-norm if $\|A\|_2 \leq 1$. Show that if A is a contraction, there exist matrices B, C, and D of order n such that

$$\begin{pmatrix} A & B \\ C & D \end{pmatrix}$$

is unitary.

THE FOLLOWING TWO EXERCISES DEFINE A UNITARY TRANSFORMATION — THE DIRECT ROTATION — THAT MAPS $\mathcal{X}$ ONTO $\mathcal{Y}$, AND THEY SHOW THAT IT IS IN SOME SENSE OPTIMAL. FOR FURTHER DETAILS SEE THE PAPER BY DAVIS AND KAHAN [53].

6. In Theorem 5.2 let the columns of X and Y form canonical bases for $\mathcal{X}$ and $\mathcal{Y}$, so that U_{11} and V_{11} are the identity. Let

$$U = Q^\mathrm{H} \begin{pmatrix} \Gamma & -\Sigma & 0 \\ \Sigma & \Gamma & 0 \\ 0 & 0 & I \end{pmatrix} Q.$$

Show that U is a unitary matrix such that $U\mathcal{X} = \mathcal{Y}$. Moreover, $\|I - U\|_2 = 2\sin\frac{\theta_1}{2}$.

7. Let V be an orthogonal matrix such that $V\mathcal{X} = \mathcal{Y}$. Show that $\|I - V\|_2 \geq 2\sin\frac{\theta_1}{2}$.

———⋄———

THE FOLLOWING EXERCISE DERIVES A SPECIAL CASE OF THE GENERALIZED SINGULAR VALUE DECOMPOSITION INTRODUCED BY VAN LOAN [247, 248]. THE FORM GIVEN HERE IS DUE TO PAIGE AND SAUNDERS [174].

8. Let

$$X = \begin{pmatrix} X_1 \\ X_2 \end{pmatrix},$$

5. PAIRS OF SUBSPACES

where X_1 and X_2 are square and X has full rank. Show that there are unitary matrices U_1 and U_2 and a nonsingular matrix, such that

$$\mathrm{diag}(U_1, U_2)^{\mathrm{H}} X = \begin{pmatrix} \Gamma \\ \Sigma \end{pmatrix} S,$$

where S is nonsingular and Γ and Σ are diagonal with $\Gamma^2 + \Sigma^2 = 1$. [Hint. Consider the CS decomposition of the orthogonal part of the QR factorization of X.]

———◇———

Chapter II
Norms and Metrics

The goal of matrix perturbation analysis is to predict or bound the changes in objects associated with a matrix when the elements of the matrix change. For example we might ask how far the eigenvalues and eigenvectors of a matrix A will change when A is replaced by a nearby matrix $A + E$. A prerequisite for answering such questions is to make precise terms like "how far" and "nearby."

For eigenvalues, which are complex numbers, the absolute value function $|\cdot| : \mathbf{C} \to \mathbf{R}$ provides a natural notion of size and distance. The usefulness of this function in analysis depends on three properties:

1. $\xi \neq 0 \implies |\xi| > 0$ (definiteness),

2. $|\alpha\xi| = |\alpha||\xi|$ (homogeneity),

3. $|\xi + \eta| \leq |\xi| + |\eta|$ (the triangle inequality).

These three properties make the function $\rho(\xi, \eta) = |\xi - \eta|$ a metric over $\mathbf{C}$, which endows it with a topology, so that we may speak of limits and continuity.

A vector norm is a generalization of the absolute value; i.e., a definite, homogeneous function on $\mathbf{C}^n$ that satisfies the triangle inequality. The first section of this chapter is devoted to the study of the elementary properties of vector norms. It is also possible to define a matrix norm to be a definite, homogeneous function on $\mathbf{C}^{m \times n}$ that satisfies the triangle inequality. However, such a definition ignores the fact that matrices can be multiplied, and it is usually augmented to take this operation into account. Section 2 is devoted to the study of matrix norms, and Section 3 is devoted to the study of a particular class of matrix norms — the UNITARILY INVARIANT NORMS — which interact nicely with the Euclidean geometry of $\mathbf{C}^n$. In some applications, the object that is perturbed is not a vector or a matrix, but a subspace. Accordingly, the chapter concludes with a discussion of measures of distance or metrics for subspaces.

This chapter, like the first, is preliminary. Unlike the first, it contains material requiring lengthy, deep proofs. The following comments are for the reader who wants to get through the chapter quickly and move on to the perturbation theory itself.

The first two sections are an elementary introduction to vector and matrix norms and may be skimmed by anyone familiar with the subject. The third section contains the most challenging material in the chapter. Theorem 3.6, which characterizes unitarily invariant norms, is the key result of the first subsection. However, its supporting lemmas and proof are not required elsewhere and may be skipped. Of the material in the last subsection only Birkhoff's theorem (Theorem 3.16) and Fan's theorem (Theorem 3.17) will be used in the sequel. The material in Section 4 is used only in Chapters V and VI. The reader may find the summary (4.11) useful in sorting out the various metrics introduced in this subsection.

1. Vector Norms

1.1. Definition

As we mentioned in the introduction, a norm on $\mathbf{C}^n$ is a generalization of the absolute value of a complex number.

Definition 1.1. *A function* $\nu : \mathbf{C}^n \to \mathbf{R}$ *is said to be a* NORM *on* $\mathbf{C}^n$ *(or a* VECTOR NORM*) if* ν *satisfies the following conditions*

1. $x \neq 0 \implies \nu(x) > 0$,
2. $\nu(\alpha x) = |\alpha|\nu(x)$,

1. VECTOR NORMS

3. $\nu(x+y) \leq \nu(x) + \nu(y)$.

Three important properties follow immediately from Definition 1.1. For any norm ν,

1. $\nu(0) = 0$,
2. $\nu(-x) = \nu(x)$,
3. $|\nu(x) - \nu(y)| \leq \nu(x-y)$.

1.2. Examples

There are an infinite number of norms on $\mathbf{C}^n$. However, three of these — the 1, 2, and ∞-norms — are most commonly used in practice. The 1-norm is defined by

$$\|x\|_1 \stackrel{\text{def}}{=} \sum_{i=1}^n |\xi_i|\,;$$

the 2-norm by

$$\|x\|_2 \stackrel{\text{def}}{=} \sqrt{\sum_{i=1}^n |\xi_i|^2}\,;$$

and the ∞-norm by

$$\|x\|_\infty \stackrel{\text{def}}{=} \max_{1 \leq i \leq n} |\xi_i|.$$

The norms $\|\cdot\|_1$ and $\|\cdot\|_2$ are special cases of the HÖLDER NORMS (or p-NORMS) defined by

$$\|x\|_p = (\sum_{i=1}^n |\xi_i|^p)^{\frac{1}{p}}, \qquad 1 \leq p < \infty. \tag{1.1}$$

(For a proof that $\|\cdot\|_p$ is indeed a norm, see Exercises 1.6–1.8.) Since

$$\|x\|_\infty = \lim_{p \to \infty} \|x\|_p, \tag{1.2}$$

the norm $\|\cdot\|_\infty$ is also regarded as a Hölder norm.

The 2-norm has the useful characterization

$$\|x\|_2^2 = x^{\mathrm{H}} x,$$

from which it immediately follows that

the 2-norm is UNITARILY INVARIANT; i.e.,

$$U \text{ unitary} \implies \|Ux\|_2 = \|x\|_2.$$

The Hölder norms have the property that $\|x\|_p = \|\,|x|\,\|_p$. They also have the property that $\|x\|_p \leq \|y\|_p$ whenever $|x| \leq |y|$. These properties will play an important role in the theory of unitarily invariant norms as well as in the structured perturbation theory of linear systems and least squares problems, and it is worth establishing that they are equivalent. We begin by introducing some terminology that we will use later.

Definition 1.2. *A vector norm ν on $\mathbf{C}^n$ is* ABSOLUTE *if $\nu(|x|) = \nu(x)$ for all $x \in \mathbf{C}^n$.*

Theorem 1.3. *A vector norm ν is absolute if and only if*

$$|x| \leq |y| \implies \nu(x) \leq \nu(y). \tag{1.3}$$

Proof. Suppose that ν is absolute. Note that this implies that if $|D| = I$ then $\nu(Dx) = \nu(x)$.

Now let $|x| \leq |y|$. It suffices to prove (1.3) for the case where the first component of $|x|$ is less than or equal to the first component of $|y|$ and the other components of x and y are equal in absolute value. By multiplying x and y by suitable diagonal matrices, we may assume without loss of generality that the first components of x and y are nonnegative and the others are equal.

Let $x = (\rho\eta_1, \eta_2, \ldots, \eta_n)^\mathrm{T}$, where $0 \leq \rho \leq 1$. If we set $\hat{y} = (-\eta_1, \eta_2, \ldots, \eta_n)^\mathrm{T}$, then

$$x = \frac{1+\rho}{2}y + \frac{1-\rho}{2}\hat{y}.$$

Since $1 - \rho \geq 0$,

$$\begin{aligned}\nu(x) &= \nu\left(\tfrac{1+\rho}{2}y + \tfrac{1-\rho}{2}\hat{y}\right) \\ &\leq \nu\left(\tfrac{1+\rho}{2}y\right) + \nu\left(\tfrac{1-\rho}{2}\hat{y}\right) \\ &= \tfrac{1+\rho}{2}\nu(y) + \tfrac{1-\rho}{2}\nu(\hat{y}) \\ &= \tfrac{1+\rho}{2}\nu(|y|) + \tfrac{1-\rho}{2}\nu(|y|) \\ &= \nu(y).\end{aligned}$$

1. Vector Norms

The converse is left as an exercise. ∎

The following theorem exhibits a technique for constructing new norms from old. Its proof is left as an exercise.

Theorem 1.4. *Let μ be a norm on $\mathbf{C}^m$, and let $A \in \mathbf{C}^{m \times n}$ have rank n. Then the function ν defined by*

$$\nu(x) = \mu(Ax)$$

is a norm on $\mathbf{C}^n$.

Corollary 1.5. *Let A be positive definite. Then the function ν defined by*

$$\nu(x) = \sqrt{x^H A x}$$

is a norm.

Proof. We have

$$\nu(x) = \|A^{\frac{1}{2}} x\|_2.$$

Hence by Theorem 1.4, ν is a norm. ∎

Norms generated by a positive definite matrix A are called ELLIPTIC NORMS and are often written $\|\cdot\|_A$. It is worth noting that these norms bear the same relation to the "inner product" $y^H A x$ that the 2-norm bears to the usual inner product $y^H x$ (see Exercises 1.15–1.17).

1.3. Equivalence and Limits

In $\mathbf{R}^2$ and $\mathbf{R}^3$, the norm $\|y - x\|_2$ is the ordinary Euclidean distance between x and y. For this reason the 2-norm on $\mathbf{C}^n$ is also called the Euclidean norm. Moreover, the function

$$\rho_2(x, y) = \|y - x\|_2$$

is a METRIC on $\mathbf{C}^n$; that is, it satisfies the conditions

1. $\rho_2(x, y) \geq 0$,
2. $\rho_2(x, y) = 0 \iff x = y$,
3. $\rho_2(x, y) = \rho_2(y, x)$,

4. $\rho_2(x,z) \leq \rho_2(x,y) + \rho_2(y,z)$.

The metric ρ_2 defines a topology on $\mathbf{C}^n$; that is, it provides $\mathbf{C}^n$ with a collection of open sets from which the notions of closed sets, compactness, limits, and continuity may be defined.

If ν is a norm on $\mathbf{C}^n$, the function

$$\rho_\nu(x,y) = \nu(y-x)$$

is also a metric and hence defines a topology for $\mathbf{C}^n$. It turns out that this topology is the same as the topology generated by the Euclidean norm. This is a consequence of the equivalence of norms on finite dimensional spaces. We will prove this result in two steps, the first of which is of independent interest.

Lemma 1.6. *Let ν be a norm on $\mathbf{C}^n$. Then ν is continuous in the Euclidean metric.*

Proof. We need only prove that for any $\epsilon > 0$ there is a $\delta > 0$ such that $|\nu(y) - \nu(x)| < \epsilon$ whenever $\|y - x\|_2 < \delta$. From Definition 1.1,

$$\begin{aligned} |\nu(y) - \nu(x)| &\leq \nu(y-x) \\ &= \nu[\textstyle\sum_{i=1}^n (\eta_i - \xi_i)\mathbf{1}_i] \\ &\leq \textstyle\sum_{i=1}^n |\eta_i - \xi_i|\nu(\mathbf{1}_i) \\ &\leq \gamma \|y - x\|_2, \end{aligned} \qquad (1.4)$$

where $\gamma = \sqrt{\sum_{i=1}^n \nu^2(\mathbf{1}_i)} > 0$ is independent of x and y. If we take $\delta = \epsilon/\gamma$, then $|\nu(y) - \nu(x)| < \epsilon$ provided that $\|y - x\|_2 < \delta$. ∎

This lemma allows us to prove the equivalence of norms.

Theorem 1.7. *Let ν and μ be norms on $\mathbf{C}^n$. Then there are positive numbers π_1 and π_2 depending only on ν and μ such that*

$$\pi_1 \nu(x) \leq \mu(x) \leq \pi_2 \nu(x), \qquad \forall x \in \mathbf{C}^n.$$

Proof. Without loss of generality, we may take $x \neq 0$. The first step is to prove the theorem for the case $\nu(\cdot) = \|\cdot\|_2$. In this case, it follows from (1.4) with $y = 0$ that we may take $\pi_2 = \sqrt{\sum_{i=1}^n \mu^2(\mathbf{1}_i)}$. Thus we need only determine π_1.

1. VECTOR NORMS

The unit sphere $\mathcal{S}_2 = \{x : \|x\|_2 = 1\}$ is closed and bounded. Since μ is continuous (Lemma 1.4), μ achieves a minimal value π_1 on $\mathcal{S}_2$. Now let $y = x/\|x\|_2$. Then $y \in \mathcal{S}_2$, and

$$\mu(x) = \mu(\|x\|_2 y) = \|x\|_2 \mu(y) \geq \pi_1 \|x\|_2. \tag{1.5}$$

Now let μ and ν be arbitrary norms. From the foregoing, there are positive numbers $\sigma_1, \sigma_2, \tau_1,$ and τ_2 such that

$$\sigma_1 \|x\|_2 \leq \mu(x) \leq \sigma_2 \|x\|_2, \tag{1.6}$$

and

$$\tau_1 \|x\|_2 \leq \nu(x) \leq \tau_2 \|x\|_2. \tag{1.7}$$

From (1.6) and (1.7),

$$0 < \pi_1 \equiv \frac{\sigma_1}{\tau_2} \leq \frac{\mu(x)}{\nu(x)} \leq \frac{\sigma_2}{\tau_1} \equiv \pi_2. \quad\blacksquare$$

The following example shows the relations between the 1, 2, and ∞-norms.

Example 1.8. For all x

1. $\|x\|_2 \leq \|x\|_1 \leq \sqrt{n}\|x\|_2$,
2. $\frac{1}{\sqrt{n}}\|x\|_2 \leq \|x\|_\infty \leq \|x\|_2$,
3. $\|x\|_\infty \leq \|x\|_1 \leq n\|x\|_\infty$.

In all cases, equality can be attained.

Theorem 1.7 shows that all norms generate the same topology on $\mathbf{C}^n$, in the sense that for any sequence $x_1, x_2, \ldots$ we have $\lim_{k\to\infty} \mu(x - x_k) = 0$ if and only if $\lim_{k\to\infty} \nu(x - x_k) = 0$. Thus we can use any norm to define the notion of a limit of a sequence of vectors.

However, it is possible to define limits without using norms. A very natural definition is the following. Let $x_k = (\xi_1^{(k)}, \ldots, \xi_n^{(k)})^{\mathrm{T}}$ ($k = 1, 2, \ldots$). If

$$\lim_{k\to\infty} \xi_i^{(k)} = \xi_i, \qquad i = 1, \ldots, n,$$

then we say that the sequence of vectors $\{x_k\}$ has the limit $x = (\xi_1, \ldots, \xi_n)^{\mathrm{T}}$, or that x_k converges to x, and we write

$$\lim_{k \to \infty} x_k = x.$$

The following theorem shows that this component-wise convergence is the same as convergence in any norm.

Theorem 1.9. *For any vector norm ν,*

$$\lim_{k \to \infty} x_k = x \iff \lim_{k \to \infty} \nu(x_k - x) = 0. \qquad (1.8)$$

Proof. The result is trivial when $\nu(\cdot) = \|\cdot\|_\infty$. Hence by the equivalence of norms, the result holds for any norm. ∎

1.4. Linear Functionals and Dual Norms

A LINEAR FUNCTIONAL on $\mathbf{C}^n$ is a continuous function $\varphi : \mathbf{C}^n \to \mathbf{C}$ that is linear. The matrix representing such a function has dimensions $1 \times n$; i.e., it is a row vector. Thus to each linear functional φ on $\mathbf{C}^n$ there corresponds a unique vector y such that

$$\varphi(x) = y^{\mathrm{H}} x \qquad (1.9)$$

for all $x \in \mathbf{C}^n$.

There is a rather natural way in which a linear functional can be given a norm. In ordinary conversation we would say that a linear functional was big if it mapped vectors of ordinary size into large ones; and conversely we would say it was small if it mapped ordinary vectors into small ones. If we make the notions of "big," "small," and "ordinary" precise by choosing a specific vector norm ν, then we can define the "size" of the functional φ by

$$\nu^*(\varphi) = \max_{\nu(x)=1} |\varphi(x)|. \qquad (1.10)$$

Note that ν^* is well defined, since $|\varphi(x)|$ is continuous, and by the equivalence of norms the ν-sphere $\mathcal{S}_\nu = \{x : \nu(x) = 1\}$ is closed and bounded. It is easy to verify that ν^* is indeed a norm.

According to (1.9) every linear functional φ on $\mathbf{C}^n$ can be identified with a vector $y \in \mathbf{C}^n$. Hence (1.10) defines a new norm on $\mathbf{C}^n$. This justifies the following definition.

1. VECTOR NORMS

Definition 1.10. *Let ν be a norm on $\mathbf{C}^n$. Then the function ν^* defined by*
$$\nu^*(y) = \max_{\nu(x)=1} |y^{\mathrm{H}} x|$$
is the DUAL NORM *of ν on $\mathbf{C}^n$.*

The dual of the 2-norm is easily seen to be itself. The dual of the 1-norm is the ∞-norm, and vice versa. More generally if $p, q > 1$ satisfy
$$\frac{1}{p} + \frac{1}{q} = 1,$$
then the Hölder norms $\|\cdot\|_p$ and $\|\cdot\|_q$ are dual.

These examples suggest that we cannot generate new norms by taking the dual of a dual norm — we simply get back the original norm. We are going to prove that this is indeed true, but to do so we must first establish an important result on the extension of linear functionals.

A linear functional $\varphi : \mathcal{X} \to \mathbf{C}$ defined on a proper subspace $\mathcal{X}$ of $\mathbf{C}^n$ has a representation in the form (1.9). However, the vector y is not unique; for example, it can be replaced by $y + z$, where z is any vector in $\mathcal{X}^\perp$. Since (1.9) defines φ on all of $\mathbf{C}^n$, another way of expressing the nonuniqueness of y is to say that there are many ways of extending the functional φ from $\mathcal{X}$ to $\mathbf{C}^n$. The following theorem shows that among these extensions is one that does not increase the norm of the functional.

Theorem 1.11 (Hahn, Banach). *Let ν be a norm on $\mathbf{C}^n$. Let $\mathcal{X}$ be a subspace of $\mathbf{C}^n$, and let $\varphi : \mathcal{X} \to \mathbf{C}$ be a linear functional satisfying*
$$\max_{\substack{x \in \mathcal{X} \\ \nu(x)=1}} |\varphi(x)| = \mu.$$
Then φ can be extended to a linear functional on $\mathbf{C}^n$ that satisfies
$$\max_{\substack{x \in \mathbf{C}^n \\ \nu(x)=1}} |\varphi(x)| = \mu.$$

Proof. Without loss of generality, we may assume that $\mu = 1$. If $\mathcal{X} = \mathbf{C}^n$, then we are through. Otherwise, there is a vector $u \neq 0$ that does not lie in $\mathcal{X}$. We shall show how to extend φ to the space $\mathcal{X}'$ spanned by $\mathcal{X}$ and u in such a way that

$$\max_{\substack{x \in \mathcal{X}' \\ \nu(x)=1}} |\varphi(x)| = 1. \tag{1.11}$$

Let x_1 and x_2 be two vectors in $\mathcal{X}$. Then

$$\varphi(x_1) - \varphi(x_2) \leq \nu(x_1 - x_2) \leq \nu(x_1 + u) + \nu(x_2 + u).$$

Hence

$$\varphi(x_1) - \nu(x_1 + u) \leq \varphi(x_2) + \nu(x_2 + u).$$

This inequality implies that any element in the set $\{\varphi(x) - \nu(x + u) : x \in \mathcal{X}\}$ is less than or equal to any element in the set $\{\varphi(x) + \nu(x+u) : x \in \mathcal{X}\}$. If we set $\varphi(-u)$ equal to any value lying between these two sets (or equal to their one common value if they intersect), then

$$\varphi(x) - \nu(x + u) \leq \varphi(-u) \leq \varphi(x) + \nu(x + u).$$

In other words

$$|\varphi(x + u)| \leq \nu(x + u).$$

Now extend φ to $\mathcal{X}'$ by linearity; i.e., for $x \in \mathcal{X}$ set

$$\varphi(x + \alpha u) = \varphi(x) + \alpha\varphi(u).$$

If $\alpha \neq 0$, we have

$$|\varphi(x + \alpha u)| \leq |\alpha||\varphi(\alpha^{-1}x + u)| \leq |\alpha|\nu(\alpha^{-1}x + u) = \nu(x + \alpha u).$$

Hence (1.11) holds.

If $\mathcal{X}' \neq \mathbf{C}^n$, we may select another vector u' not in $\mathcal{X}'$ and extend φ to the space spanned by u' and $\mathcal{X}'$ in such a way that its norm is not increased. After a finite number of such extensions, we shall have extended φ to all of $\mathbf{C}^n$. ∎

We are now in a position to prove the duality theorem.

Theorem 1.12. *Let ν be a norm on $\mathbf{C}^n$. Then $\nu^{**} = \nu$.*

1. VECTOR NORMS

Proof. From the definition of dual norm we have

$$|y^H x| \leq \nu^*(y)\nu(x). \tag{1.12}$$

Consequently

$$\nu^{**}(x) = \sup_{\nu^*(y)=1} |y^H x| \leq \nu(x).$$

To show equality define a linear functional φ on the space spanned by x by $\varphi(\alpha x) = \alpha\nu(x)$. Then (with an abuse of notation) $\nu^*(\varphi) = 1$. By the Hahn–Banach theorem, there is an extension of φ to $\mathbf{C}^n$ with $\nu^*(\varphi) = 1$. Let z be the vector representing φ. Then $\nu^*(z) = 1$ and $|z^H x| = \nu(x)$. Hence

$$\nu^{**}(x) = \sup_{\nu^*(y)=1} |y^H x| \geq \nu(x),$$

which establishes the theorem. ∎

We note for later reference that (1.12) is a generalization of the CAUCHY INEQUALITY,

$$|y^H x| \leq \|y\|_2 \|x\|_2.$$

For this reason it is sometimes called the GENERALIZED CAUCHY INEQUALITY.

Notes and References

The quantitative notion of distance or size is as old as the measuring stick. Our 2-norm in 3-space is just the Euclidean notion of size — the basis of Greek geometry — and the triangle inequality says that the length of one side of a triangle is less than the sum of the lengths of the other two sides.

There are two ways of generalizing Euclidean length to a vector space. The first, due to Minkowski [157, 1911, v.2, pp.131–229], uses convex bodies to define norms. Specifically if $\mathcal{K}$ is a compact convex set containing the origin, we may define a norm $\nu_\mathcal{K}(x)$ as the reciprocal of the number σ such that σx lies on the boundary of $\mathcal{K}$ (in this approach we must add $\alpha \geq 0$ to the homogeneity condition). Minkowski established the equivalence of his norms to the 2-norm, introduced the dual norm (he called it the polar norm), and showed that the dual of the dual was the original. Although Minkowski worked only in 3-space, it is obvious (as it must have been to him) that the approach generalizes. For a modern treatment along these lines see [121, Ch.2].

The second approach is to add the axioms for a norm to those for a vector space, which was done independently by Banach [9, 1922] and Wiener [268, 1922]. For Wiener the matter seems to have been little more than a mathematical exercise. Banach, on the other hand, developed the notion extensively and went on to apply it. In any event, normed linear spaces are known today as BANACH SPACES. It is of interest that both Banach and Wiener use the modern notation $\|\cdot\|$ to denote a norm.

Norms of the form $\sqrt{x^{\mathrm{H}}Ax}$ arise frequently in the analysis of iterative methods for linear systems [276]. In spite of the equivalence of norms, a statement made in one elliptic norm may in practice mean something entirely different from the same statement made in the Euclidean norm. We will return to this point in Section III.2, where we discuss the limitations of absolute and relative errors defined in terms of norms.

We have already pointed out that Minkowski had the concept of the dual norm. The approach taken here is due to Hahn [99, 1927] and Banach [10, 1929]. The proof given here is adapted from the one in the elegant book by Riesz and Sz.-Nagy [184, 1955]. The inequality $y^{\mathrm{T}}x \leq \|x\|_2\|y\|_2$ for real vectors is due to Cauchy [41, 1821, Note II, Théorèm XVI]. It is also associated with the names Schwarz and Bunyakovski.

Exercises

1. Let ν be a vector norm. Show that $|\nu(x) - \nu(y)| \leq \nu(x - y)$.

2. Verify directly that the functions $\|\cdot\|_p$ ($p = 1, 2, \infty$) are norms.

3. (Cauchy inequality). Show that
$$|x^{\mathrm{H}}y| \leq \|x\|_2\|y\|_2$$
with equality if and only if x and y are linearly dependent.

4. Show that up to a constant multiple the 2-norm is the only unitarily invariant norm on $\mathbf{C}^n$.

5. Show that $|y^{\mathrm{H}}x| \leq \|x\|_1\|y\|_\infty$. Conclude that $\|x\|_2^2 \leq \|x\|_1\|x\|_\infty$.

THE FOLLOWING EXERCISES SHOW THAT THE HÖLDER NORMS ARE TRULY NORMS. THEY FOLLOW BECKENBACH AND BELLMAN [20].

1. VECTOR NORMS

6. (Arithmetic-geometric mean inequality). Let $\xi > 0$. Show that

$$\xi^\alpha - \alpha\xi + \alpha - 1 \begin{cases} \geq 0 \text{ if } \alpha > 1 \text{ or } \alpha < 0, \\ \leq 0 \text{ if } 0 < \alpha < 1. \end{cases}$$

Conclude that for $0 < \alpha < 1$ and $\xi_1, \xi_2 \geq 0$,

$$\xi_1^\alpha \xi_2^{1-\alpha} \leq \alpha\xi_1 + (1-\alpha)\xi_2.$$

Equivalently if $p > 1$ and $\frac{1}{p} + \frac{1}{q} = 1$, then

$$\alpha^{\frac{1}{p}} \beta^{\frac{1}{q}} \leq \frac{\alpha}{p} + \frac{\beta}{q} \tag{1.13}$$

for all nonnegative α and β.

7. (Hölder's inequality[118, 1889]). Show that if $p > 1$ and $\frac{1}{p} + \frac{1}{q} = 1$, then

$$|y^H x| \leq \|x\|_p \|y\|_q.$$

[Hint: In (1.13) take $\alpha = \xi/\|x\|_p$ and $\beta = \eta_i/\|x\|_q$, and sum the resulting inequalities.]

8. (Minkowski's inequality [156, 1896]). Show that for $p > 1$

$$\|x + y\|_p \leq \|x\|_p + \|y\|_p.$$

[Hint: Assume without loss of generality that $x, y > 0$. Write

$$\sum_i (\xi_i + \eta_i)^p = \sum_i \xi_i(\xi_i + \eta_i)^{p-1} + \sum_i \eta_i(\xi_i + \eta_i)^{p-1},$$

and apply Hölder's inequality twice.]

9. Show that if $\frac{1}{p} + \frac{1}{q} = 1$, then the Hölder norms $\|\cdot\|_p$ and $\|\cdot\|_q$ are dual.

---◇---

10. Let μ be a norm on $\mathbf{C}^m$ and ν be a norm on $\mathbf{C}^n$. Partition any vector $x \in \mathbf{C}^{m+n}$ in the form $x = (x_1^H\ x_2^H)^H$, where $x_1 \in \mathbf{C}^m$. Show that the following functions are norms on $\mathbf{C}^{m+n}$:

1. $\mu(x_1) + \nu(x_2)$,
2. $\sqrt{\mu(x_1)^2 + \nu(x_2)^2}$,
3. $\max\{\mu(x_1), \nu(x_2)\}$.

11. Let ν be a norm on $\mathbf{C}^n$. Show that the function
$$\rho_\nu(x,y) = \nu(y-x)$$
is a metric; i.e.,

 1. $\rho_\nu(x,y) \geq 0$,
 2. $\rho_\nu(x,y) = 0 \iff x = y$,
 3. $\rho_\nu(x,y) = \rho_\nu(y,x)$,
 4. $\rho_\nu(x,z) \leq \rho_\nu(x,y) + \rho_\nu(y,z)$.

12. A real valued function on $\mathbf{C}^n \times \mathbf{C}^n$ is a PSEUDO-METRIC if it satisfies all the defining conditions for a metric except
$$\rho_2(x,y) = 0 \iff x = y.$$
Show that the relation $\rho(x,y) = 0$ is an equivalence relation on $\mathbf{C}^n$. Denote the corresponding equivalence classes by $\langle x \rangle$. Show that the function $\hat{\rho}(\langle x \rangle, \langle y \rangle) = \rho(x,y)$ is well defined and is a metric over the space of equivalence classes $\langle x \rangle$.

13. Verify the inequalities in Example 1.8.

14. Prove Theorem 1.4.

THE FOLLOWING EXERCISES CONCERN NORMS GENERATED BY POSITIVE DEFINITE MATRICES.

15. An inner product on $\mathbf{C}^n$ is a continuous mapping $(\cdot,\cdot) : \mathbf{C}^n \oplus \mathbf{C}^n \to \mathbf{R}$ that satisfies

 1. $x \neq 0 \iff (x,x) > 0$,
 2. $(\alpha x + \beta y, z) = \alpha(x,z) + \beta(y,z)$,
 3. $(y,x) = \overline{(x,y)}$.

Show that any inner product has a unique representation of the form $(x,y) = y^H A x$, where A is positive definite. Conclude that the function ν defined by $\nu^2(x) = (x,x)$ is a norm. It is the NORM GENERATED BY THE INNER PRODUCT $(\cdot,\cdot)$.

16. Let ν be a norm generated by an inner product. Show that the unit ball $\{x : \nu(x) \leq 1\}$ is an ellipsoid (hence the alternative name ELLIPTIC NORM for these norms). Describe the lengths and situations of the axes.

1. VECTOR NORMS

17. (Jordan and von Neumann [127, 1935]). Let ν be a norm on $\mathbf{C}^n$. Show that a necessary and sufficient condition for ν to be generated by an inner product is that it satisfy the RHOMBUS IDENTITY

$$\nu^2(x+y) + \nu^2(x-y) = 2[\nu^2(x) + \nu^2(y)].$$

———◇———

18. Let ν be a norm on $\mathbf{C}^n$ The sequence $x_1, x_2, \ldots$ is a CAUCHY SEQUENCE if for every $\epsilon > 0$ there is an integer N such that $\nu(x_i - x_j) \leq \epsilon$ whenever $i, j \geq N$.

1. Show that this definition is independent of the choice of norm.
2. Show that $\mathbf{C}^n$ is COMPLETE; that is, every Cauchy sequence in $\mathbf{C}^n$ converges in $\mathbf{C}^n$. [Hint: use the corresponding fact about complex numbers.]

THE FOLLOWING EXERCISES EXPLORE SOME OF THE RELATIONS OF NORMS AND CONVEXITY.

19. Let ν be a vector norm and let

$$\mathcal{B}_\nu = \{x : \nu(x) \leq 1\}$$

be the unit ν-ball. Show that $\mathcal{B}_\nu$ is closed, bounded, convex, and equilibrated ($x \in \mathcal{B}_\nu$ and $|\alpha| \leq 1 \Rightarrow \alpha x \in \mathcal{B}_\nu$). Show further that $\mathcal{B}_\nu$ contains the origin in its interior. Conversely if $\mathcal{B}$ is a closed, bounded, convex, equilibrated set containing the origin in its interior, then the function $\nu_\mathcal{B}$ defined by

$$\nu_\mathcal{B}(x) = \inf\{\sigma^{-1} > 0 : \sigma x \in \mathcal{B}\} \qquad (1.14)$$

is a vector norm.

20. Let $\mathcal{B} \subset \mathbf{R}^n$ be a closed, bounded, convex set containing the origin. Show that the function $\nu_\mathcal{B}$ defined by (1.14) is a norm in which the homogeneity condition is replaced by

$$\alpha \geq 0 \implies \nu_\mathcal{B}(x) = 0.$$

21. Show that the Hahn–Banach theorem is equivalent to the following statement. If x is a point outside a closed, bounded, convex set containing the origin in its interior, then there is a hyperplane that separates x from the set.

———◇———

22. Verify that the function defined by (1.10) is a norm.

THE PURPOSE OF THE FOLLOWING EXERCISES IS TO SHOW THAT THE EQUIVALENCE OF NORMS FAILS IN INFINITE DIMENSIONAL SPACES.

23. For $1 \le p \le \infty$ let ℓ_p be the set of all infinite sequences $x = (\xi_1, \xi_2, \ldots)$ such that $\sum_{i=1}^{\infty} |\xi_i|^p < \infty$ (for $p = \infty$ take the limit). Show that if $p_1 < p_2$ then ℓ_{p_1} is properly contained in ℓ_{p_2}.

24. Show that the function $\|\cdot\|_p$ defined by

$$\|x\|_p = (\sum_{i=1}^{\infty} |\xi_i|^p)^{\frac{1}{p}}$$

is a norm on ℓ_p. Prove that ℓ_p is complete; that is, Cauchy sequences in the p-norm converge in the p-norm. (Note: ℓ_2 is called a HILBERT SPACE.)

25. Show that although all the Hölder norms are defined on ℓ_1, they are not equivalent. In particular if $p_1 < p_2$ then there is a sequence x_k in ℓ_1 such that $\|x_k\|_{p_1} \to \infty$ while $\|x_k\|_{p_2} \to 0$.

——⋄——

2. Matrix Norms

2.1. Basic Concepts

Since the space of $m \times n$ matrices is a vector space of dimension mn, it is natural to define a matrix norm in the same way as vector norm. In the following definition, we will do this; however, as we shall see later, a matrix norm needs an additional property to be really useful.

Definition 2.1. *A function* $\nu : \mathbf{C}^{m \times n} \to \mathbf{R}$ *is a* NORM *on* $\mathbf{C}^{m \times n}$ *(or a* MATRIX NORM*) if it satisfies the following conditions:*

1. $A \ne 0 \Longrightarrow \nu(A) > 0$,

2. $\nu(\alpha A) = |\alpha|\nu(A)$,

3. $\nu(A + B) \le \nu(A) + \nu(B)$.

2. MATRIX NORMS

This definition has the consequence that all the properties of vector norms developed in the previous section remain true of matrix norms. For example, all matrix norms are equivalent and generate the same topology, in which they are all continuous functions. It makes no theoretical difference whether we define convergence of matrices elementwise or as convergence in any matrix norm.

A natural generalization of the Euclidean norm is given in the following definition.

Definition 2.2. *Let $A \in \mathbf{C}^{m \times n}$. The FROBENIUS NORM of A is the number*

$$\|A\|_\mathrm{F} \stackrel{\text{def}}{=} \sqrt{\sum_{i=1}^{m}\sum_{j=1}^{n} |\alpha_{ij}|^2} = \operatorname{trace}(A^\mathrm{H} A)^{\frac{1}{2}}. \tag{2.1}$$

Note that when $A \in \mathbf{C}^{n \times 1}$, i.e., when A is a vector, the Frobenius norm reduces to the 2-norm.

Our Definition 2.1 of matrix norm has one important defect: it makes no concession to the fact that matrices can be multiplied. What we would like is an analogue of the triangle inequality for matrix multiplication. In fact the Frobenius norm satisfies such an inequality: namely,

$$\|AB\|_\mathrm{F} \leq \|A\|_\mathrm{F} \|B\|_\mathrm{F},$$

whenever the product AB is defined (Exercise 2.1). However, not every matrix norm satisfies this kind of equality, as the following example shows.

Example 2.3. *Let us attempt to generalize the ∞-norm as we did the Euclidean norm by defining*

$$\nu_\infty(A) = \max_{i,j} |\alpha_{ij}|.$$

Clearly this is a matrix norm. However, if

$$A = \begin{pmatrix} 1 & 1 \\ 1 & 1 \end{pmatrix},$$

then $\nu_\infty(A \cdot A) = 2 > 1 = \nu_\infty(A)\nu_\infty(A)$.

The submultiplicative inequality satisfied by the Frobenius norm allows us to obtain bounds on the products of matrices in terms of the individual matrices. So important is this for matrix analysis, that norms with this property are given a special name.

Definition 2.4. Let μ, ν, and ρ be norms on $\mathbf{C}^{m\times n}$, $\mathbf{C}^{n\times k}$, and $\mathbf{C}^{m\times k}$. Then μ, ν, and ρ are CONSISTENT if

$$\rho(AB) \le \mu(A)\nu(B)$$

whenever $A \in \mathbf{C}^{m\times n}$ and $B \in \mathbf{C}^{n\times k}$. In particular, a matrix norm ν on $\mathbf{C}^{n\times n}$ is consistent if $\nu(AB) \le \nu(A)\nu(B)$ for all $A, B \in \mathbf{C}^{n\times n}$.

Since vector norms can be identified with matrix norms, Definition 2.4 includes the notion of consistency of a vector norm and a matrix norm. For example, the Frobenius norm and the vector 2-norm are consistent — that is $\|Ax\|_\mathrm{F} \le \|A\|_\mathrm{F}\|x\|_2$ — because $\|x\|_2 = \|x\|_\mathrm{F}$.

The following theorem shows that for any consistent matrix norm there is a consistent vector norm.

Theorem 2.5. Let $\|\cdot\|$ be a consistent matrix norm on $\mathbf{C}^{n\times n}$. Then there is a norm ν on $\mathbf{C}^n$ that is consistent with $\|\cdot\|$.

Proof. Chose a nonzero vector $a \in \mathbf{C}^n$ and define

$$\nu(x) = \|xa^\mathrm{T}\|.$$

It is easy to verify that ν is a vector norm. Moreover, since

$$\nu(Ax) = \|Axa^\mathrm{T}\| \le \|A\|\|xa^\mathrm{T}\| = \|A\|\nu(x),$$

ν is consistent with $\|\cdot\|$. ∎

Consistent matrix norms have an important relation to the eigenvalues of a matrix. Let us define the SPECTRAL RADIUS of a matrix A to be the number

$$\rho(A) \stackrel{\text{def}}{=} \max\{|\lambda| : \lambda \in \mathcal{L}(A)\}.$$

Then we have the following theorem.

2. MATRIX NORMS

Theorem 2.6. *Let* $\|\cdot\|$ *be a consistent matrix norm. Then for any matrix A*

$$\rho(A) \leq \|A\|. \qquad (2.2)$$

Proof. By Theorem 2.5, there is a vector norm ν that is consistent with $\|\cdot\|$. Let x be an eigenvector of A corresponding to an eigenvalue λ; i.e.,

$$Ax = \lambda x.$$

Taking norms we get

$$|\lambda|\nu(x) = \nu(\lambda x) = \nu(Ax) \leq \|A\|\nu(x).$$

Since $\nu(x) > 0$ we may divide by $\nu(x)$ to get $|\lambda| \leq \|A\|$. The result (2.2) follows from the fact that $\lambda \in \mathcal{L}(A)$ is arbitrary. ∎

2.2. Operator Norms

Recall that in Section 1.4 we defined the dual to the norm ν by

$$\nu^*(y) = \max_{\nu(x)=1} |y^H x|. \qquad (2.3)$$

An immediate consequence of this was the generalized Cauchy inequality

$$|y^H x| \leq \nu^*(y)\nu(x),$$

which may be interpreted as saying that the norms $|\cdot|$, ν^* and ν on $\mathbf{C}$, $\mathbf{C}^{1\times n}$, and $\mathbf{C}^{n\times 1}$ are consistent. It turns out that (2.3) represents a general technique for generating a class of consistent norms, which we will call OPERATOR NORMS.

Let μ be a norm on $\mathbf{C}^m$ and ν be a norm on $\mathbf{C}^n$. By the equivalence of norms, the function μ is continuous and the ν-sphere $\mathcal{S}_\nu = \{x : \nu(x) = 1\}$ is closed and bounded. Hence for any $m \times n$ matrix A, we may define the number $\|A\|_{\mu,\nu}$ by

$$\|A\|_{\mu,\nu} = \max_{\nu(x)=1} \mu(Ax). \qquad (2.4)$$

As the notation $\|A\|_{\mu,\nu}$ suggests, the function $\|\cdot\|_{\mu,\nu}$ is actually a norm.

Theorem 2.7. *Let μ and ν be as above, and let $\|\cdot\|_{\mu,\nu}$ be defined by (2.4). Then $\|\cdot\|_{\mu,\nu}$ is a norm on $\mathbf{C}^{m\times n}$, which is consistent with μ and ν.*

Proof. We first prove consistency. From (2.4),

$$\|A\|_{\mu,\nu} = \max_{x \neq 0} \frac{\mu(Ax)}{\nu(x)}.$$

Therefore,
$$\mu(Ax) \leq \|A\|_{\mu,\nu}\nu(x). \tag{2.5}$$

Now we prove that $\|\cdot\|_{\mu,\nu}$ is a matrix norm by showing that it satisfies the conditions of Definition 2.1.

1. Positive definiteness. If $A \neq 0$, there is a index i such that $A\mathbf{1}_i \neq 0$. Then from (2.5), $0 < \mu(A\mathbf{1}_i) \leq \|A\|_{\mu,\nu}\nu(\mathbf{1}_i)$, which implies that $\|A\|_{\mu,\nu} > 0$.

2. Homogeneity. For any α we have

$$\|\alpha x\|_{\mu,\nu} = \max_{\nu(x)=1} \mu(\alpha Ax) = \max_{\nu(x)=1} |\alpha|\mu(Ax) = |\alpha|\|A\|_{\mu,\nu}.$$

3. Triangle inequality. Let $A, B \in \mathbf{C}^{m \times n}$. Suppose that x satisfies $\nu(x) = 1$ and $\mu[(A+B)x] = \|A+B\|_{\mu,\nu}$. Then from (2.5),

$$\begin{aligned}\|A+B\|_{\mu,\nu} &= \mu[(A+B)x] \\ &\leq \mu(Ax) + \mu(Bx) \\ &\leq \|A\|_{\mu,\nu}\nu(x) + \|B\|_{\mu,\nu}\nu(x) \\ &= \|A\|_{\mu,\nu} + \|B\|_{\mu,\nu}. \quad \blacksquare\end{aligned}$$

Theorem 2.7 justifies the following definition.

Definition 2.8. *Let μ and ν be norms on $\mathbf{C}^m$ and $\mathbf{C}^n$. The norm $\|\cdot\|_{\mu,\nu}$ defined by (2.4) is called an* OPERATOR NORM *on $\mathbf{C}^{m \times n}$. It is also said to be the norm* SUBORDINATE *to the vector norms μ and ν.*

Equation (2.5) shows that an operator norm subordinate to two vector norms is consistent with them. The following theorem shows that under appropriate conditions, operator norms are consistent with themselves. Its proof is left as an exercise.

Theorem 2.9. *Let μ, ν, and ρ be norms on $\mathbf{C}^m, \mathbf{C}^n$, and $\mathbf{C}^k$, and let $\|\cdot\|_{\mu,\nu}, \|\cdot\|_{\nu,\rho}$, and $\|\cdot\|_{\mu,\rho}$ be the subordinate operator norms. Then*

$$\|AB\|_{\mu,\rho} \leq \|A\|_{\mu,\nu}\|B\|_{\nu,\rho}.$$

2. Matrix Norms

We saw in Example 2.3 that generalizing the vector ∞-norm by extending its algebraic definition to matrices failed to produce a consistent matrix norm. The notion of an operator norm provides a means of extending the definition of the Hölder norms to all matrices. Specifically, if we define

$$\|A\|_p = \max_{\|x\|_p=1} \|Ax\|_p, \tag{2.6}$$

then by Theorem 2.9

$$\|AB\|_p \leq \|A\|_p \|B\|_p \tag{2.7}$$

whenever the product AB is defined. We say that these Hölder matrix norms form a CONSISTENT FAMILY OF MATRIX NORMS. Another consistent family is the Frobenius norm defined by (2.1).

For $p = 1, 2, \infty$, the Hölder matrix norms have explicit characterizations.

Theorem 2.10. Let $A = (\alpha_{ij}) \in \mathbf{C}^{m \times n}$. Then

$$\|A\|_1 = \max_{1 \leq j \leq n} \sum_{i=1}^{m} |\alpha_{ij}|, \tag{2.8}$$

$$\|A\|_\infty = \max_{1 \leq i \leq m} \sum_{j=1}^{n} |\alpha_{ij}|, \tag{2.9}$$

and

$$\|A\|_2 = \sqrt{\lambda_{\max}(A^H A)} = \sigma_{\max}(A), \tag{2.10}$$

where $\lambda_{\max}(A^H A)$ is the largest eigenvalue of $A^H A$ and $\sigma_{\max}(A)$ is the largest singular value of A.

Proof. To prove (2.8) let $A = (a_1, \cdots, a_n)$ be partitioned by columns. For any $x \neq 0$ we have

$$\|Ax\|_1 = \left\| \sum_{j=1}^{n} \xi_j a_j \right\|_1 \leq \sum_{j=1}^{n} |\xi_j| \|a_j\|_1 \leq \max_{1 \leq j \leq n} \|a_j\|_1 \|x\|_1.$$

Hence

$$\|A\|_1 \leq \max_{1 \leq j \leq n} \|a_j\|_1. \tag{2.11}$$

On the other hand, if $\max_{1\leq j\leq n} \|a_j\|_1 = \|a_k\|_1$, then

$$\frac{\|A1_k\|_1}{\|1_k\|_1} = \|a_k\|_1 = \max_{1\leq j\leq n} \|a_j\|_1.$$

Hence

$$\|A\|_1 \geq \max_{1\leq j\leq n} \|a_j\|_1. \tag{2.12}$$

The two inequalities (2.11) and (2.12) together imply (2.8).

The characterization (2.9) is proved similarly. Equation (2.10) follows from Theorem I.4.3. ∎

In view of (2.8)–(2.10), the norm $\|\cdot\|_1$ is sometimes called the COLUMN SUM NORM; the norm $\|\cdot\|_\infty$, the ROW SUM NORM; and the norm $\|\cdot\|_2$, the SPECTRAL NORM.

There is no computationally convenient characterization of the spectral norm, but it does have a number of nice properties that make it useful for theoretical purposes. Some of these properties are contained in the following theorem.

Theorem 2.11. *For any matrix A*

1. $\|A\|_2 = \max\limits_{\substack{\|x\|_2=1 \\ \|y\|_2=1}} |y^{\mathrm{H}} A x|$,

2. $\|A^{\mathrm{H}}\|_2 = \|A^{\mathrm{T}}\|_2 = \|A\|_2$,

3. $\|A^{\mathrm{H}} A\|_2 = \|A\|_2^2$,

4. *If U and V are unitary, $\|U^{\mathrm{H}} A V\|_2 = \|A\|_2$,*

5. $\|A\|_2^2 \leq \|A\|_1 \|A\|_\infty$.

Proof. The first four items follow directly from the singular value decomposition, and their proofs are left as an exercise. For the last item, note that from (2.7) and Theorem 2.6

$$\|A\|_2^2 = \lambda_{\max}(A^{\mathrm{H}} A) \leq \|A^{\mathrm{H}} A\|_1 \leq \|A^{\mathrm{H}}\|_1 \|A\|_1 = \|A\|_\infty \|A\|_1. \blacksquare$$

Notes and References

The spectral norm was introduced by Peano [177, 1888], who established its basic properties. The Frobenius norm appears briefly in the paper of Peano just cited, but only as a bound for the spectral norm. Schur [193, 1909] uses it in an important bound on the eigenvalues of a matrix. Frobenius [76, 77, 1911] appears to be first to regard the function $\|\cdot\|_F$ as what we would call a matrix norm. Actually he worked with the quantity $\|\cdot\|_F^2$, which he called the *Spannung* of a matrix, and established the equivalent of the triangle inequality, consistency, invariance under orthogonal transformations.

Although norms of operators in infinite dimensional spaces were a staple of functional analysis almost from the beginning, they seem to have percolated more slowly into matrix theory. Applications in numerical analysis played a large role in the process, thanks in large part to a series of conferences in Gatlinburg, Tennessee, hosted by A. S. Householder, which combined both the theoretical and the numerical aspects of matrix theory. Any list of the more influential works would include von Neumann [254, 1937],[256, 1947], Faddeeva [68, 1959], Mirsky [158, 1960], Householder [121, 1964], and Wilkinson [269, 1965].

Exercises

1. Show that $\|AB\|_F \leq \|A\|_F \|B\|_F$ whenever the product AB is defined.

2. Prove Theorem 2.9.

3. Establish the first four items in Theorem 2.11.

4. Show that $\|A\|_2^2 \leq \|A\|_1 \|A\|_\infty$.

5. Let $A = (a_1 \ldots a_n)$, where $\|a_i\|_2 = 1$ ($i = 1, \ldots, n$). Show that $\|A\|_2 \leq \sqrt{n}$.

6. Let ν be defined on $\mathbf{C}^{n \times n}$ by $\nu(A) = n \max_{i,j} |\alpha_{ij}|$. Show that ν is a consistent matrix norm.

7. (Gastinel [121, p.61]). Show that if ν is a matrix norm on $\mathbf{C}^{n \times n}$, there is a constant τ such that the function $A \mapsto \tau \nu(A)$ is a consistent matrix norm.

8. Let ν be a consistent matrix norm on $\mathbf{C}^n$ and let X be nonsingular. Show that the function ν_X defined by

$$\nu_X(A) = \nu(X^{-1}AX)$$

is a consistent matrix norm.

9. Let ν be a norm on $\mathbf{C}^{n\times n}$ and let B be nonsingular. Let the norm μ be defined by $\mu(x) = \nu(Bx)$. Show that the operator norms $\|\cdot\|_\mu$ and $\|\cdot\|_\nu$ subordinate to μ and ν are related by the equation $\mu(A) = \nu(BAB^{-1})$.

10. Let $\|\cdot\|$ be the operator norm subordinate to an absolute vector norm. Show that
$$\|\text{diag}(\delta_1, \delta_2, \ldots, \delta_n)\| = \max_i |\delta_i|. \tag{2.13}$$
Conversely, if (2.13) is satisfied for all diagonal matrices by an operator norm, then it is generated by an absolute norm.

11. (Mirsky [121, p.61]). Show that
$$\inf_{X \text{ nonsingular}} \|X^{-1}AX\|_\text{F} = \sum_{\lambda \in \mathcal{L}(A)} |\lambda|^2,$$
with equality for some particular X if and only if A is diagonalizable.

12. Show that the numbers ρ_{pq} in the following table satisfy $\|A\|_p \le \rho_{pq}\|A\|_q$, where $A \in \mathbf{C}^{n\times n}$. Show that equality can be attained.

| | \multicolumn{4}{c}{q} | | | |
p	1	2	∞	F
1	1	$\sqrt{n}$	n	$\sqrt{n}$
2	$\sqrt{n}$	1	$\sqrt{n}$	1
∞	n	$\sqrt{n}$	1	$\sqrt{n}$
F	$\sqrt{n}$	$\sqrt{n}$	$\sqrt{n}$	1

13. Build the table in Exercise 2.12 for the same norms over $\mathbf{C}^{m\times n}$.

14. Let $A \in \mathbf{C}^{n\times n}$ and let $\epsilon > 0$. Show that there is a consistent matrix norm $\|\cdot\|_{A,\epsilon}$ such that
$$\|A\|_{A,\epsilon} = \rho(A) + \epsilon.$$
[Hint: Reduce A to Schur form. Use a diagonal similarity transformation to reduce the off-diagonal elements so that the infinity norm is less than $\rho(A) + \epsilon$. Finally, use Exercise 2.8 to undo the transformations.]

15. Show that the spectral norm and the Frobenius norm are UNITARILY INVARIANT; that is,
$$U, V \text{ unitary} \implies \|U^\text{H} A V\|_\text{p} = \|A\|_\text{p}, \quad p = 2, \text{F}.$$

2. Matrix Norms

16. Let A be square. Show that there is a consistent norm $\|\cdot\|$ such that $\|A\| = \rho(A)$ if and only if every eigenvalue $\lambda \in \mathcal{L}(A)$ with $|\lambda| = \rho(A)$ is nondefective.

THE FOLLOWING EXERCISES INVESTIGATE THE PROPERTIES OF THE POWERS OF A MATRIX $A \in \mathbf{C}^{n \times n}$.

17. Show that $\lim_{k \to \infty} A^k = 0$ if and only if $\rho(A) < 1$.

18. (Neumann series). Let $\rho(A) < 1$. Show that $I - A$ is nonsingular and $\sum_{k=0}^{\infty} A^k = (I - A)^{-1}$.

19. Show that if ν is a consistent matrix norm, then $\lim_{k \to \infty} \nu(A^k)^{\frac{1}{k}} = \rho(A)$.

———⋄———

20. Let the infinite series $\phi(\zeta) = \sum_{k=0}^{\infty} \gamma_k \zeta^k$ have radius of convergence σ. Show that if $\rho(A) < \sigma$ then $\sum_{k=0}^{\infty} \gamma_k A^k$ converges. We write $\phi(A)$ for its limit.

21. In the last exercise, let $|\lambda| < \sigma$. Show that

$$\phi[J_k(\lambda)] = \begin{pmatrix} \phi(\lambda) & \phi'(\lambda) & \phi''(\lambda)/2 & \phi^{(3)}(\lambda)/3! & \cdots & \\ 0 & \phi(\lambda) & \phi'(\lambda) & \phi''(\lambda)/2 & \phi^{(3)}(\lambda)/3! & \cdots \\ 0 & 0 & \phi(\lambda) & \phi'(\lambda) & \phi''(\lambda)/2 & \phi^{(3)}(\lambda)/3! & \cdots \\ \vdots & \vdots & \vdots & \vdots & \vdots & \vdots \end{pmatrix}.$$

22. For any $A \in \mathbf{C}^{n \times n}$ define

$$e^A = \sum_{k=0}^{\infty} \frac{A^k}{k!}.$$

Show the following.

1. If $AB = BA$, then $e^{A+B} = e^A e^B$.
2. $\det(e^A) = e^{\text{trace}(A)}$.
3. $de^{\tau A}/d\tau = A e^{\tau A}$.

23. Show that $\rho(A) < 1$ if and only if there is a positive definite matrix Q such that $Q - AQA^{\mathrm{H}}$ is positive definite.

24. A matrix is STABLE if all of its eigenvalues have negative real parts. Show that A is stable if and only if $\lim_{t\to+\infty} e^{tA} = 0$.

25. Show that $-A$ is stable if and only if there is a positive definite matrix M such that $AM + MA^{\mathrm{H}}$ is positive definite. [Hint: Use Exercise 2.23.]

3. Unitarily Invariant Norms

3.1. Von Neumann's Theory

An important property of Euclidean space is that shapes and distances do not change under rotations. In particular for any vector x and for any unitary matrix U we have

$$\|Ux\|_2 = \|x\|_2.$$

An analogous property is shared by the spectral and Frobenius norms: namely, for any unitary matrices U and V for which the product $U^{\mathrm{H}}AV$ is defined,

$$\|U^{\mathrm{H}}AV\|_p = \|A\|_p, \qquad p = 2, \mathrm{F}.$$

These examples suggest the following definition.

Definition 3.1. A norm $\|\cdot\|$ on $\mathcal{C}^{m\times n}$ is UNITARILY INVARIANT if it satisfies

$$\|U^{\mathrm{H}}AV\| = \|A\|$$

for all unitary U and V. It is NORMALIZED if

$$\|A\| = \|A\|_2 \tag{3.1}$$

whenever A is of rank one.

We have observed that the spectral and Frobenius norms are unitarily invariant. However, not all norms are unitarily invariant, as the following example shows.

Example 3.2. Let

$$A = \begin{pmatrix} 1 & 1 \\ 1 & 1 \end{pmatrix},$$

3. Unitarily Invariant Norms

so that $\|A\|_\infty = 2$. Let

$$U = \begin{pmatrix} \frac{1}{\sqrt{2}} & \frac{1}{\sqrt{2}} \\ -\frac{1}{\sqrt{2}} & \frac{1}{\sqrt{2}} \end{pmatrix}$$

Then

$$\|UA\|_\infty = \left\| \begin{pmatrix} \frac{2}{\sqrt{2}} & \frac{2}{\sqrt{2}} \\ 0 & 0 \end{pmatrix} \right\|_\infty = \frac{4}{\sqrt{2}}.$$

Since not all matrix norms are unitarily invariant, we may ask which ones are. One purpose of this section is to establish a characterization, due to von Neumann, in terms of certain vector norms, which, in this connection, are called SYMMETRIC GAUGE FUNCTIONS.

In one direction the connection is easy to establish. Let A be of order n, and let $U^H A V = \Sigma$ be the singular value decomposition of A. Let $\|\cdot\|$ be a unitarily invariant norm. Since U and V are unitary,

$$\|A\| = \|\Sigma\|.$$

Thus $\|A\|$ is a function Φ of the singular values of A.

Since $\|\cdot\|$ is a norm, the function Φ, regarded as a mapping from $\mathbf{R}^n$ to $\mathbf{R}$, is also a norm. Since by interchanging columns of U and V we can make the singular values of A appear in any order, Φ must be symmetric in its arguments. Since by multiplying a column of U by -1, we can change the sign of the corresponding singular value, Φ must depend only on the absolute values of its argument. Moreover, if $\|\cdot\|$ is normalized and A has rank one, then $\|A\| = \Phi(\sigma_1 \mathbf{1}_1) = \sigma_1$. All this suggests the following definition.

Definition 3.3. *A function* $\Phi : \mathbf{R}^n \to \mathbf{R}$ *is a* SYMMETRIC GAUGE FUNCTION *if it satisfies the following conditions.*

1. $x \neq 0 \implies \Phi(x) > 0$.

2. $\Phi(\rho x) = |\rho| \Phi(x)$.

3. $\Phi(x + y) \leq \Phi(x) + \Phi(y)$.

4. *For any permutation matrix* P *we have* $\Phi(Px) = \Phi(x)$.

5. $\Phi(|x|) = \Phi(x)$.

The function Φ is NORMALIZED if

$$\Phi(\mathbf{1}_1) = 1.$$

In the language of norms, a symmetric gauge function is an absolute norm that is invariant under permutation transformations. If $x = (\xi_1, \cdots, \xi_n)^{\mathrm{T}}$, we will often write $\Phi(\xi_1, \cdots, \xi_n)$ for $\Phi(x)$.

We have just sketched a proof that every unitarily invariant norm is a symmetric gauge function of the singular values of its argument. The converse is also true: if Φ is a symmetric gauge function and if $\|\cdot\|_\Phi$ is defined by

$$\|A\|_\Phi = \Phi(\sigma_1, \cdots, \sigma_n), \qquad (3.2)$$

where $\sigma_1, \cdots, \sigma_n$ are the singular values of A, then $\|\cdot\|_\Phi$ is a unitarily invariant norm. It is easy to see that $\|A\|_\Phi$ is a definite, homogeneous function. However, to show that it satisfies the triangle inequality requires more work. We begin with an inequality, due to von Neumann, relating the singular values of two matrices with the trace of their product.

Lemma 3.4. *Let A and B have singular values $\sigma_1 \geq \sigma_2 \geq \cdots \geq \sigma_n$ and $\tau_1 \geq \tau_2 \geq \cdots \geq \tau_n$. Then*

$$\max_{U,V \text{ unitary}} \Re \operatorname{trace}(AUBV^{\mathrm{H}}) = \sum_{i=1}^n \sigma_i \tau_i. \qquad (3.3)$$

Proof. It is sufficient to prove the theorem for the case where the σ_i are positive and distinct (otherwise perturb the σ_i so they are positive and distinct and take the limit in (3.3) as the perturbation approaches zero).

Let $\Sigma = \operatorname{diag}(\sigma_1, \ldots, \sigma_n)$ and $T = \operatorname{diag}(\tau_1, \ldots, \tau_n)$. By passing to the singular value decompositions of A and B, we see that (3.3) is equivalent to

$$\sup_{U,V \text{ unitary}} \Re \operatorname{trace}(\Sigma U T V^{\mathrm{H}}) \leq \sum_{i=1}^n \sigma_i \tau_i. \qquad (3.4)$$

3. UNITARILY INVARIANT NORMS

Since the set of unitary matrices is closed and bounded, there are unitary matrices U_0 and V_0 for which the supremum in (3.4) is attained. Let $C = U_0 T V_0^H$.

We claim that C is a diagonal matrix with nonnegative diagonal elements. To see that the diagonals of C are nonnegative, let us suppose, say, that $\gamma_{11} \neq 0$ is not positive. Then by multiplying the first row of C by $\bar{\gamma}_{11}/|\gamma_{11}|$ (n.b., this is a unitary transformation) we increase $\Re \operatorname{trace}(\Sigma C)$, contrary to the optimality of C.

To show that the off-diagonal elements of C are zero, let us suppose, say, that $\gamma_{12} \neq 0$. By multiplying the first row of C by $\bar{\gamma}_{12}/|\gamma_{12}|$ and dividing the first column by the same number — a unitary transformation that does not change the trace — we may take $\gamma_{12} > 0$.
Let
$$R_\theta = \begin{pmatrix} \cos\theta & -\sin\theta \\ \sin\theta & \cos\theta \end{pmatrix}$$
and let $C_\theta = C \operatorname{diag}(R_\theta, I_{n-2})$. Then

$\Re \operatorname{trace}(\Sigma C_\theta) = \sigma_1(\gamma_{11}\cos\theta$
$\qquad + \gamma_{12}\sin\theta) + \sigma_2(\gamma_{22}\cos\theta - \Re\gamma_{21}\sin\theta) + \sum_{i=3}^{n}\sigma_i\gamma_{ii}$,

and
$$\left.\frac{d\Re\operatorname{trace}(C_\theta)}{d\theta}\right|_{\theta=0} = \sigma_1\gamma_{12} - \sigma_2\Re\gamma_{21}. \tag{3.5}$$

If this derivative is nonzero, then $\Re\operatorname{trace}(C_\theta) > \Re\operatorname{trace}(C)$ for sufficiently small θ (positive or negative depending on the sign of the derivative). Otherwise, let $\hat{C}_\theta = \operatorname{diag}(R, I_{n-2})C$. Then
$$\left.\frac{d\Re\operatorname{trace}(\hat{C}_\theta)}{d\theta}\right|_{\theta=0} = \sigma_2\gamma_{12} - \sigma_1\Re\gamma_{21}.$$

Since $\sigma_1 > \sigma_2$, $\gamma_{12} > 0$, and the derivative (3.5) is zero, this latter derivative cannot be zero, and a small change in θ will increase $\Re\operatorname{trace}(\Sigma\hat{C}_\theta)$. Either case is a contradiction of the optimality of C.

Since C is nonnegative and diagonal, its diagonal elements must be the singular values of B; i.e., $C = \operatorname{diag}(\tau_{\pi(1)}, \ldots, \tau_{\pi(n)})$ for some permutation π of $\{1, 2, \ldots, n\}$. Hence

$$\sup_{U,V \text{ unitary}} \Re\operatorname{trace}(\Sigma U T V^H) \leq \sum_{i=1}^{n}\sigma_i\tau_{\pi(i)} \leq \sum_{i=1}^{n}\sigma_i\tau_i,$$

the last inequality following from the fact that the σ_i and the τ_i are nonincreasing. ∎

Since Φ is itself a norm, it has a dual norm Φ^* whose dual is Φ (Theorem 1.12). It turns out that we can characterize $\|A\|_\Phi$ in terms of the dual norm, and this characterization is just what we need to establish the triangle inequality for $\|\cdot\|_\Phi$.

Lemma 3.5. *Let Φ be a symmetric gauge function and let Φ^* be its dual. Let the function $\|\cdot\|_\Phi$ be defined by (3.2). Then*

$$\|A\|_\Phi = \max_{\|X\|_{\Phi^*}=1} \Re\operatorname{trace}(X^{\mathrm{H}}A).$$

Proof. Let $\sigma(A) = \{\alpha_1, \ldots, \alpha_n\}$, and let $\sigma(X) = \{\xi_1, \ldots, \xi_n\}$ for any X. Then

$$\begin{aligned}
\max_{\|X\|_{\Phi^*}=1} \Re\operatorname{trace}(X^{\mathrm{H}}A) &= \max_{\substack{\|X\|_{\Phi^*}=1 \\ U,V \text{ unitary}}} \Re\operatorname{trace}(VX^{\mathrm{H}}UA) \\
&= \max_{\Phi^*(\xi_1,\ldots,\xi_n)=1} \sum_{i=1}^n \xi_i \sigma_i \quad \text{(Lemma 3.4)} \\
&= \Phi(\sigma_1, \ldots, \sigma_n) \quad \text{(by duality)} \\
&= \|A\|_\Phi \quad \text{(by (3.2))}. \quad \blacksquare
\end{aligned}$$

We now use Lemma 3.5 to prove that every symmetric gauge function generates a unitarily invariant norm.

Theorem 3.6 (von Neumann). *Let Φ be a symmetric gauge function on $\mathbf{R}^n$ and let $\|\cdot\|_\Phi$ be defined by (3.2). Then $\|\cdot\|_\Phi$ is a unitarily invariant norm on $\mathbf{C}^{n\times n}$. Conversely, if $\|\cdot\|$ is a unitarily invariant norm on $\mathbf{C}^{n\times n}$, then there is a symmetric gauge function Φ on $\mathbf{R}^n$ such that $\|A\| = \|A\|_\Phi$.*

Proof. We need only prove that $\|\cdot\|_\Phi$ satisfies the triangle inequality. From Lemma 3.5,

$$\begin{aligned}
\|A + B\|_\Phi &= \max_{\|X\|_{\Phi^*}=1} \Re\operatorname{trace}[X^{\mathrm{H}}(A+B)] \\
&\leq \max_{\|X\|_{\Phi^*}=1} \Re\operatorname{trace}(X^{\mathrm{H}}A) + \max_{\|X\|_{\Phi^*}=1} \Re\operatorname{trace}(X^{\mathrm{H}}B) \\
&= \|A\|_\Phi + \|B\|_\Phi. \quad \blacksquare
\end{aligned}$$

3.2. Properties of Unitarily Invariant Norms

The correspondence between symmetric gauge functions and unitarily invariant norms allows us to transfer results about the former to the latter. This subsection is devoted to establishing the basic properties of unitarily invariant norms.

The first item of business is to extend the definition of unitarily invariant norms to rectangular matrices. Suppose that Φ is a symmetric gauge function on $\mathbf{R}^n$. Then if $m = \min\{k, l\} \leq n$, a unitarily invariant norm on $\mathbf{C}^{k \times l}$ may be defined by

$$\|A\|_\Phi = \Phi(\sigma_1, \sigma_2, \ldots, \sigma_m, 0, \ldots, 0),$$

where $\sigma_1, \sigma_2, \ldots, \sigma_m$ are the singular values of A. We shall call these norms THE FAMILY OF NORMS GENERATED BY Φ. This convention allows us to use what is essentially the same norm on matrices of varying dimensions.

In the most important cases, the gauge function Φ can be regarded as defined for all infinite sequences $\xi_1, \xi_2, \ldots$ with only a finite number of nonzero elements, in which case the corresponding norm is defined for matrices of all dimensions. For example, the function

$$\Phi_2(\xi_1, \xi_2, \ldots) = \max_i |\xi_i|$$

generates the spectral norm, and

$$\Phi_F(\xi_1, \xi_2, \ldots) = \sqrt{\sum_i |\xi_i|^2}$$

generates the Frobenius norm.

The fact that a symmetric gauge function is an absolute norm has the following important consequences for unitarily invariant norms.

Theorem 3.7. *Let A and B have singular values $\sigma_1 \geq \cdots \geq \sigma_n$ and $\tau_1 \geq \cdots \geq \tau_n$. If $\sigma_i \leq \tau_i$ ($i = 1, \ldots, n$), then for every unitarily invariant norm $\|\cdot\|$, we have $\|A\| \leq \|B\|$.*

Proof. Let Φ be the symmetric gauge function that generates $\|\cdot\|$. Then since Φ is absolute,

$$\|A\| = \Phi(\sigma_1, \ldots, \sigma_n) \leq \Phi(\tau_1, \ldots, \tau_n) = \|B\|. \blacksquare$$

The following useful result is a corollary of this theorem and Theorem I.4.4.

Corollary 3.8. *Let $\|\cdot\|$ be a unitarily invariant norm, and let the matrix A be partitioned in the form*

$$A = \begin{pmatrix} A_{11} & A_{12} \\ A_{21} & A_{22} \end{pmatrix}.$$

Then $\|A_{11}\| \leq \|A\|$.

The 2-norm plays a special role in the theory of unitarily invariant norms as the following theorem shows.

Theorem 3.9. *Let $\|\cdot\|$ be a family of unitarily invariant norms. Then*

$$\|AB\| \leq \|A\|\|B\|_2 \qquad (3.6)$$

and

$$\|AB\| \leq \|A\|_2\|B\|. \qquad (3.7)$$

Also

$$\|AB\| \geq \|A\|\mathrm{inf}_2(B) \qquad (3.8)$$

and

$$\|AB\| \geq \mathrm{inf}_2(A)\|B\|. \qquad (3.9)$$

Moreover, if $\|\cdot\|$ is normalized, then

$$\|A\|_2 \leq \|A\|. \qquad (3.10)$$

Proof. The inequalities (3.6) and (3.7) are immediate corollaries of Theorem 3.7 and Theorem I.4.5. The inequalities (3.8) and (3.9) are corollaries of Exercise I.4.6.

To establish (3.10), let $\sigma_1 \geq \cdots \geq \sigma_n$ be the singular values of A. Then since Φ is absolute,

$$\|A\| = \Phi(\sigma_1, \sigma_2, \ldots, \sigma_n) \geq \Phi(\sigma_1, 0, \ldots, 0) = \sigma_1 = \|A\|_2. \blacksquare$$

Combining (3.6), (3.7), and (3.10) we have the following corollary.

Corollary 3.10. *A family of normalized, unitarily invariant norms is consistent.*

3.3. Doubly Stochastic Matrices and Fan's Theorem

As far as unitarily invariant norms are concerned, the principal result of this subsection is a theorem of Ky Fan, which gives conditions under which one matrix dominates another in any unitarily invariant norm. However, to establish it we must first prove some important theorems on doubly stochastic matrices, one of which will be used later in this book.

Definition 3.11. *Let A be a matrix with nonnegative elements. Then A is* STOCHASTIC *if $A\mathbf{1} = \mathbf{1}$. A stochastic matrix A is* DOUBLY STOCHASTIC *if A^{T} is also stochastic.*

Since the elements of a row of a stochastic matrix sum to one, they may be regarded as probabilities — hence the name stochastic. A doubly stochastic matrix is one whose rows and columns sum to one.

The first theorem gives necessary and sufficient conditions for two vectors to be related by a doubly stochastic matrix. A little notation will be helpful in stating and proving it. Let $x = (\xi_1, \ldots, \xi_n)^{\mathrm{T}}$ and $y = (\eta_1, \ldots, \eta_n)^{\mathrm{T}}$ be real vectors with

$$\begin{aligned}\xi_1 \geq \cdots \geq \xi_n,\\ \eta_1 \geq \cdots \geq \eta_n.\end{aligned} \quad (3.11)$$

We shall write

$$x \succ y$$

if

$$\xi_1 + \cdots + \xi_k \geq \eta_1 + \cdots + \eta_k, \quad k = 1, \ldots, n-1.$$

and

$$\xi_1 + \cdots + \xi_n = \eta_1 + \cdots + \eta_n.$$

We say that x MAJORIZES y.

Theorem 3.12 (Hardy–Littlewood–Pólya). *Let the vectors x and y satisfy (3.11). Then a necessary and sufficient condition for there to exist a doubly stochastic matrix S with*

$$y = Sx \quad (3.12)$$

is that

$$x \succ y \quad (3.13)$$

Proof. The necessity of the condition is left as an exercise.

For the sufficiency, note that (3.13) implies that if all the x_i's are equal to some number, then the y_i's are equal to the same number, and we may take $S = I$. Moreover, if $x \succ y$ and we add a constant to the elements of x and y then we still have $x \succ y$. Consequently we may assume without loss of generality that

$$\xi_1 > 0 > \xi_n$$

and

$$\sum_{i=1}^{n} \xi_i = 0.$$

The proof is by induction. The result is trivial for $n = 1$. For $n = 2$, the most general form of a doubly stochastic matrix is

$$\begin{pmatrix} \alpha & 1-\alpha \\ 1-\alpha & \alpha \end{pmatrix},$$

where $0 \le \alpha \le 1$. Thus (3.12) requires that

$$\eta_1 = \alpha \xi_1 + (1-\alpha)\xi_2$$

and

$$\eta_2 = (1-\alpha)\xi_1 + \alpha \xi_2.$$

But the hypotheses of the theorem imply that $\xi_1 \ge \eta_1 \ge \xi_2$, which in turn implies that the first of these equalities can be satisfied for a unique $a \in [0,1]$. Summing the two equalities, we see that the second is equivalent to $\xi_1 + \xi_2 = \eta_1 + \eta_2$, which is the equality in (3.13).

For the general case, let us first note that if equality occurs in any of the inequalities (3.13), then x and y can be partitioned in the form

$$x = \begin{pmatrix} x_1 \\ x_2 \end{pmatrix} \text{ and } y = \begin{pmatrix} y_1 \\ y_2 \end{pmatrix},$$

where the pairs x_1, y_1 and x_2, y_2 satisfy the hypotheses of the theorem. Thus there are doubly stochastic matrices S_1 and S_2 such that $y_i = S_i x_i$ ($i = 1, 2$). It follows that $S = \mathrm{diag}(S_1, S_2)$ is the matrix required by the theorem.

3. UNITARILY INVARIANT NORMS

Let k be the index of the smallest positive ξ_i and l be the index of the largest negative ξ_i; i.e.,

$$\xi_k > 0 = \xi_{k+1} = \cdots = \xi_{l-1} > \xi_l.$$

Let x' be the vector obtained by replacing ξ_k by $\xi_k - \alpha$ and ξ_l by $\xi_l + \alpha$. For sufficiently small α, we have $x \succ x' \succ y$.

Now consider the following three cases.

$k > 1$: Here we have equality in the first of the relations $x \succ x'$.

$l < n$: Here we have equality in the next to last of the relations $x \succ x'$.

$k = 1, l = n$: This is equivalent to the 2×2 case.

In all three cases, there is a doubly stochastic matrix S' such that $x' = S'x$. Now let us increase α from zero until one of two things happens.

1. We obtain an equality in the relations $x' \succ y$.

2. x'_k or x'_l becomes zero.

In the first case, there is a doubly stochastic matrix S_y such that $y = S_y x'$, in which case $S = S_y S'$ is the matrix required by the theorem. In the second case, x' has one more zero component than x. We may then repeat the above construction to obtain a new vector $x'' \preceq x'$ and a double stochastic matrix S'' satisfying $x'' = S''x'$. Again we either have equality in the relation $x'' \succ y$, in which case the theorem is proved, or x'' has one more zero element than x'. Ultimately this reduction must furnish the required doubly stochastic matrix or produce a zero vector $x^{(m)}$. This latter implies that $y = 0$ and the matrix $S = S^{(m)} \cdots S''S'$ is the required doubly stochastic matrix. ∎

The second theorem is a characterization of doubly stochastic matrices which says that they are the convex hull of all permutation matrices. The proof requires a theorem of independent interest. We begin with a definition.

Definition 3.13. *Let $T \in \mathbf{C}^{m \times n}$, where $m \leq n$. If π is any permutation of $\{1, \ldots, n\}$ and $1 \leq j_1 < \cdots < j_m \leq n$, then the vector*

$$\left(\tau_{\pi(1), j_1}, \ldots, \tau_{\pi(m), j_m}\right)$$

is called a PERMUTATION VECTOR *of T.*

Theorem 3.14 (Hall). *Let $T \in \mathbf{C}^{m \times n}$ ($m \leq n$). Then there are permutation matrices P and Q such that PTQ has a $p \times q$ zero submatrix with $p + q > n$ if and only if every permutation vector of T contains a zero component.*

Proof. We will first show that if T can be permuted to the form

$$T = \begin{array}{c} p \\ m-p \end{array} \overset{\begin{array}{cc} q & n-q \end{array}}{\begin{pmatrix} 0 & T_{12} \\ T_{21} & T_{22} \end{pmatrix}}, \qquad (3.14)$$

where $p + q > n$, then any permutation vector must have zero components. Suppose to the contrary that $(\tau_{i_1,j_1}, \ldots, \tau_{i_m,j_m})$ is a permutation vector with no zero elements. Then from (3.14), first $m-(n-q)$ integers $i_1, \ldots, i_{m-(n-q)}$ must be distinct and lie between $p+1$ and m (inclusive). It follows that $m - p \geq m - (n - q)$ or $n \geq p + q$, a contradiction.

The proof of the converse is by induction. It is trivial for $m = 1$ or $n = 1$. Therefore assume that $m, n > 1$ and that every permutation vector of T has a zero component. Without loss of generality we may suppose that $\tau_{mn} \neq 0$. Then every permutation vector of the $(m-1) \times (n-1)$ leading principle submatrix must have a zero component. For any permutation vector not having a zero component could be combined with τ_{mn} to give a permutation vector of T that does not have a zero component — a contradiction.

By the induction hypothesis, T can be permuted to the form (3.14), where now $p + q = n$. It follows that T_{12} is square and T_{21} has at least as many columns as rows. Now at least one of the matrices T_{21} or T_{12} must have all its permutation vectors with zero components; for otherwise we could piece together permutation vectors from T_{21} and T_{12} having nonzero components to form a permutation vector for T whose components are nonzero.

Assume that all the permutation vectors of T_{21} have zero components. By the induction hypothesis, we can permute the rows and columns of T so that it has the form

$$\begin{array}{c} p \\ r \\ m-p-r \end{array} \overset{\begin{array}{cc} s & n-r \end{array}}{\begin{pmatrix} 0 & \hat{T}_{12} \\ 0 & \hat{T}_{22} \\ T_{31} & \hat{T}_{22} \end{pmatrix}}, \qquad (3.15)$$

3. UNITARILY INVARIANT NORMS

where $r + s > q$. Then $(p+r) + s > p + q = n$, which shows that (3.15) is the required matrix. ∎

From Theorem 3.14 we get the following corollary.

Corollary 3.15. *If $T \in \mathbf{R}^{n \times n}$ is a nonzero multiple of a doubly stochastic matrix, then T has a permutation vector consisting of positive elements.*

Proof. The proof is by contradiction. Without loss of generality, we may assume that T is stochastic. If every permutation vector of T contains a zero element, then from Theorem 3.14 the matrix T can be permuted to have a zero $p \times q$ submatrix with $p + q = n + 1$. Since the property of being doubly stochastic is invariant under permutations, we may assume that the zero submatrix is located in the upper left corner of T; i.e.,

$$T = \begin{matrix} \\ p \\ n-p \end{matrix} \begin{pmatrix} \overset{q}{0} & \overset{n-q}{T_{12}} \\ T_{21} & T_{22} \end{pmatrix}.$$

Now the sum of all elements of T_{12} is p and the sum of all elements of T_{21} is q. But $p + q > n$ which is greater than the sum of all elements of T. The contradiction establishes the corollary. ∎

We may now establish our characterization of double stochastic matrices.

Theorem 3.16 (Birkhoff). *The set of all doubly stochastic matrices of order n is the convex hull of all permutation matrices of order n; that is, any doubly stochastic matrix S can be expressed as a convex combination of the permutation matrices P_i ($i = 1, \ldots, n!$) :*

$$S = \sum_{i=1}^{n!} \sigma_i P_i, \quad \sum_{i=1}^{n!} \sigma_i = 1, \sigma_i \geq 0, i = 1, \ldots, n!. \tag{3.16}$$

Proof. Any matrix of the form (3.16) is clearly doubly stochastic. It therefore remains to show that any doubly stochastic matrix $S = (\sigma_{ij})$ of order n has the form (3.16).

By Corollary 3.15, S has a regular set $\sigma_{1i_1}, \ldots, \sigma_{ni_n}$ in which $\sigma_{ki_k} > 0$. Let $\sigma_1 = \min_{1 \leq k \leq n}\{\sigma_{ki_k}\}$, and let P_1 be the permutation matrix with 1 in its $(1, i_1), \ldots, (n, i_n)$ elements. Let $S_1 = S - \sigma_1 P_1$.

Clearly the matrix S_1 has the following properties:

1. S_1 is nonnegative;

2. The sum of all elements in each row and the sum of all elements in each column are equal to $1 - \sigma_1 \geq 0$;

3. The number of zero elements of S_1 is greater than that of S by at least one.

If $1 - \sigma_1 = 0$, then $S_1 = 0$; i.e., $S = \sigma_1 P_1$ and the theorem is proved. If $1 - \sigma_1 > 0$, then by Corollary 3.15 the matrix S_1 has a regular set consisting of positive elements. Repeating the above argument, we get a matrix $S_2 = S - \sigma_1 P_1 - \sigma_2 P_2$ in which $\sigma_2 > 0$ and P_2 is a permutation matrix. As above, the matrix S_2 is nonnegative. The sum of the elements in each row and the sum of the elements in each column are equal to $1 - \sigma_1 - \sigma_2 \geq 0$. Finally the number of zero elements of S_2 is greater than that of S by least two.

Continuing in this way we may produce a sequence $P_1, P_2, \ldots$ of permutations and multipliers $\sigma_1, \sigma_2, \ldots$, with $0 < \sigma_i \leq 1$. The sequence terminates when $1 - \sigma_1 - \cdots - \sigma_m = 0$, at which point

$$S = \sum_{i=1}^{m} \sigma_i P_i$$

is the required convex combination. ∎

We may now prove Fan's theorem.

Theorem 3.17 (Fan). *Let $x, y \in \mathbf{R}^n$ satisfy*

$$\xi_1 \geq \ldots \geq \xi_n \geq 0,$$
$$\eta_1 \geq \ldots \geq \eta_n \geq 0. \qquad (3.17)$$

Then
$$\xi_1 + \cdots + \xi_k \geq \eta_1 + \cdots + \eta_k, \quad k = 1, \ldots, n. \qquad (3.18)$$

is a necessary and sufficient condition for

$$\Phi(x) \geq \Phi(y)$$

to hold for all symmetric gauge functions Φ.

3. Unitarily Invariant Norms

Proof. For the necessity consider the symmetric gauge functions

$$\Phi_k(x) \stackrel{\text{def}}{=} \max_{1 \leq i_1 < \cdots < i_k \leq n} \{|\xi_{i_1}| + \cdots + |\xi_{i_k}|\}. \tag{3.19}$$

If x and y satisfy (3.17) then $\Phi_k(x) \geq \Phi_k(y)$ ($k = 1, \ldots, n$) is equivalent to (3.18).

For sufficiency, note that by successively reducing ξ_n, then ξ_{n-1}, and so on, we can obtain a vector $\hat{x} \leq x$ such that $\hat{x} \succ y$. By the Hardy–Littlewood–Pólya theorem, there is a doubly stochastic matrix S such that $y = S\hat{x} \leq Sx$. By Birkhoff's theorem we can write S as a convex combination of permutation matrices:

$$S = \sum_{i=1}^{n!} \sigma_i P_i, \qquad \sum_{i=1}^{n!} \sigma_i = 1, \sigma_i \geq 0, i = 1, \ldots, n!.$$

It then follows that for any symmetric gauge function Φ,

$$\Phi(y) \leq \Phi(Sx) = \Phi\left(\sum_{i=1}^{n!} \sigma_i P_i x\right) \leq \sum_{i=1}^{n!} \sigma_i \Phi(P_i x) = \Phi(x). \blacksquare$$

The symmetric gauge functions Φ_k defined by (3.19) have associated unitarily invariant norms $\|\cdot\|_{\Phi_k}$. When Fan's theorem is recast in terms of these norms it takes the following form.

Corollary 3.18. In order for

$$\|A\| \leq \|B\|$$

for every unitarily invariant norm it is necessary and sufficient that

$$\|A\|_{\Phi_k} \leq \|B\|_{\Phi_k}, \qquad k = 1, \ldots, n.$$

Notes and References

The characterization of unitarily invariant norms as symmetric gauge functions of singular values is due to von Neumann [254, 1932]. Lemma 3.4 is of independent interest, since it links the eigenvalues and the singular values of a matrix (see Exercise 3.2). The proof given here is new.

For surveys with extensive bibliographies of the subjects of doubly stochastic matrices and majorization see [159, 5]. For the Hardy–Littlewood–Pólya theorem see [103, 1934]. The proof given here is due to Ostrowski [168, 1952]. For Birkhoff's theorem see [33, 1946], and for Fan's see [70, 1951]. Fan's paper is the culmination of a flury of results, initiated by Weyl [266, 1949] on inequalities bounding eigenvalues in terms of singular values. As Fan points out, these results can be used to establish von Neumann's characterization of unitarily invariant norms in terms of singular values.

Theorem 3.14, which we have called Hall's theorem, is associated with the names König and Frobenius. Hall [100, 1935] actually proved a set theoretic version of the theorem given here and noted that it was a generalization of a graph theoretic theorem of König [136, 1916], which van der Warden [246, 1927] had recast in a set theoretic form. The association with Frobenius appears to be spurious. The earliest reference we know that mentions him in this connection is by Dulmage and Halperin [62, 1955], from which the proof given here is adapted. This paper cites Frobenius's famous paper on nonnegative matrices [78, 1912]. However, the theorem does not appear there, at least not in an obvious form.

Exercises

1. Show that for any unitarily invariant norm $\|\cdot\|$,

$$\left\|\begin{pmatrix} A_{11} & 0 \\ 0 & A_{22} \end{pmatrix}\right\| \leq \left\|\begin{pmatrix} A_{11} & A_{21} \\ A_{12} & A_{22} \end{pmatrix}\right\|.$$

2. Let $\mathcal{L}(A) = \{\lambda_1, \ldots, \lambda_n\}$ and $\mathcal{S}(A) = \{\sigma_1, \ldots, \sigma_n\}$. Show that

$$|\lambda_1| + \cdots + |\lambda_n| \leq \sigma_1 + \cdots + \sigma_n.$$

3. (Fan and Hoffman [71]). Let $\|\cdot\|$ be a unitarily invariant norm.

    1. Show that if H is Hermitian and U is unitary then

    $$\|H - I\| \leq \|H - U\| \leq \|H + I\|.$$

    2. Show that for any Hermitian matrix H,

    $$\left\|A - \frac{A + A^{\mathrm{H}}}{2}\right\| \leq \|A - H\|.$$

4. For any real vector x let $x^+ = (x + |x|)/2$ be the result of setting the negative components of x to zero. Show that $x \succ y$ if and only if $\mathbf{1}^T(x - \tau\mathbf{1})^+ \geq \mathbf{1}^T(y - \tau\mathbf{1})^+$ for all real τ.

5. Show that $x \succ y$ if and only if y is a convex combination of all vectors of the form Px where P is a permutation matrix.

6. (Hall's theorem, the original version [100]). Let $\mathcal{A}$ be a set consisting of n elements. For $m \leq n$ suppose that $\mathcal{A} = \bigcup_{i=1}^m \mathcal{B}_i$. Show that there exist m distinct elements $a_1, \ldots, a_m$ of $\mathcal{A}$ such that $a_i \in \mathcal{B}_i$ ($i = 1, \ldots, m$) if and only if every union of k of the sets $\mathcal{B}_i$ contains at least k distinct elements of $\mathcal{A}$.

4. Metrics on Subspaces of $\mathbf{C}^n$

A difficulty in framing a workable perturbation theory for eigenvectors is that there may be no unique eigenvector corresponding to a multiple eigenvalue. For example, any nonzero vector in the space spanned by the unit vectors $\mathbf{1}_1$ and $\mathbf{1}_2$ is an eigenvector of the matrix

$$A = \begin{pmatrix} 1 & 0 & 0 \\ 0 & 1 & 0 \\ 0 & 0 & 2 \end{pmatrix}.$$

When A is perturbed, two distinct eigenvectors can precipitate from this subspace, and for different perturbations these eigenvectors can be quite different, even when the perturbations are small. For example, when $\epsilon > 0$ the matrix

$$\begin{pmatrix} 1 & 0 & 0 \\ 0 & 1+\epsilon & 0 \\ 0 & 0 & 2 \end{pmatrix}$$

has eigenvectors $\mathbf{1}_1$ and $\mathbf{1}_2$, whereas the matrix

$$\begin{pmatrix} 1 & \epsilon & 0 \\ \epsilon & 1 & 0 \\ 0 & 0 & 2 \end{pmatrix}$$

has the very different eigenvectors $(1\ 1\ 0)^{\mathrm{T}}$ and $(1\ -1\ 0)^{\mathrm{T}}$. In spite of such differences, the two eigenvectors will always span a space that is very near to the space spanned by $\mathbf{1}_1$ and $\mathbf{1}_2$. Thus it makes more sense to derive perturbation bounds for the subspace, which is stable, rather than for the eigenvectors, which are not.

In order to derive such a theory, we must first specify what we mean by the distance between subspaces. Since we will be comparing only subspaces of the same dimensions, we can restrict ourselves to the problem of introducing a notion of distance on the set of all l dimensional subspaces of $\mathbf{C}^n$, a set that we will denote by $\mathbf{C}^n_l$ (or $\mathbf{R}^n_l$ in the real case). Unfortunately, this can be done in a number of ways, not all of which turn out as we might hope.

Example 4.1. *Consider the space $\mathbf{R}^2_1$ of all infinitely extended lines in the plane that pass through the origin. Given any subspace $\mathcal{X} \in \mathbf{R}^2_1$, there is a unique vector $x(\mathcal{X}) = (\xi_1\ \xi_2)^{\mathrm{T}}$ in $\mathcal{X}$ that lies in the half open semicircle $\{x : \|x\|_2 = 1, 0 \leq \xi_1, -1 < \xi_2 \leq 1\}$. It is easy to verify that the function*

$$\rho(\mathcal{X}, \mathcal{Y}) = \angle[x(\mathcal{X}), x(\mathcal{Y})]$$

is a metric on $\mathbf{R}^2_1$. However, the lines along the directions $(0, -1)$ and $(\epsilon, -1)$ are nearly π apart in this metric, even though the lines themselves approach one another as $\epsilon \to 0$.

In the above example we gave $\mathbf{R}^2_1$ the topology of a semicircle open at one end, which allows lines that are near in the usual sense of the word to be far apart in the sense of the ρ metric. This shows that we have to take some care in defining distances and metrics on $\mathbf{C}^n_l$. In the first subsection we will introduce one widely used distance function — the gap — and derive its properties. In the second subsection we will consider metrics that are unitarily invariant.

4.1. The Gap

In defining a notion of distance between subspaces, it is natural to begin with the distance between a point and a subspace.

4. METRICS ON SUBSPACES OF $\mathbf{C}^n$

Definition 4.2. *Let $\mathcal{X}$ be a subspace of $\mathbf{C}^n$ and let $y \in \mathbf{C}^n$. If ν is a norm on $\mathbf{C}^n$, then the ν-DISTANCE between y AND $\mathcal{X}$ IS THE FUNCTION*

$$\delta_\nu(y, \mathcal{X}) \stackrel{\text{def}}{=} \min_{x \in \mathcal{X}} \nu(y - x). \tag{4.1}$$

An elementary compactness argument shows that δ_ν is well defined. When ν is the 2-norm, it follows from Theorem I.2.5 that

$$\delta_2(y, \mathcal{X}) = \|(I - P_\mathcal{X})y\|_2. \tag{4.2}$$

In other words, the 2-distance between y and $\mathcal{X}$ is the distance between y and its projection onto $\mathcal{X}$. For this reason, a minimizing vector x in (4.1) is sometimes called a ν-projection of y onto $\mathcal{X}$.

We are now in a position to define a distance between subspaces.

Definition 4.3. *Let $\mathcal{X}, \mathcal{Y} \in \mathbf{C}_l^n$ and let ν be a norm on $\mathbf{C}^n$. Then the ν-GAP BETWEEN $\mathcal{X}$ AND $\mathcal{Y}$ is the number*

$$\rho_{g,\nu}(\mathcal{X}, \mathcal{Y}) \stackrel{\text{def}}{=} \max \left\{ \max_{\substack{x \in \mathcal{X} \\ \nu(x)=1}} \delta_\nu(x, \mathcal{Y}), \max_{\substack{y \in \mathcal{Y} \\ \nu(y)=1}} \delta_\nu(y, \mathcal{X}) \right\}.$$

Thus the gap is the largest distance from $\mathcal{Y}$ of a vector of length one lying in $\mathcal{X}$, or vice versa, whichever is greater.

The definition of the gap function satisfies our intuitive notions of what a distance between subspaces should be. In Example 4.1 it gives $\mathbf{R}_1^2$ a topology in which lines that intersect at small angles have a small distance from one another. However, the gap function need not be a metric, and this means that we must establish from first principles that a gap function actually generates a topology. We shall do this in two stages. First we shall prove that all gap functions are equivalent, in the same sense that all norms are equivalent. We will then show that one gap function, $\rho_{g,2}$ is a metric. It will then follow from their equivalence that all gap functions generate the same topology.

Theorem 4.4. *Let μ and ν be norms on $\mathbf{C}^n$, and let*

$$\alpha \mu(x) \leq \nu(x) \leq \beta \mu(x), \quad \alpha, \beta > 0.$$

Then

$$\frac{\alpha}{\beta} \rho_{g,\mu}(\mathcal{X}, \mathcal{Y}) \leq \rho_{g,\nu}(\mathcal{X}, \mathcal{Y}) \leq \frac{\beta}{\alpha} \rho_{g,\mu}(\mathcal{X}, \mathcal{Y}). \tag{4.3}$$

Proof. First we will establish the second inequality in (4.3). Let $x \in \mathcal{X}$ with $\nu(x) = 1$ and let $y \in \mathcal{Y}$. Let $x' = x/\mu(x)$, so that $\mu(x') = 1$, and let $y' = y/\mu(x)$. Then

$$\nu(x-y) = \mu(x)\nu(x'-y') \leq \frac{1}{\alpha}\nu(x'-y') \leq \frac{\beta}{\alpha}\mu(x'-y').$$

From this it follows that

$$\delta_\nu(x,\mathcal{Y}) \leq \frac{\beta}{\alpha}\delta_\mu(x',\mathcal{Y}).$$

As x ranges over all vectors in $\mathcal{X}$ with $\nu(x) = 1$, x' ranges over all vectors in $\mathcal{X}$ with $\mu(x') = 1$. Hence

$$\max_{\substack{x \in \mathcal{X} \\ \nu(x)=1}} \delta_\nu(x,\mathcal{Y}) \leq \frac{\beta}{\alpha} \max_{\substack{x' \in \mathcal{X} \\ \mu(x')=1}} \delta_\mu(x',\mathcal{Y}).$$

Similarly,

$$\max_{\substack{y \in \mathcal{Y} \\ \nu(y)=1}} \delta_\nu(y,\mathcal{X}) \leq \frac{\beta}{\alpha} \max_{\substack{y' \in \mathcal{Y} \\ \mu(y')=1}} \delta_\mu(y',\mathcal{X}),$$

and the second inequality follows from the definition of the gap.

The first inequality in (4.3) follows from the second and from the fact that

$$\beta^{-1}\nu(x) \leq \mu(x) \leq \alpha^{-1}\nu(x). \quad \blacksquare$$

We shall now prove that $\rho_{g,2}$ is a metric. The proof is based on the following characterization of $\rho_{g,2}(\mathcal{X},\mathcal{Y})$ in terms of the canonical angles between $\mathcal{X}$ and $\mathcal{Y}$.

Theorem 4.5. *Let $\mathcal{X}, \mathcal{Y} \in \mathbf{C}_l^n$, and let $\Theta = \operatorname{diag}(\theta_1, \ldots, \theta_l)$, where $\theta_1 \geq \cdots \geq \theta_l$ are the canonical angles between $\mathcal{X}$ and $\mathcal{Y}$. Then*

$$\rho_{g,2}(\mathcal{X},\mathcal{Y}) = \sin\theta_1 = \|\sin\Theta\|_2 \qquad (4.4)$$

Proof. By (4.2),

$$\delta_2(x,\mathcal{Y}) = \|(I - P_\mathcal{Y})x\|_2.$$

4. METRICS ON SUBSPACES OF $\mathbf{C}^n$

Hence
$$\max_{\substack{x \in \mathcal{X} \\ \|x\|_2=1}} \delta_2(x, \mathcal{Y}) = \max_{\substack{x \in \mathcal{X} \\ \|x\|_2=1}} \|(I - P_\mathcal{Y})x\|_2$$
$$= \max_{\substack{x \in \mathcal{X} \\ \|x\|_2 \le 1}} \|(I - P_\mathcal{Y})x\|_2$$
$$= \max_{\|r\|_2=1} \|(I - P_\mathcal{Y})P_\mathcal{X} r\|_2$$
$$= \|(I - P_\mathcal{Y})P_\mathcal{X}\|_2.$$

Thus by Theorem I.5.5,
$$\max_{\substack{x \in \mathcal{X} \\ \|x\|_2=1}} \delta_2(x, \mathcal{Y}) = \sin \theta_1.$$

Similarly,
$$\max_{\substack{y \in \mathcal{Y} \\ \|y\|_2=1}} \delta_2(y, \mathcal{X}) = \sin \theta_1,$$

and the result follows from the definition of the gap. ∎

Since by Theorem I.5.5 the singular values of $P_\mathcal{X} - P_\mathcal{Y}$ are the sines of the singular values of the canonical angles between $\mathcal{X}$ and $\mathcal{Y}$, we have the following corollary.

Corollary 4.6. *In the 2-norm,*
$$\rho_{g,2}(\mathcal{X}, \mathcal{Y}) = \|P_\mathcal{X} - P_\mathcal{Y}\|_2. \tag{4.5}$$

It immediately follows from Corollary 4.6 that $\rho_{g,2}$ is a metric on $\mathbf{C}_l^n$ [e.g., to get the triangle inequality write $\rho_{g,2}(\mathcal{X}, \mathcal{Z}) = \|P_\mathcal{X} - P_\mathcal{Z}\|_2 \le \|P_\mathcal{X} - P_\mathcal{Y}\|_2 + \|P_\mathcal{Y} - P_\mathcal{Z}\|_2 = \rho_{g,2}(\mathcal{X}, \mathcal{Y}) + \rho_{g,2}(\mathcal{Y}, \mathcal{Z})$]. Consequently, $\rho_{g,2}$ induces a topology on $\mathbf{C}_l^n$, and by Theorem 4.4 all gap functions induce the same topology, which we will call the GAP TOPOLOGY.

Equation (4.5) does not hold in general; if we replace the 2-norm by another norm, the equality may fail. However, the right-hand side, regarded as a function of $\mathcal{X}$ and $\mathcal{Y}$, remains a metric on $\mathbf{C}_l^n$. We leave the proof of the following theorem as an exercise (the subscript p in (4.6) stands for projection).

Theorem 4.7. *Let ν be a matrix norm on $\mathbf{C}^{n \times n}$. Then the function*
$$\rho_{p,\nu}(\mathcal{X}, \mathcal{Y}) \stackrel{\text{def}}{=} \nu(P_\mathcal{X} - P_\mathcal{Y}) \tag{4.6}$$
is a metric on $\mathbf{C}_l^n$, which generates the gap topology.

4.2. Unitarily Invariant Metrics

A metric ρ on $\mathbf{C}_l^n$ is UNITARILY INVARIANT if $\rho(\mathcal{X},\mathcal{Y}) = \rho(U\mathcal{X}, U\mathcal{Y})$ for all unitary matrices U. In this subsection we will be interested in unitarily invariant metrics that generate the gap topology. Now the metric of Example 4.1 is not unitarily invariant and does not generate the gap topology. It is therefore appealing to conjecture that any unitarily invariant metric must generate the gap topology. Unfortunately, the conjecture is not true, as the following example shows.

Example 4.8. Let $\mathcal{X}, \mathcal{Y} \in \mathbf{R}_1^2$ and let $\theta(\mathcal{X}, \mathcal{Y})$ be the canonical angle between $\mathcal{X}$ and $\mathcal{Y}$. Define $\rho(\mathcal{X}, \mathcal{Y})$ by

$$\rho(\mathcal{X},\mathcal{Y}) = \begin{cases} \theta(\mathcal{X},\mathcal{Y}), & \text{if } \theta(\mathcal{X},\mathcal{Y}) \text{ is rational,} \\ 1, & \text{if } \theta(\mathcal{X},\mathcal{Y}) \text{ is irrational.} \end{cases}$$

Then it is easily verified that ρ is a unitarily invariant metric. But two subspaces with an irrational canonical angle are at a distance of unity, no matter how near they are in the gap topology.

Fortunately, there are many unitarily invariant metrics that generate the gap topology. One of them is the gap function $\rho_{g,2}$, since $\|\cdot\|_2$ is unitarily invariant. However, this metric effectively exhausts class of unitarily invariant metrics that can be generated by gap functions, since up to a constant multiple the 2-norm is the only unitarily invariant vector norm on $\mathbf{C}^n$. But there are many unitarily invariant *matrix* norms, which can be used in the definition (4.6) to give unitarily invariant metrics generating the gap topology.

Theorem 4.9. *If ν is a unitarily invariant matrix norm, then $\rho_{p,\nu}$ is a unitarily invariant metric on $\mathbf{C}_l^n$.*

By (4.4), $\rho_{g,2}(\mathcal{X},\mathcal{Y}) = \|\sin\Theta(\mathcal{X},\mathcal{Y})\|_2$, and it is natural to ask if we can find new unitarily invariant metrics of the form $\nu[\sin\Theta(\mathcal{X},\mathcal{Y})]$, where ν is a unitarily invariant matrix norm. Unfortunately, these metrics are just the $\rho_{p,\nu}$ metrics in disguise.

Theorem 4.10. *Let ν be a unitarily invariant matrix norm on $\mathbf{C}^{l\times l}$. Then there is a unitarily invariant matrix norm ν' such that*

$$\nu[\sin\Theta(\mathcal{X},\mathcal{Y})] = \rho_{p,\nu'}(\mathcal{X},\mathcal{Y})$$

for all $\mathcal{X}, \mathcal{Y} \in \mathbf{C}_l^n$.

4. Metrics on Subspaces of $\mathbf{C}^n$

Proof. We will treat only the case $2l \leq n$, leaving the case $2l > n$ as an exercise. Let $\sigma_1 \geq \cdots \geq \sigma_l$ be the sines of the canonical angles between $\mathcal{X}$ and $\mathcal{Y}$. By Theorem I.5.5, we know that the singular values of $P_\mathcal{X} - P_\mathcal{Y}$ are $\sigma_1, \sigma_1, \ldots, \sigma_l, \sigma_l, 0, \ldots, 0$.

Let Φ be the symmetric gauge function that generates ν, and define $\Phi' : \mathbf{R}^n \to \mathbf{R}$ as follows. For any vector $x \in \mathbf{R}^n$ with $|\xi_{i_1}| \geq |\xi_{i_2}| \geq \cdots \geq |\xi_{i_n}|$ let

$$\Phi'(x) = \Phi\left(\frac{|\xi_{i_1}| + |\xi_{i_2}|}{2}, \frac{|\xi_{i_3}| + |\xi_{i_4}|}{2}, \ldots, \frac{|\xi_{i_{2l-1}}| + |\xi_{i_{2l}}|}{2}\right).$$

It is easily verified that Φ' is a symmetric gauge function. Let ν' be the unitarily invariant matrix norm generated by Φ'. Then

$$\nu'(P_\mathcal{X} - P_\mathcal{Y}) = \Phi'(\sigma_1, \sigma_1, \ldots, \sigma_l, \sigma_l, 0, \ldots, 0)$$
$$= \Phi(\sigma_1, \ldots, \sigma_l)$$
$$= \nu[\sin \Theta(\mathcal{X}, \mathcal{Y})]. \quad \blacksquare$$

We now turn to two different metrics on $\mathbf{C}_l^n$. To motivate the first, let $\mathcal{X}, \mathcal{Y} \in \mathbf{C}_l^n$ and let the columns of X and Y form orthonormal bases for $\mathcal{X}$ and $\mathcal{Y}$. If $\mathcal{X} = \mathcal{Y}$, then there is an unitary matrix $Q = Y^H X$ such that $X = YQ$ or equivalently $\|X - YQ\|_F = 0$. This suggests that we use the number

$$\rho_b(\mathcal{X}, \mathcal{Y}) \stackrel{\text{def}}{=} \min_{Q \text{ unitary}} \|X - YQ\|_F$$

as a measure of the distance between $\mathcal{X}$ and $\mathcal{Y}$ (the subscript b stands for basis).

Theorem 4.11. *The function ρ_b is a unitarily invariant metric on $\mathbf{C}_l^n$. If γ_i is the cosine of the ith canonical angle between $\mathcal{X}$ and $\mathcal{Y}$, then*

$$\rho_b(\mathcal{X}, \mathcal{Y}) = \sqrt{2\sum_i (1 - \gamma_i)}. \quad (4.7)$$

Proof. It is easy to see that ρ_b is a unitarily invariant, nonnegative function that is zero if and only if its arguments are equal. We will now show that it satisfies the triangle inequality.

Let the columns of X, Y, Z form orthonormal bases for $\mathcal{X}, \mathcal{Y}, \mathcal{Z} \in \mathbf{C}_l^n$. Let
$$\rho_{\mathrm{b}}(\mathcal{X}, \mathcal{Y}) = \|X - YQ_{X,Y}\|_{\mathrm{F}},$$
$$\rho_{\mathrm{b}}(\mathcal{Y}, \mathcal{Z}) = \|Y - ZQ_{Y,Z}\|_{\mathrm{F}},$$
$$\rho_{\mathrm{b}}(\mathcal{X}, \mathcal{Z}) = \|X - ZQ_{X,Z}\|_{\mathrm{F}}.$$
Then
$$\rho_{\mathrm{b}}(\mathcal{X}, \mathcal{Z}) \leq \|X - ZQ_{Y,Z}Q_{X,Y}\|_{\mathrm{F}}$$
$$= \|X - YQ_{X,Y} + YQ_{X,Y} - ZQ_{Y,Z}Q_{X,Y}\|_{\mathrm{F}}$$
$$\leq \|X - YQ_{X,Y}\|_{\mathrm{F}} + \|Y - ZQ_{Y,Z}\|_{\mathrm{F}}$$
$$= \rho_{\mathrm{b}}(\mathcal{X}, \mathcal{Y}) + \rho_{\mathrm{b}}(\mathcal{Y}, \mathcal{Z}).$$

We will establish (4.7) for the case $2l \leq n$. Without loss of generality, we may assume that X and Y are canonical bases for $\mathcal{X}$ and $\mathcal{Y}$. Thus we must find a unitary matrix Q that minimizes
$$\left\| \begin{pmatrix} I \\ 0 \\ 0 \end{pmatrix} - \begin{pmatrix} \Gamma \\ \Sigma \\ 0 \end{pmatrix} Q \right\|_{\mathrm{F}}^2 = \|I - \Gamma Q\|_{\mathrm{F}}^2 + \|\Sigma\|_{\mathrm{F}}^2.$$

The second term on the right-hand side of the above equation is independent of Q. Hence Q must minimize
$$\|I - \Gamma Q\|_{\mathrm{F}}^2 = \mathrm{trace}(I - Q^{\mathrm{H}}\Gamma - \Gamma Q + \Gamma^2).$$

This quantity is minimized when the diagonals of Q are one, and since Q is unitary, $Q = I$. Hence
$$\|I - \Gamma Q\|_{\mathrm{F}}^2 + \|\Sigma\|_{\mathrm{F}}^2 = \mathrm{trace}(I - 2\Gamma + \Gamma^2 + \Sigma^2)$$
$$= 2\mathrm{trace}(I - \Gamma) = 2\sum_i(1 - \gamma_i). \quad \blacksquare$$

The second metric is defined by the formula
$$\rho_\theta(\mathcal{X}, \mathcal{Y}) = \arccos \frac{|\det(X^{\mathrm{H}}Y)|}{\sqrt{\det(X^{\mathrm{H}}X)\det(Y^{\mathrm{H}}Y)}},$$
where the columns of X and Y form bases for $\mathcal{X}$ and $\mathcal{Y}$. We will also consider the closely related metric
$$\rho_{\mathrm{s}}(\mathcal{X}, \mathcal{Y}) = \sin \rho_\theta(\mathcal{X}, \mathcal{Y}).$$

4. Metrics on Subspaces of $\mathbf{C}^n$

It is easily seen that these functions are unitarily invariant and independent of the choice of X and Y. By choosing canonical bases we see that
$$\rho_\theta(\mathcal{X}, \mathcal{Y}) = \arccos \prod_i \gamma_i, \qquad (4.8)$$
where, as usual, the γ_i's are the cosines of the canonical angles between $\mathcal{X}$ and $\mathcal{Y}$. The proof of the following theorem shows that they are metrics on $\mathbf{C}^n_l$.

Theorem 4.12. *The functions ρ_θ and ρ_s are unitarily invariant metrics on $\mathbf{C}^n_l$.*

Proof. The fact that ρ_s is a metric follows immediately from the fact that ρ_θ is a metric, which we will now establish. From (4.8) we have

$$\rho_\theta(\mathcal{X}, \mathcal{Y}) = 0 \iff \prod_i \gamma_i = 1$$
$$\iff \gamma_1 = \gamma_2 = \cdots = 1$$
$$\iff \mathcal{X} = \mathcal{Y}.$$

Thus it remains only to show that ρ_θ satisfies the triangle inequality.

Let the columns of X, Y, Z form orthonormal bases for $\mathcal{X}, \mathcal{Y}, \mathcal{Z} \in \mathbf{C}^n_l$. Then we must show that

$$\arccos |\det(X^\mathrm{H} Z)| \leq \arccos |\det(X^\mathrm{H} Y)| + \arccos |\det(Y^\mathrm{H} Z)|.$$

For $S = X, Y, Z$, let $\delta^{(S)}_{i_1 \ldots i_l}$ denote the determinant of the matrix formed from the i_1th, ..., i_lth rows of S. By the Binet-Cauchy formula (see, e.g., [81, V.I, p.9]) we have

$$\det(X^\mathrm{H} Y) = \sum_{1 \leq i_1 < \ldots < i_l \leq n} \delta^{(\bar{X})}_{i_1 \ldots i_l} \delta^{(Y)}_{i_1 \ldots i_l}, \qquad (4.9)$$

with similar formulas for $\det(X^\mathrm{H} Z)$ and $\det(Y^\mathrm{H} Z)$. For $S = X, Y, Z$, let the components of v_S be the numbers $\delta^{(S)}_{i_1 \ldots i_l}$ taken in some fixed order. Then
$$\det(X^\mathrm{H} Y) = v_X^\mathrm{H} v_Y,$$
with similar formulas for $\det(X^\mathrm{H} Z)$ and $\det(Y^\mathrm{H} Z)$. Thus our problem reduces to showing that

$$\arccos |v_X^\mathrm{H} v_Z| \leq \arccos |v_X^\mathrm{H} v_Y| + \arccos |v_Y^\mathrm{H} v_Z|. \qquad (4.10)$$

The QR decomposition of the matrix $(v_Y \; v_X \; v_Z)$ gives a unitary matrix U such that

$$v_Y = U(1, 0, \ldots, 0)^\mathrm{T},$$
$$v_X = U(\alpha_1, \alpha_2, 0, \ldots, 0)^\mathrm{T}, \qquad |\alpha_1|^2 + |\alpha_2|^2 = 1,$$
$$v_Z = U(\beta_1, \beta_2, \beta_3, 0, \ldots, 0)^\mathrm{T}, \qquad |\beta_1|^2 + |\beta_2|^2 + |\beta_3|^2 = 1.$$

Since

$$|\bar{\alpha}_1 \beta_1 + \bar{\alpha}_2 \beta_2| \geq |\alpha_1||\beta_1| - |\alpha_2||\beta_2|$$
$$= |\alpha_1||\beta_1| - \sqrt{1 - |\alpha_1|^2}\sqrt{1 - |\beta_1|^2 - |\beta_3|^2}$$
$$\geq |\alpha_1||\beta_1| - \sqrt{1 - |\alpha_1|^2}\sqrt{1 - |\beta_1|^2}$$
$$= \cos(\arccos|\alpha_1| + \arccos|\beta_1|),$$

we have

$$\arccos|\bar{\alpha}_1 \beta_1 + \bar{\alpha}_2 \beta_2| \leq \arccos|\alpha_1| + \arccos|\beta_1|$$

which is the required inequality (4.10). ∎

Since most perturbation bounds deal with small quantities, it is instructive to examine the asymptotic behavior of the metrics introduced above when the canonical angles θ_i are small. Specifically, we have

$$\begin{aligned}
&1. \; \rho_{\mathrm{g},2} = \sin\theta_1 & &\cong \|\Theta\|_2, \\
&2. \; \rho_{\mathrm{p,F}} = \sqrt{2\sum \sin^2\theta_i} & &\cong \sqrt{2}\|\Theta\|_\mathrm{F}, \\
&3. \; \rho_\mathrm{b} = \sqrt{2\sum(1 - \cos\theta_i)} & &\cong \|\Theta\|_\mathrm{F}, \\
&4. \; \rho_\theta = \arccos(\prod \cos\theta_i) & &\cong \|\Theta\|_\mathrm{F}.
\end{aligned} \qquad (4.11)$$

In particular, comparing (4.11.3) and (4.11.4) with (4.11.2) we see that the metrics ρ_b and ρ_θ also generate the gap topology.

Notes and References

The original impetus for comparing subspaces came from functional analysis. According to Berkson [27], the gap or "opening" $\rho_{g,2}$ was first defined on Hilbert space by Krein and Krasnoselski [137, 1947]. Krein, Krasnoselski, and Milman later extended the notion to an arbitrary Banach space [138, 1948]. For more on the gap and its applications see Kato [135].

4. Metrics on Subspaces of $\mathbf{C}^n$

The metric ρ_b was introduced by Paige [173, 1984] and the metric ρ_θ y Lu [149, 1963]. At this writing there is no simple characterization of unitarily invariant metrics for subspaces. One reason is that these metrics — unlike norms, which are defined for one object on a space with a linear structure — are relations between two objects in a space with a complicated structure. However, the survey in this section suggests that any reasonable approach will yield something that can be expressed in terms of canonical angles — at least asymptotically [cf. (4.11)].

Exercises

In the following exercises $\mathcal{X}$ and $\mathcal{Y}$ are subspaces of $\mathbf{C}^n$ and ν is a norm on $\mathbf{C}^n$. Unless otherwise stated $\mathcal{X}$ and $\mathcal{Y}$ have the same dimension.

1. Show that if $\dim(\mathcal{X}) > \dim(\mathcal{Y})$ then there is a point $x \in \mathcal{X}$ such that $\delta_\nu(x, \mathcal{Y}) = \nu(x)$. [Note: This theorem is nontrivial. For a proof and further references see [135, Ch.IV,Sec.2], from which some of the following exercises were excavated.]

2. Show that $\rho_{g,\nu}(\mathcal{X}, \mathcal{Y}) \leq 1$, with equality if $\dim(\mathcal{X}) \neq \dim(\mathcal{Y})$.

3. Let
$$\hat{\delta}_\nu(y, \mathcal{X}) = \min_{\substack{x \in \mathcal{X} \\ \nu(x)=1}} \nu(y - x)$$
and
$$\hat{\rho}_{g,\nu}(\mathcal{X}, \mathcal{Y}) = \max\left\{ \max_{\substack{x \in \mathcal{X} \\ \nu(x)=1}} \hat{\delta}_\nu(x, \mathcal{Y}),\ \max_{\substack{y \in \mathcal{Y} \\ \nu(y)=1}} \hat{\delta}_\nu(y, \mathcal{X}) \right\}.$$
Show that $\hat{\rho}_{g,\nu}$ is a metric.

4. Show that
$$\rho_{g,\nu}(\mathcal{X}, \mathcal{Y}) \leq \hat{\rho}_{g,\nu}(\mathcal{X}, \mathcal{Y}) \leq 2\rho_{g,\nu}(\mathcal{X}, \mathcal{Y}).$$

5. Show that if $\{\mathcal{X}_k\}$ is a Cauchy sequence in $\hat{\rho}_{g,\nu}$ then there is a subspace $\mathcal{Y}$ such that $\mathcal{X}_k \to \mathcal{Y}$ in $\hat{\rho}_{g,\nu}$ (i.e., the space of all subspaces is complete).

6. (Schäffer [153]) . Let
$$\pi(\mathcal{X}, \mathcal{Y}) = \max\left\{ \min\{\nu(I - C) : C\mathcal{X} = \mathcal{Y}\}, \min\{\nu(I - C) : C\mathcal{Y} = \mathcal{X}\} \right\}$$

and
$$\rho_s(\mathcal{X}, \mathcal{Y}) = \log[1 + \pi(\mathcal{X}, \mathcal{Y})].$$
Show that ρ_s is a metric.

Chapter III
Linear Systems and Least Squares Problems

In this chapter we will be concerned with the solution of the linear system
$$Ax = b,$$
where A is a nonsingular matrix, and with the closely related least squares problem
$$\text{minimize} \quad \|b - Ax\|_2, \qquad (1)$$
where A is a general $m \times n$ matrix. A solution of the latter problem is given by $A^\dagger b$, where $A^\dagger$ is the pseudo-inverse of A to be defined below. When A is nonsingular, $A^\dagger = A^{-1}$, so that the pseudo-inverse solves the first problem as well.

This chapter begins with an introductory section, after which we will treat the perturbation of matrix inverses and pseudo-inverses and in consequence the solution of linear systems and linear least squares problems. As we have noted, the former is a special case of the latter, and in principle we could approach the subject by developing the perturbation theory of pseudo-inverses and least squares problems and note what happens when $m = n$. However, the perturbation of pseudo-inverses is complicated by the fact that the pseudo-inverse need not be a continuous function of its elements. We will therefore begin with the simpler case of matrix inverses and linear systems. This approach also has the advantage of presenting some of the key ideas of perturbation theory in a comparatively simple setting.

1. The Pseudo-Inverse and Least Squares

1.1. Generalized Inverses and the Pseudo-Inverse

Let $A \in \mathbf{C}^{n \times n}$. It is well known that if A is nonsingular then there is a unique matrix X such that

$$AX = XA = I. \tag{1.1}$$

The matrix X is called the inverse of A and is denoted by A^{-1}. In this case the linear system $Ax = b$ has a unique solution $x = A^{-1}b$.

It is natural to attempt to generalize the idea of an inverse to the case where A is singular or even fails to be square. This can be done by requiring X to satisfy conditions that are less restrictive than (1.1). By varying the conditions, we can obtain many different "generalized inverses," each suited to its own application. In this book we will be particularly concerned with the geometry of $\mathbf{C}^n$, and the following PENROSE CONDITIONS are the most appropriate:

$$\begin{aligned}&1.\quad AXA = A,\\&2.\quad XAX = X,\\&3.\quad (AX)^{\mathrm{H}} = AX,\\&4.\quad (XA)^{\mathrm{H}} = XA.\end{aligned} \tag{1.2}$$

Note that the first condition alone implies that $X = A^{-1}$ when A is nonsingular. However, it does not define X uniquely when A is singular. We are therefore free to impose additional conditions. The conditions 2–4 have geometric implications, which we will explore in the next section.

It is customary to denote an "inverse" satisfying a subset of the Penrose conditions, say conditions i, j, and k, by writing $A^{(i,j,k)}$. Thus $A^{(1)}$ satisfies only the first condition. The matrix $A^{(1,2,3,4)}$, which satisfies all four, is written $A^\dagger$, and is called the MOORE-PENROSE GENERALIZED INVERSE or the PSEUDO-INVERSE of A.

1. THE PSEUDO-INVERSE AND LEAST SQUARES

There are explicit formulas for some of the generalized inverses generated from the Penrose conditions. Let

$$A = U \begin{pmatrix} \Sigma_+ & 0 \\ 0 & 0 \end{pmatrix} V^H \qquad (1.3)$$

be the singular value decomposition of A. Let us seek $A^{(1)}$ in the form

$$A^{(1)} = V \begin{pmatrix} T & K \\ L & M \end{pmatrix} U^H.$$

By direct multiplication

$$AA^{(1)}A = U \begin{pmatrix} \Sigma_+ T \Sigma_+ & 0 \\ 0 & 0 \end{pmatrix} V^H,$$

and it follows from the first Penrose condition that $T = \Sigma_+^{-1}$. Thus any (1)-inverse has the form

$$A^{(1)} = V \begin{pmatrix} \Sigma_+^{-1} & K \\ L & M \end{pmatrix} U^H,$$

where K, L, and M, are arbitrary.

If we now seek a $(1,2)$-inverse in the form

$$A^{(1,2)} = V \begin{pmatrix} \Sigma_+^{-1} & K \\ L & M \end{pmatrix} U^H,$$

then by the second Penrose condition

$$A^{(1,2)} = A^{(1,2)}AA^{(1,2)} = V \begin{pmatrix} \Sigma_+ & 0 \\ 0 & L\Sigma_+ K \end{pmatrix} U^H.$$

Thus any $(1,2)$-inverse has the form

$$A^{(1,2)} = V \begin{pmatrix} \Sigma_+^{-1} & K \\ L & M \end{pmatrix} U^H,$$

where K and L are arbitrary and $L\Sigma_+ K = M$.
In the same way we can prove that any $(1,2,3)$-inverse has the form

$$A^{(1,2,3)} = V \begin{pmatrix} \Sigma_+^{-1} & 0 \\ L & 0 \end{pmatrix} U^{\mathrm{H}},$$

with L arbitrary; and any $(1,2,4)$-inverse has the form

$$A^{(1,2,4)} = V \begin{pmatrix} \Sigma_+^{-1} & K \\ 0 & 0 \end{pmatrix} U^{\mathrm{H}},$$

with K arbitrary. Finally the pseudo-inverse has the form

$$A^{(1,2,3,4)} = V \begin{pmatrix} \Sigma_+^{-1} & 0 \\ 0 & 0 \end{pmatrix} U^{\mathrm{H}}. \qquad (1.4)$$

These representations are independent of the choice of singular vectors (see the comments following Theorem I.4.1). In particular, since there is nothing arbitrary about (1.4), we have established the existence and uniqueness of the pseudo-inverse.

Theorem 1.1. *Let $A \in \mathbf{C}^{m \times n}$. Then there is a unique matrix $X \in \mathcal{C}^{n \times m}$ that satisfies the Penrose conditions (1.2).*

The following properties of the pseudo-inverse are easily established from (1.2) or (1.4)

Theorem 1.2. *For any matrix A the following hold.*

1. $(A^\dagger)^\dagger = A$.
2. $(\bar{A})^\dagger = \overline{(A^\dagger)}$.
3. $(A^{\mathrm{T}})^\dagger = (A^\dagger)^{\mathrm{T}}$.
4. $\operatorname{rank}(A) = \operatorname{rank}(A^\dagger) = \operatorname{rank}(AA^\dagger) = \operatorname{rank}(A^\dagger A)$.
5. $(AA^{\mathrm{H}})^\dagger = A^{\mathrm{H}\dagger} A^\dagger$, $(A^{\mathrm{H}} A)^\dagger = A^\dagger A^{\mathrm{H}\dagger}$.
6. $(AA^{\mathrm{H}})^\dagger AA^{\mathrm{H}} = AA^\dagger$, $(A^{\mathrm{H}} A)^\dagger A^{\mathrm{H}} A = A^\dagger A$.

1. THE PSEUDO-INVERSE AND LEAST SQUARES

7. If $A \in \mathbf{C}^{m \times n}$ has rank n, then $A^\dagger = (A^H A)^{-1} A^H$ and $A^\dagger A = I^{(n)}$.

8. If $A \in \mathbf{C}^{m \times n}$ has rank m, then $A^\dagger = A^H (A A^H)^{-1}$ and $A A^\dagger = I^{(m)}$.

9. If A has the full rank factorization $A = F G^H$, where $\operatorname{rank}(F) = \operatorname{rank}(G) = \operatorname{rank}(A)$, then
$$A^\dagger = G(F^H A G)^{-1} F^H$$
and
$$A^\dagger = (G^\dagger)^H F^\dagger.$$
In particular, for the singular value factorization $A = U_1 \Sigma_+ V_1$,
$$A^\dagger = V_1 \Sigma_+^{-1} U_1^H.$$

10. If U, V are unitary matrices, then
$$(UAV)^\dagger = V^H A^\dagger U^H.$$

11. If
$$A = \begin{pmatrix} D & 0 \\ 0 & 0 \end{pmatrix} \in \mathbf{C}_r^{m \times n},$$
with $D = \operatorname{diag}(\delta_1, \ldots, \delta_r)$ and $\delta_i \neq 0$ for $i = 1, \ldots, r$, then
$$A^\dagger = \begin{pmatrix} D^{-1} & 0 \\ 0 & 0 \end{pmatrix} \in \mathbf{C}_r^{n \times m}.$$

Theorem 1.2 shows that the pseudo-inverse has many properties in common with the ordinary inverse. However, it fails to share other properties. It is left as an exercise to construct examples to show that

1. $(AB)^\dagger$ is not necessarily the same as $B^\dagger A^\dagger$,

2. $AA^\dagger$ is not necessarily the same as $A^\dagger A$,

3. $(A^k)^\dagger$ is not necessarily the same as $(A^\dagger)^k$,

4. The nonzero eigenvalues of $A^\dagger$ are not necessarily the reciprocals of the nonzero eigenvalues of A.

1.2. Projections and Least Squares

As we saw in Section I.2, any solution of the problem (1) of minimizing $\|b-Ax\|_2$ must satisfy $Ax = P_A b$, where P_A is the orthogonal projection onto the column space of A. It turns out that this projection can be expressed in terms of the pseudo-inverse of A.

Theorem 1.3. *For any matrix A,*

1. $P_A = AA^\dagger$ *is the orthogonal projector onto* $\mathcal{R}(A)$,

2. $P_{A^H} = A^\dagger A$ *is the orthogonal projector onto* $\mathcal{R}(A^H)$,

3. $I - P_{A^H}$ *is the orthogonal projector onto* $\mathcal{N}(A)$.

Proof. From the Penrose conditions, $AA^\dagger$ is a Hermitian idempotent and hence is the orthogonal projector onto

$$\mathcal{R}(AA^\dagger) \subset \mathcal{R}(A).$$

Since $A = (AA^\dagger)A$, we have $\operatorname{rank}(AA^\dagger) \geq \operatorname{rank}(A)$. Hence $\mathcal{R}(AA^\dagger) = \mathcal{R}(A)$, which establishes Part 1.

Part 2 follows from Part 1 on observing

$$\begin{aligned} P_{A^H} &= A^H(A^H)^\dagger \\ &= [A^H(A^H)^\dagger]^H \\ &= A^\dagger A. \end{aligned}$$

To establish Part 3, note that by Part 2

$$A(I - P_{A^H}) = A - AA^\dagger A = A - A = 0,$$

so that $I - P_{A^H}$ is an orthogonal projector into $\mathcal{N}(A)$. But if $Ax = 0$, then

$$(I - P_{A^H})x = x - A^\dagger Ax = x,$$

so that $I - P_{A^H}$ projects onto $\mathcal{N}(A)$. ∎

The characterization $Ax = P_A b$ of the solution of the least squares problem (1) is not sufficient to determine x when A is not of full column rank. The following theorem gives a complete characterization.

1. THE PSEUDO-INVERSE AND LEAST SQUARES

Theorem 1.4 (Penrose). *The solutions of the problem (1) have the general form*
$$x = A^\dagger b + (I - P_{A^H})z, \qquad (1.5)$$
where z is arbitrary. Of all the solutions, $A^\dagger b$ has the smallest 2-norm.

Proof. By Theorem I.2.5, any minimizing vector Ax must satisfy $Ax = P_A b$. Since $A(A^\dagger b) = P_A b$, the vector $x = A^\dagger b$ is one solution.
Let us now seek a general solution in the form
$$x = A^\dagger b + y. \qquad (1.6)$$
Since
$$Ay = A(x - A^\dagger b) = Ax - P_A b = 0,$$
we have $y \in \mathcal{N}(A)$. By Theorem 1.3,
$$y = (I - P_{A^H})z,$$
where z is arbitrary. This establishes (1.5).
Since $A^\dagger b$ is orthogonal to $(I - P_{A^H})z$, from
$$\|x\|_2^2 = \|A^\dagger b\|_2^2 + \|(I - P_{A^H})z\|_2^2$$
we see that $\|x\|$ is minimal when $z = 0$. ∎

When A has full column rank, then $P_{A^H} = I$, and the solution of (1) is unique.

We conclude this subsection with two sets of equations that are always satisfied by least squares solutions. The first are the classic NORMAL EQUATIONS. The second, which involves both the solution and the residual, are called the EXPANDED EQUATIONS and are useful in a number of applications. The proof of the following theorem is left as an exercise.

Theorem 1.5. *Let x be a solution of the least squares problem (1), and let $r = b - Ax$ be the associated* RESIDUAL VECTOR. *Then*
$$A^H A x = A^H b$$
and
$$\begin{pmatrix} I & A \\ A^H & 0 \end{pmatrix} \begin{pmatrix} r \\ x \end{pmatrix} = \begin{pmatrix} b \\ 0 \end{pmatrix}. \qquad (1.7)$$

Notes and References

Although Gauss [84, 1821] exhibited the rows of the pseudo-inverse to prove his celebrated minimum variance theorem (see Exercise 1.9), he had no concept of the pseudo-inverse as a matrix or an operator, and it would be mistaken to impute the notion to him. The true fathers of the pseudo-inverse are Moore [162, 1920], Bjerhammer [34, 1951] (the full rank case), Penrose [178, 179, 1955, 1956], and to a lesser extent Bergman et al. [26, 1950], who introduced the (1,2)-inverse of a symmetric matrix and specialized it to the pseudo-inverse. Penrose — perhaps because of his elegant algebraic characterization of the pseudo-inverse — touched off a vogue in the subject of generalized inverses. By 1976, Nashed and Rall [163] were able to compile a bicentennial bibliography running to 1776 entries, in which they all but say that a number of these papers contributed more to the promotion of their authors than to the promotion of science.

Things have settled down since then, and now is a good time to sift the residuum.

Any attempt to use a generalized inverse in the case where the matrix is not of full rank must come to grips with the fact that under such circumstances the generalized inverse is not a continuous function of its elements. This observation was first made by Penrose [178] for the pseudo-inverse, but it is true of any (1)-inverse.* Thus, to use a pseudo-inverse in practice, one must determine the rank of the matrix and project the errors appropriately. Unfortunately, determining the rank of a matrix in the presence of errors is a very difficult problem (e.g., see [211]). For this reason, papers about the applications of generalized inverses have a certain theoretical air about them: they leave you all dressed up with nowhere to go.

The clear winner in the generalized inverse sweepstakes is the pseudo-inverse applied to full rank problems. It is unique, continuous, and computable (although one seldom has need of an explicit pseudo-inverse). Moreover, its connection with orthogonal projections makes it useful in discussing the geometry of n-space. A distant second is the Drazin generalized-inverse [61, 1958] (see Exercise 1.23), which enters into the perturbation theory of eigenvalues and eigenvectors, one of the few cases in which we know the rank *a priori* (see Example V.2.10).

The principle of least squares was used by Gauss in astronomical calculations

---

*This follows from the fact that $\text{rank}(A) = \text{trace}(A^{(1)}A)$ and that the trace of a matrix, being an integer, is discontinuous unless the matrix is of full rank.

1. THE PSEUDO-INVERSE AND LEAST SQUARES

in the 1790s. However, Legendre [143, 1805] first published the method, and Gauss's subsequent claim to it [82, 83, 1809] sparked a famous priority dispute (for a discussion and further references see [219]). The applications of least squares are far too many to survey here. For the statistician's point of view, see [196]. [Warning: the notation is quite different. A statistician does a "regression analysis," which is usually written in the form $\mathbf{y} = \mathbf{X}\boldsymbol{\beta} + \mathbf{e}$, where $\mathbf{X}$ has n rows and typically p columns] For the numerical analyst's point of view, see [38, 142].

The least squares problem can be generalized in a number of ways. One is to replace the 2-norm by an elliptic norm (See Corollary II.1.5 and Exercise II.1.16). Actually this can be done in two ways, since we can generalize Theorem 1.4 by requiring that x be the solution of minimal T-norm that minimizes $\|b - Ax\|_S$. The theory of these problems can be approached through elliptic pseudo-inverses (see Exercises 1.13–1.15). However, the numerical solution of such problems requires a different approach (see [171, 172]).

Another generalization is to require that the solution satisfy a linear equality constraint. This problem can also be approach via generalized inverses, but the returns on this approach are not yet in. See [64] for details and further references.

Exercises

1. (Moore's characterization [162]). Show that if

   1. $AXA = A$,
   2. $XAX = X$,
   3. $\mathcal{R}(A) = \mathcal{R}(X^H)$,
   4. $\mathcal{R}(A^H) = \mathcal{R}(X)$,

   then $X = A^\dagger$.

2. (Bjerhammer's characterization [34]). Let A have full column rank and let B be any matrix such that $A^H B = 0$ and $(A\ B)$ is nonsingular. If

$$\begin{pmatrix} X^H \\ Y^H \end{pmatrix} = (A\ B)^{-1},$$

then $X^H = A^\dagger$.

3. Show that if $A^{(1)}$ is a 1-inverse, then $\operatorname{rank}(A) \leq \operatorname{rank}(A^{(1)})$. Moreover $AA^{(1)}$ is an (oblique) projection onto the column space of A. If in addition $A^{(1)}$ is a 3-inverse, then the projection is orthogonal.

4. Let A have the singular value decomposition (1.3). Determine conditions on T, K, L, and M such that

$$X = V \begin{pmatrix} T & K \\ L & M \end{pmatrix} U^{\mathrm{H}}$$

is a (1,3)-inverse.

5. Establish the existence and uniqueness of the pseudo-inverse directly from Penrose's conditions

6. Show that if A has full column rank and $A = QR$ is the QR factorization of A, then $A^\dagger = R^{-1} Q^{\mathrm{H}}$.

7. Show that if A is of full rank then $\inf_2(A) = \|A^\dagger\|_2^{-1}$. What is $\|A^\dagger\|_2^{-1}$ when A is not of full rank?

8. Prove Theorem 1.2.

9. (Gauss [84, 1821]). Let A have full rank, and let $a_i^{(\dagger)}$ denote the ith row of $A^\dagger$. Show that

$$\|a_1^{(\dagger)}\|_2 = \min_{z^{\mathrm{H}} A = \mathbf{1}_i^{\mathrm{T}}} \|z\|_2.$$

Conclude that among all matrices Z satisfying $Z^{\mathrm{H}} A = I$ (i.e., among all RIGHT INVERSES of A) the pseudo-inverse has minimal Frobenius norm. [Note: Gauss's application was the following. If $b = Ax + e$, where e is a vector of uncorrelated random variables with mean zero and variance σ^2, then any vector satisfying $z^{\mathrm{H}} A = \mathbf{1}_i^{\mathrm{T}}$ yields an unbiased estimate $z^{\mathrm{H}} b$ of the first component of x. Since the variance of $z^{\mathrm{H}} b$ is $\sigma^2 \|z\|_2^2$, *the least squares estimate is the minimum variance estimate.* This result is sometimes called the Gauss-Markov theorem, although the attribution to Markov is spurious.]

10. (Penrose [178]). Show that the equation $AXB = C$ has a solution if and only if $AA^{(1)} C B^{(1)} B = C$, in which case the most general form of the solution is

$$X = A^{(1)} C B^{(1)} + Y - A A^{(1)} Y B^{(1)} B,$$

where Y is arbitrary.

11. Show that if A is normal, then $(A^n)^\dagger = (A^\dagger)^n$.

1. THE PSEUDO-INVERSE AND LEAST SQUARES

12. Show that

$$A^\dagger = \lim_{\tau \to 0}(\tau I + A^H A)^{-1} A^H = \lim_{\tau \to 0} A^H (\tau I + AA^H)^{-1}.$$

THE FOLLOWING EXERCISES SHOW HOW TO GENERALIZE THE NOTION OF PSEUDO-INVERSE TO THE CASE OF ELLIPTIC NORMS.

13. Let S and T be positive definite matrices of orders m and n, and let the last two Penrose conditions be replaced by

    3'. SAX is Hermitian,

    4'. XAT is Hermitian.

Show that AX is the projection onto $\mathcal{R}(A)$ that is orthogonal with respect to the inner product $(x, y)_S = y^H S x$. What is XA? The matrix $X \equiv A^{(S,T)}$ is called an ELLIPTIC or WEIGHTED PSEUDO-INVERSE.

14. Show that the most general solution of the problem of minimizing $\|b - Ax\|_S$ is $x = A^{(S,T)}b + (I - A^{(S,T)}A)z$, where z is arbitrary. In particular $A^{(S,T)}b$ has minimal T-norm.

15. Show that if A is of full column rank, then

    $$A^{(S,T)} = (A^H S A)^{-1} A^H S.$$

    In this case we write $A^{(S,T)} = A^{(S)}$.

16. (Paige [171]) . Let W be positive definite and let $W = LL^H$. Show that the problem of minimizing $\|b - Ax\|_{W^{-1}}$ is equivalent to the problem

    minimize $\|v\|_2$,

    subject to $b = Ax + Lv$.

The importance of this result is that it works when W is singular: simply let $W = LL^H$ be a full rank factorization of W.

——◇——

WHEN S AND T ARE DIAGONAL THE ELLIPTIC PSEUDO-INVERSES ARE CALLED SCALED PSEUDO-INVERSES. IN THE FOLLOWING EXERCISES WE WILL ASSUME THAT A IS OF FULL RANK AND $D \in \mathcal{D}_+$, THE SET OF ALL DIAGONAL MATRICES WITH POSITIVE DIAGONAL ELEMENTS.

17. (Stewart [216]). Let
$$\mathcal{X} = \{x \in \mathcal{R}(A) : \|x\| = 1\}$$
and
$$\mathcal{Y} = \{y : \exists D \in \mathcal{D}_+ \text{ such that } Y^\mathrm{T} Dy = 0\}.$$
Let
$$\rho = \inf_{\substack{y \in \mathcal{Y} \\ x \in \mathcal{X}}} \|y - x\|.$$
Show that
$$\sup_{D \in \mathcal{D}_+} \|AA^{(D)}\|_2 \le \rho$$
and
$$\sup_{D \in \mathcal{D}_+} \|A^{(D)}\| \le \rho \|A^\dagger\|.$$

18. (Stewart [216]). For any matrix X let $\inf_+(X)$ be the smallest nonzero singular value of X. Let the columns of U form an orthonormal basis for $\mathcal{R}(A)$. Let
$$\hat{\rho} = \min \inf_+(U_I),$$
where U_I denotes any submatrix formed from a set of rows of U. Show that $\rho \ge \hat{\rho}$.

19. (O'Leary [165]). Show that
$$\rho \le \hat{\rho}.$$

20. Devise an example to show that the above results do not hold when D ranges over the space of positive definite matrices.

———◇———

THE FOLLOWING EXERCISES TREAT THE THE LEAST SQUARES PROBLEM WITH LINEAR EQUALITY CONSTRAINTS:

$$\begin{aligned} \text{minimize} \quad & \|b_2 - A_2 x\|_2, \\ \text{subject to} \quad & A_1 x = b_1. \end{aligned} \qquad (1.8)$$

21. Let A_1 have full row rank and A_2 have full column rank. Let
$$A_\tau = \begin{pmatrix} \tau A_1 \\ A_2 \end{pmatrix} \quad \text{and} \quad b = \begin{pmatrix} b_1 \\ b_2 \end{pmatrix}.$$

1. THE PSEUDO-INVERSE AND LEAST SQUARES

Show that
$$x = \lim_{\tau \to \infty} A_\tau^\dagger b$$
is the solution of (1.8). [Note: This result suggests that we can solve a constrained least squares problem by taking τ large enough and solving an ordinary least squares problem. In essence this is true, but precautions must be taken against the effect of rounding errors. See [249] for more.]

22. (Wedin [262]). Show that if A_1 has full row rank, then the the constrained problem (1.8) has a unique solution which satisfies

$$\begin{pmatrix} 0 & 0 & A_1 \\ 0 & I & A_2 \\ A_1^H & A_2^H & 0 \end{pmatrix} \begin{pmatrix} \ell \\ r_2 \\ x \end{pmatrix} = \begin{pmatrix} b_1 \\ b_2 \\ 0 \end{pmatrix},$$

where ℓ is a vector and

$$\begin{pmatrix} 0 \\ r_2 \end{pmatrix} = \begin{pmatrix} b_1 \\ b_2 \end{pmatrix} - \begin{pmatrix} A_1 \\ A_2 \end{pmatrix} x.$$

———◇———

THE FOLLOWING EXERCISES DEVELOP THE DRAZIN GENERALIZED INVERSE [61]. DRAZIN ORIGINALLY DEFINED HIS INVERSE FOR ELEMENTS OF RINGS AND SEMIGROUPS. HERE WE APPROACH THE DRAZIN INVERSE THROUGH THE JORDAN CANONICAL FORM.

23. Let the Jordan form of A be written

$$A = X_1 J_{k_1}(\lambda_1) Y_1^H + X_2 J_{k_2}(\lambda_2) Y_2^H + \cdots + X_l J_{k_l}(\lambda_l) Y_l^H \qquad (1.9)$$

(see the subsection on invariant subspaces in Section I.3). Define

$$J_k(\lambda)^\# = \begin{cases} J_k(\lambda)^{-1} & \text{if } \lambda \neq 0 \\ 0 & \text{if } \lambda = 0 \end{cases}$$

and

$$A^\# = X_1 J_{k_1}(\lambda_1)^\# Y_1^H + X_2 J_{k_2}(\lambda_2)^\# Y_2^H + \cdots + X_l J_{k_l}(\lambda_l)^\# Y_l^H.$$

Show that $A^\#$ is the unique matrix satisfying

1. $A^\# A = AA^\#$,
2. $A^\# = (A^\#)^2 A$,
3. there is an integer m such that $A^m = A^{m+1} A^\#$.

The matrix $A^\#$ is called the DRAZIN GENERALIZED INVERSE of A.

24. Let A have the Jordan form (1.9) and let λ be an eigenvalue of A. The invariant subspace corresponding to λ is the space

$$\bigoplus_{\lambda_i = \lambda} \mathcal{R}(X_i).$$

Show that $P_\lambda = I - (\lambda I - A)(\lambda I - A)^\#$ is a (generally oblique) projection onto the invariant subspace associated with λ. What subspace does it project along? The matrix P_λ is called the SPECTRAL PROJECTION of λ.

—⋄—

2. Inverses and Linear Systems

In this section we will be concerned with two related problems. Let $A \in \mathbf{C}^{n \times n}$ be nonsingular and let $\tilde{A} = A + E$ be a perturbation of A. The first problem is to determine under what conditions $\tilde{A}$ is nonsingular and to bound a norm of $\tilde{A}^{-1} - A^{-1}$. The second problem is to derive bounds on some norm of $\tilde{x} - x$, where x is a solution of the linear system

$$Ax = b$$

and $\tilde{x}$ is the solution of

$$\tilde{A}\tilde{x} = b.$$

The close relation between the two problems is due to the fact that $\tilde{x} - x = (\tilde{A}^{-1} - A^{-1})b$.

In the statement of these problems, note the use of a tilde to denote a perturbed quantity. We will use this notational convention throughout the book:

> A symbol with a tilde over it always denotes a perturbed quantity. The unperturbed quantity is denoted by the same symbol without a tilde.

2. Inverses and Linear Systems

Here we must distinguish between primary perturbations and derived perturbations. An example of the former is $\tilde{A}$, which is obtained from A by the addition of an explicit perturbation E. The latter is represented by $\tilde{x}$, which is not explicitly perturbed but depends on the primary perturbation E. In perturbation analysis our goal is usually to obtain bounds on derived perturbations in terms of an explicit perturbation; e.g., to bound some norm of $\tilde{x} - x$ in terms of some norm of E. Generally, we will represent explicit perturbations by the letters E, F, G, H or their lower-case and Greek analogues. Usually, but not always, these letters will carry the implication that the perturbation is in some sense small.

2.1. Absolute and Relative Errors

Before we can proceed to the perturbation theory of linear systems, we must first discuss how errors are to be represented. For scalars there are two representations in general use.

Definition 2.1. *Let* $\alpha, \tilde{\alpha} \in \mathbf{C}$. *The* ABSOLUTE ERROR *or simply the* ERROR *in* $\tilde{\alpha}$ *regarded as an approximation to* α *is the number*

$$\mathrm{ae}(\alpha, \tilde{\alpha}) = |\tilde{\alpha} - \alpha|.$$

If $\alpha \neq 0$, *then the* RELATIVE ERROR *in* $\tilde{\alpha}$ *is*

$$\mathrm{re}(\alpha, \tilde{\alpha}) = \frac{|\tilde{\alpha} - \alpha|}{|\alpha|}.$$

The absolute and relative errors have a number of elementary properties, which we list here without proof.

Theorem 2.2. *In the notation of Definition 2.1:*

1. *There is a number ϵ with $|\epsilon| = \mathrm{ae}(\alpha, \tilde{\alpha})$ such that $\tilde{\alpha} = \alpha + \epsilon$.*

2. *There is a number ρ with $|\rho| = \mathrm{re}(\alpha, \tilde{\alpha})$ such that $\tilde{\alpha} = \alpha(1 + \rho)$.*

3. *If* $\mathrm{re}(\alpha, \tilde{\alpha}) < 1$, *then*

$$\frac{\mathrm{re}(\alpha, \tilde{\alpha})}{1 + \mathrm{re}(\alpha, \tilde{\alpha})} \leq \mathrm{re}(\tilde{\alpha}, \alpha) \leq \frac{\mathrm{re}(\alpha, \tilde{\alpha})}{1 - \mathrm{re}(\alpha, \tilde{\alpha})}.$$

4. If $\text{re}(\alpha, \tilde{\alpha}) = 10^{-t}$, then α and $\tilde{\alpha}$ agree to approximately t decimal digits.

The first item in the theorem says that if the absolute error is small, then $\tilde{\alpha}$ may be obtained from α by adding a number near zero. Similarly, if the relative error is small, then $\tilde{\alpha}$ may be obtained from α by multiplying by a number near one. The third item says that if the relative error is small, then for practical purposes $\text{re}(\alpha, \tilde{\alpha})$ and $\text{re}(\tilde{\alpha}, \alpha)$ are the same. The last item gives a quick way of estimating relative error; e.g., we see from it that 3.14159, regarded as an approximation to π, has a relative error of about 10^{-6}.

It is natural to attempt to generalize absolute and relative error to vectors and matrices by replacing the absolute value by a norm. The result is the following definition, which, however, is not without its difficulties.

Definition 2.3. Let $A, \tilde{A} \in \mathbf{C}^{m \times n}$, and let $\|\cdot\|$ be a norm on $\mathbf{C}^{m \times n}$. The ABSOLUTE ERROR or simply the ERROR in $\tilde{A}$ regarded as an approximation to A is the number

$$\text{ae}(A, \tilde{A}) = \|\tilde{A} - A\|.$$

If $A \neq 0$, then the RELATIVE ERROR in $\tilde{A}$ is

$$\text{re}(A, \tilde{A}) = \frac{\|\tilde{A} - A\|}{\|A\|}.$$

The following theorem list the analogues of the items in Theorem 2.2.

Theorem 2.4. In the notation of Definition 2.3:

1. There is a matrix E with $\|E\| = \text{ae}(A, \tilde{A})$ such that $\tilde{A} = A + E$.

2. If $\|\cdot\|$ is consistent and there is a matrix R such that $\tilde{A} = A(I + R)$, then $\text{re}(A, \tilde{A}) \leq \|R\|$.

3. If $\text{re}(A, \tilde{A}) < 1$, then

$$\frac{\text{re}(A, \tilde{A})}{1 + \text{re}(A, \tilde{A})} \leq \text{re}(\tilde{A}, A) \leq \frac{\text{re}(A, \tilde{A})}{1 - \text{re}(A, \tilde{A})}.$$

The difficulty with the definitions 2.3 is that they attempt to characterize the errors in the mn elements of $\tilde{A}$ by a single number. Some information has to be lost in the process, and that information may be important. This is illustrated by comparing the second items in Theorems 2.2 and 2.4. For scalars, the statement $\rho = \text{re}(\alpha, \tilde{\alpha})$ means that $\tilde{\alpha}$ is obtained from α by multiplying by a number within ρ of one. For matrices, we would like to say that $\rho = \text{re}(A, \tilde{A})$ means that $\tilde{A} = A(I + R)$, where $\|R\| \leq \rho$. However, if A is singular such an R may not exist, and if A is nonsingular the most we can say about R is that $\|R\| \leq \rho \|A\| \|A^{-1}\|$ (see Exercise 2.1). Thus to report the fact that $\tilde{A} = A(I + R)$ by saying that the relative error in $\tilde{A}$ is $\|R\|$ is to give away information.

The absence of a fourth item from Theorem 2.4 illustrates another loss of information. For example, suppose that with respect to the ∞-norm we have

$$\text{re}_\infty(x, \tilde{x}) = 10^{-t}.$$

Then we know that the largest components of x and $\tilde{x}$ agree to roughly t decimal digits. But the best thing we can say about the smaller components is that if $|\xi_i| = 10^{-k}\|x\|_\infty$, then $|\xi_i|$ and $|\tilde{\xi}_i|$ agree to about $t - k$ significant figures. In particular, if $k > t$ then the relative error says nothing at all about the accuracy of $\tilde{\xi}_i$.

As we shall see, structured perturbation theorems, such as Theorem 2.14, provide partial relief from these problems, as do componentwise bounds (Theorem 2.12). Nonetheless, we will cast many of our results in terms of absolute and relative errors of vectors and matrices. In the first place, bounds of this form do say something about the larger components. Moreover, the use of absolute and relative errors gives perturbation bounds a simplicity that makes them easier to interpret. Finally, in applications we can often scale the problem so that the components of the quantity being bounded are approximately equal, in which case a bound on the relative error is as good as bounds on the individual components.

2.2. The Inverse Matrix

The fundamental results on matrix inverses are contained in the following theorem.

Theorem 2.5. Let $A \in \mathbf{C}^{n \times n}$ be nonsingular and let $\tilde{A} = A + E$ be a perturbation of A. Let $\| \cdot \|$ be a consistent matrix norm. If $\tilde{A}$ is nonsingular, then

$$\frac{\|\tilde{A}^{-1} - A^{-1}\|}{\|\tilde{A}^{-1}\|} \leq \|A^{-1}E\|. \tag{2.1}$$

Alternatively, if

$$\|A^{-1}E\| < 1,$$

then $\tilde{A}$ is perforce nonsingular and

$$\|\tilde{A}^{-1}\| \leq \frac{\|A^{-1}\|}{1 - \|A^{-1}E\|}. \tag{2.2}$$

Moreover

$$\frac{\|\tilde{A}^{-1} - A^{-1}\|}{\|A^{-1}\|} \leq \frac{\|A^{-1}E\|}{1 - \|A^{-1}E\|}. \tag{2.3}$$

Remark 2.6. The theorem remains valid if $A^{-1}E$ is everywhere replaced by EA^{-1}.

Proof. Since $\tilde{A}$ is nonsingular, $\tilde{A}\tilde{A}^{-1} = (A + E)\tilde{A}^{-1} = I$, or $(I + A^{-1}E)\tilde{A}^{-1} = A^{-1}$. Hence

$$\tilde{A}^{-1} - A^{-1} = -A^{-1}E\tilde{A}^{-1}, \tag{2.4}$$

and (2.1) follows on taking norms in (2.4).

If $\|A^{-1}E\| < 1$, the spectral radius of $A^{-1}E$ is less than one and $I + A^{-1}E$ is nonsingular. From $\tilde{A} = A(I + A^{-1}E)$ it follows that $\tilde{A}$ is nonsingular. Moreover, from (2.4) we also have $\|\tilde{A}^{-1}\| \leq \|A^{-1}\| + \|A^{-1}E\|\|\tilde{A}^{-1}\|$, and (2.2) follows on solving this inequality for $\|\tilde{A}^{-1}\|$. Finally, (2.2) follows from (2.1) and the third item in Theorem 2.4. ∎

This theorem establishes four things: 1. a bound on $\text{re}(\tilde{A}^{-1}, A^{-1})$ which holds whenever $\tilde{A}$ is nonsingular; 2. a condition under which $\tilde{A}$ is nonsingular; 3. a bound on $\tilde{A}^{-1}$; 4. a bound on $\text{re}(A^{-1}, \tilde{A}^{-1})$. All these are cast in terms of the number $\|A^{-1}E\|$ (or, by Remark 2.6, the number $\|EA^{-1}\|$). However, in many applications we will not know E explicitly, only an estimate of $\|E\|$. The following corollary, which also defines the condition number of a matrix, answers to these applications. It is proved by replacing $\|A^{-1}E\|$ by the upper bound $\|A^{-1}\|\|E\|$ in the conclusions of Theorem 2.5.

2. Inverses and Linear Systems

Corollary 2.7. Let
$$\kappa(A) = \|A\|\|A^{-1}\|$$
be the CONDITION NUMBER of A. If $\tilde{A}$ is nonsingular, then
$$\frac{\|\tilde{A}^{-1} - A^{-1}\|}{\|\tilde{A}^{-1}\|} \leq \kappa(A)\frac{\|E\|}{\|A\|}.$$

If in addition
$$\kappa(A)\frac{\|E\|}{\|A\|} < 1,$$
then $\tilde{A}$ is perforce nonsingular and
$$\|\tilde{A}^{-1}\| \leq \frac{\|A^{-1}\|}{1 - \kappa(A)\frac{\|E\|}{\|A\|}}. \tag{2.5}$$

Moreover
$$\frac{\|\tilde{A}^{-1} - A^{-1}\|}{\|A^{-1}\|} \leq \frac{\kappa(A)\frac{\|E\|}{\|A\|}}{1 - \kappa(A)\frac{\|E\|}{\|A\|}}. \tag{2.6}$$

It is instructive to look at the inequality (2.6) in greater detail. The left-hand side is the relative error (with respect to the norm $\|\cdot\|$) of $\tilde{A}^{-1}$ regarded as an approximation to A^{-1}. Its bound on the right-hand side consists of two factors. The first is the relative error $\|E\|/\|A\|$ in $\tilde{A}$ regarded as an approximation to A. The second is the quotient
$$\frac{\kappa(A)}{1 - \kappa(A)\frac{\|E\|}{\|A\|}}.$$

If $\kappa(A)\frac{\|E\|}{\|A\|}$ is much less than one, as it will be when the error E is small enough, the denominator has negligible effect, and the second factor is essentially $\kappa(A)$. Thus, the inequality (2.6) says that

> a relative error in the matrix A may be magnified by as much as $\kappa(A)$ in its inverse.

The word "magnify" is appropriate here because

$$1 \leq \|I\| = \|AA^{-1}\| \leq \|A\|\|A^{-1}\| = \kappa(A).$$

It is difficult to overstate the insight that one gets from this bound. Its precursor, the inequality (2.3), suggests that the inverse of a matrix will be WELL CONDITIONED — that is, insensitive to perturbations — provided its inverse is sufficiently small. But it does not say what "sufficiently small" is. The bound (2.6), on the other hand, makes it clear that a well conditioned matrix is one with a small condition number. And small is well defined, since the condition number is bounded below by unity. For example, to the extent that Item 4 in Theorem 2.2 holds for relative errors in matrices, we have the following rule of thumb:

If a matrix has a condition number of 10^k and its elements are perturbed in their t-th digits, then the elements of its inverse will be perturbed in their $(t - k)$-th digits.

People often say that ill-conditioned matrices are nearly singular. The following theorem gives substance to this way of speaking.

Theorem 2.8. *Let A be a nonsingular matrix. Let $\kappa(A) = \|A\|\|A^{-1}\|$ be the condition number of A with respect to a consistent matrix norm $\|\cdot\|$. Then for any matrix E,*

$$A + E \text{ is singular} \implies \frac{\|E\|}{\|A\|} \geq \kappa^{-1}(A). \tag{2.7}$$

Moreover, if the norm $\|\cdot\|$ is subordinate to a vector norm (also written $\|\cdot\|$), then there is a matrix E with $\|E\|/\|A\| = \kappa^{-1}(A)$ such that $A + E$ is singular.

Proof. If $A + E$ is singular, then by Theorem 2.5

$$1 \leq \|A^{-1}E\| \leq \|A^{-1}\|\|E\|,$$

which with a little manipulation is seen to be equivalent to the implicand in (2.7).

Now suppose that $\|\cdot\|$ is subordinate to the vector norm $\|\cdot\|$, and let x be a vector of norm one such that $\|A^{-1}x\| = \|A^{-1}\|$. Let $y = A^{-1}x/\|A^{-1}\|$, so that

$$Ay = \frac{x}{\|A^{-1}\|}.$$

Let $\|\cdot\|_*$ be the norm dual to the vector norm $\|\cdot\|$, and let z be a vector such that $\|z\|_* = 1$ and

$$z^H y = \max_{\|u\|_*=1} |u^H y|.$$

It follows from Theorem II.1.12 that

$$z^H y = \|y\| = 1.$$

Let

$$E = -\frac{xz^H}{\|A^{-1}\|}.$$

Then

$$(A+E)y = Ay - \frac{xz^H}{\|A^{-1}\|}y = \frac{x}{\|A^{-1}\|} - \frac{x}{\|A^{-1}\|} = 0,$$

so that $A + E$ is singular. The theorem will be proved if we can show that $\|E\| = \|A^{-1}\|^{-1}$. But

$$\|E\| = \max_{\|v\|=1} \frac{\|(xz^H)v\|}{\|A^{-1}\|} = \frac{\|x\|}{\|A^{-1}\|} \max_{\|v\|=1} |z^H v| = \|x\| \|z\|_* \|A^{-1}\|^{-1},$$

and the result follows from the fact that x and z have norm one. ∎

The first part of this theorem states that to make a matrix A singular we must introduce a relative perturbation of at least $\kappa^{-1}(A)$. Thus, well-conditioned matrices are not nearly singular. The second part says that for a broad class of norms, which includes the Hölder norms, we can make A singular by introducing a relative perturbation of $\kappa^{-1}(A)$. In these norms, the larger the condition number, the nearer to singularity is the matrix.

In general, the condition number is not easy to characterize. However, for the spectral norm, the condition number can be expressed in terms of the singular values. The proof of the following theorem is left as an exercise.

Theorem 2.9. Let A have singular values $\sigma_1 \geq \sigma_2 \geq \cdots \geq \sigma_n > 0$. Then

$$\kappa_2(A) = \frac{\sigma_1}{\sigma_n}.$$

As modern numerical linear algebra began to develop in the late 1940s, positive definite matrices were very much to the fore. Since the singular values and the eigenvalues of a positive definite matrix are the same, the condition number was sometimes defined as the ratio of the largest to the smallest eigenvalue. Certainly if this ratio is large, the matrix must be ill conditioned (see Exercise 2.4). However, the following example shows that it can fail completely to indicate ill-conditioning.

Example 2.10. Let A_n be the matrix illustrated below for $n = 5$:

$$A_5 = \begin{pmatrix} 1 & -1 & -1 & -1 & -1 \\ 0 & 1 & -1 & -1 & -1 \\ 0 & 0 & 1 & -1 & -1 \\ 0 & 0 & 0 & 1 & -1 \\ 0 & 0 & 0 & 0 & 1 \end{pmatrix}.$$

The eigenvalues of A_n are all one; hence the ratio of the largest to the smallest eigenvalue is one. On the other hand it is easily seen from the equation

$$\begin{pmatrix} 1 & -1 & -1 & -1 & -1 \\ 0 & 1 & -1 & -1 & -1 \\ 0 & 0 & 1 & -1 & -1 \\ 0 & 0 & 0 & 1 & -1 \\ -2^{-3} & 0 & 0 & 0 & 1 \end{pmatrix} \begin{pmatrix} 1 \\ \frac{1}{2} \\ \frac{1}{4} \\ \frac{1}{8} \\ \frac{1}{8} \end{pmatrix} = 0$$

that replacing the $(n,1)$-element of A_n by -2^{2-n} makes A exactly singular. Since $\|A_n\|_2 \geq 1$, it follows from Theorem 2.8 that $\kappa_2(A_n) \geq 2^{n-2}$.

This example also shows that the determinant is not a good measure of singularity, since the determinant of A_n is always one.

Corollary 2.7 shows that the condition number will never underestimate the sensitivity of a matrix to perturbations. Theorem 2.8 shows that for subordinate norms the condition number is sharp in that it truly estimates the distance to singularity. Nonetheless, when the condition number is used in practice, it often overestimates the actual error. The phenomenon is known as ARTIFICIAL ILL-CONDITIONING, and it is instructive to see how it comes about.

2. INVERSES AND LINEAR SYSTEMS 123

Consider the matrix

$$A_\eta = \begin{pmatrix} 1 & \eta \\ 1 & -\eta \end{pmatrix}, \qquad \eta > 0.$$

Its inverse is

$$A_\eta^{-1} = \frac{1}{2}\begin{pmatrix} 1 & 1 \\ \eta^{-1} & -\eta^{-1} \end{pmatrix}.$$

From this it is seen that if $\eta < 1$

$$\kappa_\infty(A_\eta) = 1 + \frac{1}{\eta},$$

and A_η becomes increasingly ill-conditioned as η approaches zero.

Now in the sense of being nearly singular, A_η is truly ill conditioned when η is small, for we can construct a small perturbation that will render A singular. In fact if

$$E_\eta = \begin{pmatrix} 0 & -\eta \\ 0 & \eta \end{pmatrix}, \qquad (2.8)$$

then $\|E_\eta\|_\infty = \eta$ and $A + E$ is singular. However, in applications we do not construct the error in A; rather it is given as part of the problem. For example, suppose that η represents a column scaling of an original matrix A_1 (extreme scaling can result from changes of units; e.g., years to seconds). This matrix is presumed to have an error E in it — say its elements are bounded by ϵ, where ϵ is small. Now when A_1 is scaled so that it becomes A_η, E inherits that scaling, so that its elements are bounded by the elements of the matrix

$$\begin{pmatrix} \epsilon & \eta\epsilon \\ \epsilon & \eta\epsilon \end{pmatrix}.$$

This says that a perturbation like E_η in (2.8) cannot occur. In other words, although A_η becomes increasingly ill conditioned, the nature of the underlying problem insures that perturbations exhibiting this ill-conditioning are forbidden. It is this restriction on the range of the perturbations that makes the ill-conditioning artificial.

It should be stressed that artificial ill-conditioning in no way represents a failure of our perturbation *theorems*: they were designed to predict the behavior of the inverse matrix under arbitrary perturbations, and they handle the worst cases well. It does, however, represent a failure of our perturbation *theory*, which gives no way of incorporating the structure of an error into the its bounds. In the next subsection we will see that componentwise bounds can alleviate the situation. But they are not a cure-all, and at the present state of our knowledge we must cope by using ad-hoc methods, usually some form of rescaling. The practitioner who encounters a large condition number should examine his data for artificial ill-conditioning before concluding that his results are inaccurate.

2.3. Linear Systems

In this subsection we will be concerned with perturbations of the system $Ax = b$. Since the solution of the perturbed system $\tilde{A}\tilde{x} = b$ satisfies $\tilde{x} - x = (\tilde{A}^{-1} - A^{-1})b$, we immediately obtain from (2.3)

$$\|\tilde{x} - x\| \leq \frac{\|A^{-1}E\|\|A^{-1}\|\|b\|}{1 - \|A^{-1}E\|}. \qquad (2.9)$$

However, this bound illustrates one of the ironies of matrix perturbation theory: a general result often does not give the best bound when applied in a special case. In particular the factor $\|A^{-1}\|\|b\|$ may be replaced by $\|x\|$, as the following theorem shows.

Theorem 2.11. *Let A be nonsingular and let $\tilde{A} = A + E$ be a perturbation of A. For $b \in \mathbf{C}^n$ let*

$$Ax = b.$$

Let $\|\cdot\|$ be a consistent matrix norm that is also consistent with the vector norm $\|\cdot\|$. If there is a vector $\tilde{x}$ such that

$$\tilde{A}\tilde{x} = b, \qquad (2.10)$$

then

$$\frac{\|\tilde{x} - x\|}{\|\tilde{x}\|} \leq \|A^{-1}E\|.$$

2. Inverses and Linear Systems

If in addition
$$\|A^{-1}E\| < 1,$$
then (2.10) always has a unique solution, which satisfies
$$\frac{\|\tilde{x}-x\|}{\|x\|} \leq \frac{\|A^{-1}E\|}{1-\|A^{-1}E\|}.$$
If $\kappa(A) = \|A\|\|A^{-1}\|$ is the condition number of A and
$$\kappa(A)\frac{\|E\|}{\|A\|} < 1,$$
then
$$\frac{\|\tilde{x}-x\|}{\|x\|} \leq \frac{\kappa(A)\dfrac{\|E\|}{\|A\|}}{1-\kappa(A)\dfrac{\|E\|}{\|A\|}}.$$

Proof. The proof is *mutatis mutandis* the same as that of Theorem 2.5 and its corollary. The key is the relation $\tilde{x} + A^{-1}E\tilde{x} = x$. ∎

Most of the observations to be made about this theorem have already been made in the preceding subsection. Here, as there, the condition number determines the relative perturbation of the solution. Artificial ill-conditioning can occur with linear systems, just as with inverses. Perhaps the most interesting feature of the bounds is their independence of the right-hand side b of the linear system.

One way of dealing with artificial ill-conditioning is to examine the effects of special perturbations on individual components. We will illustrate the technique for perturbations in a single column of the matrix of a linear system.

Theorem 2.12. *Let A be nonsingular, and let $\tilde{A} = A + E$, where*
$$E = e\mathbf{1}_j^{\mathrm{T}}, \tag{2.11}$$
For $b \neq 0$ let $Ax = b$, and suppose $\tilde{A}\tilde{x} = b$. Then
$$|\tilde{\xi}_i - \xi_i| \leq \|a_i^{(-1)}\|_*\|e\||\tilde{\xi}_j|, \tag{2.12}$$
where $a_i^{(-1)\mathrm{H}}$ is the ith row of A^{-1} and $\|\cdot\|$ and $\|\cdot\|_*$ are dual.

Proof. From the relation
$$\tilde{x} - x = -A^{-1}E\tilde{x},$$
we have upon multiplying by 1_i^T and substituting the right-hand side of (2.11) for E
$$\tilde{\xi}_i - \xi_i = -a_i^{(-1)H}e\tilde{\xi}_j.$$
The theorem now follows on taking norms. ∎

The advantage of this result over more general bounds is that it is invariant under column scaling. For example, if we replace A by AD, where $D = \mathrm{diag}(\delta_1, \delta_2, \ldots, \delta_n)$ is nonsingular, then we must also make the following substitutions in (2.12):

1. $\xi_i \leftarrow \delta_i^{-1}\xi_i$,

2. $\xi_j \leftarrow \delta_j^{-1}\xi_j$,

3. $e \leftarrow \delta_j e$,

4. $a_i^{(-1)H} \leftarrow \delta_i^{-1}a_i^{(-1)H}$.

The effect of these substitutions is simply to multiply both sides of (2.12) by δ_i^{-1}, which does not represent a change in the bound.

Let us now turn to the problem of perturbations in the right-hand side of the system $Ax = b$. In order to describe what is actually going on, it is necessary to introduce some additional notation. Specifically, let

$$\eta = \frac{\|A\|\|x\|}{\|b\|}. \qquad (2.13)$$

Since $\|b\| \leq \|A\|\|x\|$, we have $\eta \geq 1$. On the other hand, since $\|x\| \leq \|A^{-1}\|\|b\|$, we have $\eta \leq \kappa(A)$. When η is near $\kappa(A)$, we say that the solution of the system $Ax = b$ REFLECTS THE CONDITION of A. In other words, when η is near $\kappa(A)$, the size of the solution x is nearly as big as we might predict on the basis of the condition number alone.

With these preliminaries, we can state the following theorem.

Theorem 2.13. Let $A \in \mathbf{C}^{n \times n}$ be nonsingular. For $b \neq 0$, let $Ax = b$, and let $A\tilde{x} = b + e$. Then

$$\frac{\|\tilde{x} - x\|}{\|x\|} \leq \frac{\kappa(A)}{\eta}\frac{\|e\|}{\|b\|}. \qquad (2.14)$$

2. Inverses and Linear Systems

Proof. We have $\tilde{x} - x = A^{-1}e$, from which

$$\|\tilde{x} - x\| \leq \|A^{-1}\|\|e\|.$$

The result now follows on dividing by $\|x\| = \eta\|b\|/\|A\|$. ∎

The left-hand side of the bound (2.14) is the relative error in $\tilde{x}$. The factor $\|e\|/\|b\|$ is the relative error in b. The factor $\kappa(A)/\eta$ is the condition number of the problem. Whatever the condition of the matrix A, if x reflects that condition, then x is insensitive to perturbations in b, whatever they may be. This is in contrast to the sensitivity of x to perturbations in A, where it is $\kappa(A)$ alone that predicts the worst case.

Theorem 2.12 shows that by manipulating the relation $\tilde{x} - x = -A^{-1}E\tilde{x}$ we can obtain bounds that to some extent circumvent the problems of artificial ill-conditioning. The focus there was the components of x. The following STRUCTURED PERTURBATION THEOREM homes in on the components of A.

Theorem 2.14 (Bauer, Skeel). *Let A be nonsingular, Let $Ax = b$, and $(A + E)\tilde{x} = b + e$. Let $\|\cdot\|$ be an absolute vector norm, and let $\|\cdot\|$ also denote a consistent matrix norm (e.g., the subordinate operator norm). If for some nonnegative S, s, and ϵ*

$$|E| \leq \epsilon S \quad \text{and} \quad |e| \leq \epsilon s$$

and in addition

$$\epsilon\| |A^{-1}| S\| < 1,$$

then

$$\|\tilde{x} - x\| \leq \frac{\epsilon\| |A^{-1}|(S|x| + s)\|}{1 - \epsilon\| |A^{-1}| S\|}.$$

Proof. From the identity

$$\tilde{x} - x = A^{-1}Ex + A^{-1}e + A^{-1}E(\tilde{x} - x),$$

it follows that

$$|\tilde{x} - x| \leq \epsilon|A^{-1}|S|x| + \epsilon|A^{-1}|s + \epsilon|A^{-1}|S|\tilde{x} - x|.$$

The theorem now follows on taking norms and solving for $\|\tilde{x} - x\|$. ∎

The point of this theorem is that that it gives us control over the form of the errors. By choosing the STRUCTURE MATRICES S and s appropriately we may model the behavior of the error. For example, if we wish to consider only errors in the matrix A, we may set $s = 0$. Again, if we wish to consider only relative errors in the elements of A, we may take $S = |A|$. These substitutions and a little manipulation yield the following corollary.

Corollary 2.15. *Let $e = 0$ and $|E| \leq \epsilon |A|$. Set*

$$\kappa_{\mathrm{BS}}(A) = \| |A^{-1}||A| \|.$$

If $\epsilon\kappa_{\mathrm{BS}}(A) < 1$, then

$$\frac{\|\tilde{x} - x\|}{\|x\|} \leq \frac{\epsilon\kappa_{\mathrm{BS}}(A)}{1 - \epsilon\kappa_{\mathrm{BS}}(A)}. \tag{2.15}$$

The number $\kappa_{\mathrm{BS}}(A)$ is the BAUER–SKEEL CONDITION NUMBER of A. It has the property that it is independent of row scaling, which therefore cannot be a source of artificial ill-conditioning in (2.15). Unfortunately, the bound can be quite sensitive to column scaling.

We conclude this subsection with the topic of RESIDUAL BOUNDS. . Generally speaking, most problems in matrix computations can be cast in the form of solving an equation $r(x) = 0$. For example, $r(x) = b - Ax$ for the linear system $Ax = b$. The result of the computation will not be the exact solution but an approximate solution $\tilde{x}$, usually one for which the RESIDUAL $r(\tilde{x})$ is small. The problem of residual bounds is to construct a bound on $\tilde{x} - x$ in terms of $r(\tilde{x})$.

There are many possible solutions to this problem; but one — the METHOD OF BACKWARD PERTURBATIONS — is particularly fruitful. The technique is to show that the computed solution is the exact solution of a problem with slightly perturbed data and to bound the perturbation. Conventional perturbation theory can be then used to determine the accuracy of the purported solution $\tilde{x}$.

This technique is illustrated by the following theorem and its proof.

Theorem 2.16 (Rigal–Gaches). *Let $A \in \mathbf{C}^{n \times n}$. Let $\|\cdot\|$ denote a vector norm and its subordinate matrix norm. For any $\tilde{x} \neq 0$, let $r = b - A\tilde{x}$. Then there is a matrix E satisfying*

$$\|E\| = \frac{\|r\|}{\|\tilde{x}\|}$$

2. INVERSES AND LINEAR SYSTEMS

such that
$$(A + E)\tilde{x} = b,$$
and E is the smallest matrix with this property. Hence, if A is nonsingular and $Ax = b$, we have

$$\frac{\|\tilde{x} - x\|}{\|\tilde{x}\|} \leq \kappa(A) \frac{\|r\|}{\|A\| \|\tilde{x}\|}. \quad (2.16)$$

Proof. Let $\|\cdot\|_*$ be the norm dual to $\|\cdot\|$. Let y be a vector satisfying $\|y\|_* = 1$ and $y^H x = \|x\|$. Set

$$E = \frac{ry^H}{\|x\|}. \quad (2.17)$$

Then
$$b - (A + E)x = r - Ex = r - \frac{ry^H x}{\|x\|} = r - r = 0.$$

The norm of E is

$$\|E\| = \max_{\|z\|=1} \frac{\|ry^H z\|}{\|x\|} = \frac{\|r\|}{\|x\|} \max_{\|z\|=1} |y^H z| = \frac{\|r\|}{\|x\|},$$

the last inequality following from the fact that $\|y\|_* = 1$. Moreover, if E is any matrix satisfying $(A + E)x = b$, then $Ex = r$. Hence $\|E\| \|x\| \geq \|r\|$ or $\|E\| \geq \|r\|/\|x\|$. The bound (2.16) follows directly from Theorem 2.11. ∎

Provided we are willing to sacrifice optimality, there is nothing sacred about the choice (2.17) of E. For example, if $\tilde{\xi}_i \neq 0$, the choice

$$E = \frac{r \mathbf{1}_i^T}{\|\tilde{\xi}_i \mathbf{1}_i\|}$$

is also a backward perturbation for the problem. It has the property that it concentrates the error in the ith column of A. Other choices might place the error entirely on nonzero elements of A, an important consideration in dealing with sparse matrices.

There is a structured backward perturbation theorem, the analogue of Theorem (2.16) in the spirit of the Bauer–Skeel theorem.

Theorem 2.17 (Oettli–Prager). Let $r = b - A\tilde{x}$. Let S and s be nonnegative and set

$$\epsilon = \max_i \frac{|\rho_i|}{(S|\tilde{x}| + s)_i} \qquad (2.18)$$

(here $0/0 = 0$ and otherwise $\rho/0 = \infty$). If $\epsilon \neq \infty$, there is a matrix E and a vector e with

$$|E| \leq \epsilon S \quad \text{and} \quad |e| \leq \epsilon s \qquad (2.19)$$

such that
$$(A + E)\tilde{x} = b + e. \qquad (2.20)$$

Moreover, ϵ is the smallest number for which such matrices exist.

Proof. From (2.18) we have

$$|\rho_i| \leq \epsilon (S|\tilde{x}| + s)_i.$$

This in turn implies that $r = D(S|\tilde{x}| + s)$, where $|D| \leq \epsilon I$. It is then easily verified that $E = DS \operatorname{diag}(\operatorname{sign}(\xi_1), \ldots, \operatorname{sign}(\xi_n))$ and $e = -Ds$ are the required backwards perturbations.

On the other hand, given perturbations E and e satisfying (2.19) and (2.20) for some ϵ, we have

$$|r| = |b - A\tilde{x}| = |E\tilde{x} - e| \leq \epsilon(S|\tilde{x}| + s).$$

Hence $\epsilon \geq |\rho_i|/(S\tilde{x} + s)_i$, which shows that the ϵ defined by (2.18) is optimal. ∎

2.4. Asymptotic Forms and Derivatives

Throughout this section we have made use of the formula

$$\tilde{A}^{-1} = A^{-1} - A^{-1}E\tilde{A}^{-1}.$$

The main drawback to this formula is the presence of $\tilde{A}^{-1}$ on both sides of the equality sign. There is little we can do about this without passing to infinite series, e.g.,

$$\tilde{A}^{-1} = A^{-1} - (A^{-1}E)A^{-1} + (A^{-1}E)^2 A^{-1} - \cdots,$$

2. INVERSES AND LINEAR SYSTEMS

or inverses, e.g.,

$$\tilde{A}^{-1} = (I - A^{-1}E)^{-1}A^{-1},$$

either of which destroys the simplicity of the relation. However, if we are willing to make do with an approximation, we can write

$$\tilde{A}^{-1} \cong A^{-1} - A^{-1}EA^{-1}.$$

Since A^{-1} is a differentiable function of its elements, this FIRST ORDER APPROXIMATION is accurate up to terms of order $\|E\|^2$.

First order approximations occupy an important place in perturbation theory. They furnish computable approximations to the perturbed objects; in fact, in many applications the term "perturbation theory" amounts to little more than the construction of first and higher order approximations. Moreover, first order approximations are often easier to work with than their exact equivalents. The following example, which requires a smattering of probability theory, illustrates this point.

Example 2.18. *Let us suppose that the elements of E are independent random variables with mean zero and standard deviation σ. Then $\|\tilde{A}^{-1} - A^{-1}\|_F$ is a random variable, whose distribution gives information about the sensitivity of A^{-1} to perturbations in A. Unfortunately, its distribution is not tractable. However, we can easily compute the number*

$$\mathbf{E}(\|A^{-1}EA\|_F^2)^{\frac{1}{2}} = \sigma\|A^{-1}\|_F^2,$$

which is the root mean square of the Frobenius norm of the linearized error. This number is analogous to our error bounds, being proportional to an error term σ and the square of the inverse. However, unlike our bounds, it is an exact equality, so that if we can ignore second order terms and higher, it is a better estimate of the actual error.

An important application of first order approximations is to calculate derivatives. For example, to compute $\partial A^{-1}/\partial \alpha_{ij}$ note that when $E = \epsilon \mathbf{1}_i \mathbf{1}_j^T$ the matrix $\tilde{A}$ is just A with its (i,j)-element perturbed by ϵ. Hence for this choice of E,

$$\frac{\partial A^{-1}}{\partial \alpha_{ij}} = \lim_{\epsilon \to 0} \frac{\tilde{A}^{-1} - A^{-1}}{\epsilon}.$$

Since $\tilde{A}^{-1} - A^{-1} = -A^{-1}EA + O(\epsilon^2)$, we have

$$\frac{\partial A^{-1}}{\partial \alpha_{ij}} = \lim_{\epsilon \to 0} \frac{-\epsilon A^{-1}\mathbf{1}_i \mathbf{1}_j^T A^{-1}}{\epsilon} = -a_{(-1)}^i a_j^{(-1)\mathrm{H}},$$

where $a_{(-1)}^i$ is the ith column of A^{-1} and $a_j^{(-1)\mathrm{H}}$ is its jth row.

Similar results hold for the solution of the linear system $Ax = b$. In the notation of the last subsection,

$$\tilde{x} \cong x - A^{-1}Ex.$$

Moreover

$$\frac{\partial x}{\alpha_{ij}} = \xi_j a_{(-1)}^i,$$

that is, the derivative is proportional to the ith column of A^{-1}.

Notes and References

The perturbation of matrix inverses and linear systems finds applications in many areas. Since the theory is simple, expositors tend to develop as much or as little as they need in the notation of their fields. We cannot begin to survey this body of literature, but to see how the theory appears to at least two numerical optimizers see [166, Section 2.3].

The theory is presented here as it is found in the numerical analysis literature, where it has become highly polished. The reason is a style of rounding-error analysis — BACKWARD ROUNDING-ERROR ANALYSIS — in which the effects of rounding error are thrown back on the original data. To cite the most famous example, if Gaussian elimination with partial pivoting is used to solve the linear system $Ax = b$, then under certain conditions, which need not concern us here, the computed solution $\tilde{x}$ satisfies $(A + E)\tilde{x} = b$, where $\|E\|/\|A\|$ is a modest multiple of the rounding unit. It remains to evaluate the effect of this error on the solution, and this necessity has made numerical analysts keen perturbation theorists. An unfortunate side effect is that the outsider must search a desert of rounding-error analysis to find little nuggets of perturbation theory. For historical comments and references on the subject of rounding error, see [270, 113].

Actually, a bound in the classic style was developed earlier by Wittmeyer [274, 1936], who showed (in our notation) that

$$\|\tilde{x} - x\|_2 = \frac{\|A^{-1}\|_2\|e\|_2 + \|b\|_2\|A^{-1}\|_2^2\|E\|_2}{1 - \|A^{-1}\|_2\|E\|_2},$$

2. INVERSES AND LINEAR SYSTEMS 133

which is equivalent to (2.9) extended to account for errors in the right-hand side. It is not clear what influence this paper exerted, since it is only sporadically referenced (e.g., in [240, 121] but not in [269].)

The notion of condition number was introduced by Turing [242, 1948] to quantify "the expression 'ill-condition' [which] is sometimes used merely as a term of abuse applicable to matrices or equations, ..." The fact that the condition number appears in the error bounds lead to attempts to find the row and column scaling of A that gives the smallest condition number, of which the most penetrating investigation was given by Bauer [13, 1963]. Not much is heard about the topic now, partly because of a better understanding of what the condition number actually means and partly because of a remarkable theorem of van der Sluis [244, 1969], which says that the condition number of a matrix is nearly optimal in the spectral norm if its columns have the same 2-norm. The present discussion of artificial ill-conditioning owes a great deal to Wilkinson's common-sense approach to the subject [269, Ch.2, pp.192–193].

Kahan [129] attributes Theorem 2.8 to Gastinel but cites no reference. The connection between singularity and condition is not an accident. Kahan [131, 1972] shows that for a number of problems the condition is related to the distance from degeneracy. Demmel [55, 1987] gives a uniform treatment of this phenomenon via differential inequalities.

The condition number $\kappa(A)$ has the drawback that it depends on A^{-1}, which seldom needs to be calculated in application. This has given rise to CONDITION ESTIMATORS, which attempt to approximate some norm of A^{-1} from a factorization of A. The first of these was suggested by Gragg and Stewart [96, 1976], and an improved version [45, 1979] was incorporated into the LINPACK codes [59, 1979]. Since then there have been many variations and improvements in the technique. See [111] for a survey.

The approach to perturbation theory through structured errors, as in Theorem 2.14, is due to Bauer [14, 1966] in the forward sense and Oettli and Prager [164, 1964] in the backward sense. Skeel [197, 1979] combined this approach with rounding-error analysis to arrive at important and original observations on the numerical solution of linear systems. The number κ_{BS} is sometimes called the Skeel condition number, but it was introduced by Bauer. A variant of Theorem 2.14 for matrix inverses was first established by Bauer [14]. The specialization to linear systems is due to Skeel [197], and the present statement is a variant of one by Higham [113].

Backward perturbation theory in the style of Rigal and Gauches [185, 1967] has important applications to quasi-Newton methods for the numerical solution of nonlinear equations and optimization problems, where the perturbation is used to update approximate Jacobians and Hessians (e.g., see [57]). For the case of Hessians, where the update must be symmetric, see Exercise 2.10.

The use of first order expansions to determine the properties of functions of random variables goes back at least to Gauss [84, 1821, Section 16]. For a systematic application of the idea to matrix perturbation theory, see [215].

Exercises

UNLESS OTHERWISE STATED, A IS A NONSINGULAR MATRIX OF ORDER n. THE SYMBOL $\|\cdot\|$ DENOTES A VECTOR NORM AND ITS SUBORDINATE MATRIX NORM.

1. Show that if $\mathrm{re}(A, \tilde{A}) = \rho$, then there exists a matrix R satisfying $\|R\| \leq \rho \kappa(A)$ such that $\tilde{A} = A(I + R)$. Show that this is the best possible result.

2. Give an example of a matrix X such that $\|I - XA\|/\|A\|\|X\|$ is small while $\|I - AX\|/\|A\|\|X\|$ is large (i.e., a matrix that is a good left inverse but a poor right inverse.)

3. Show that for any Hölder norm, if $\mathrm{re}(x, \tilde{x}) \leq \rho$ then the relative error in any component $\xi_i \neq 0$ of x is bounded by $\rho \|x\|/|\xi_i|$.

4. Show that for any consistent norm

$$\kappa(A) \geq \frac{|\lambda_{\max}(A)|}{|\lambda_{\min}(A)|},$$

where $\lambda_{\max}(A)$ is the eigenvalue of A of greatest absolute value and $\lambda_{\min}(A)$ is the eigenvalue of least absolute value. Show that for the spectral norm equality is attained whenever A is normal.

5. What is $\kappa_2(0.1I_n)$? What is $\det(0.1I_n)$?

6. Show that the inverse of the matrix A_n in Example 2.10 has the form

2. INVERSES AND LINEAR SYSTEMS

illustrated below for $n = 5$:

$$\begin{pmatrix} 1 & 1 & 2 & 4 & 8 \\ 0 & 1 & 1 & 2 & 4 \\ 0 & 0 & 1 & 1 & 2 \\ 0 & 0 & 0 & 1 & 1 \\ 0 & 0 & 0 & 0 & 1 \end{pmatrix}.$$

Generalize and derive an upper bound for $\kappa(A_n)$ in your favorite norm.

7. Let B_n be the matrix illustrated below for $n = 5$:

$$B_5 = \begin{pmatrix} 1 & 1 & 1 & 1 & 1 \\ 0 & 1 & 1 & 1 & 1 \\ 0 & 0 & 1 & 1 & 1 \\ 0 & 0 & 0 & 1 & 1 \\ 0 & 0 & 0 & 0 & 1 \end{pmatrix}.$$

What is B_n^{-1}?

8. It is sometimes objected that the matrix in Example 2.10 is not scaled properly. If we normalize the columns so that they have 2-norm one, to give a matrix $\hat{A}_n$, the $\det(\hat{A}_n) = 1/\sqrt{n!}$, which reveals the ill-conditioning of A. Comment.

9. (van der Sluis [244]). Let $\mathcal{D}_+$ be the set of all diagonal matrices with positive diagonal elements. Let

$$\kappa_{2,\text{opt}}(A) = \inf_{D \in \mathcal{D}_+} \kappa_2(AD),$$

and let

$$\hat{A} = A \operatorname{diag}(\|a_1\|_2, \ldots, \|a_n\|_2)^{-1}$$

(i.e., A scaled so that its columns have 2-norm one). Show that

$$\kappa(\hat{A}) \leq \sqrt{n} \kappa_{2,\text{opt}}(A).$$

10. (Dennis and Moré [56]). Let A be symmetric and let $r = b - A\tilde{x}$. Show that

$$E = \frac{r\tilde{x}^{\mathrm{T}} + \tilde{x}r^{\mathrm{T}}}{\|\tilde{x}\|_2} - \frac{\tilde{x}^{\mathrm{T}}r}{\|\tilde{x}\|_2^2}xx^{\mathrm{T}}$$

is the smallest symmetric matrix in the Frobenius norm for which the vector x satisfies $(A + E)\tilde{x} = b$.

11. (Bunch, Demmel, and Van Loan [40]). In the last exercise, show that if F is any matrix for which $(A + F)\tilde{x} = b$, then $\|E\|_p \leq 3\|F\|_p$ $(p = 2, \text{F})$.

3. The Pseudo-Inverse

In this section we will derive perturbation bounds for the pseudo-inverse. The task is complicated by the fact that the pseudo-inverse, unlike the inverse, is not a continuous function of its elements. The qualitative result is that

$$\lim_{\tilde{A} \to A} \tilde{A}^\dagger = A^\dagger \iff \lim_{\tilde{A} \to A} \text{rank}(\tilde{A}) = \text{rank}(A). \qquad (3.1)$$

In spite of this discontinuity, we can obtain informative bounds on $A^\dagger$, even when $\text{rank}(\tilde{A}) \neq \text{rank}(A)$.

A second complicating feature is that the bounds depend on where the perturbations lie. For example, if A is of full column rank, then a perturbation in the orthogonal complement of the column space of A, no matter how large, can have only bounded effect. Accounting for this makes our bounds more intricate than the corresponding bounds for inverses.

We will begin in the first subsection by establishing a uniform notation and nomenclature. Here we will also introduce the notion of an acute perturbation which will play a central role in the theory. The second subsection is devoted to general results, and the third to acute perturbations. In the last subsection we give derivatives and asymptotic forms.

The complexity of our results makes it imperative to have a consistent notation.

> Throughout the next three sections A will denote an $m \times n$ matrix with $m \geq n$. The projection onto the column space of A will be denoted by P, and the projection onto the row space $\mathcal{R}(A^H)$ by R. The complementary projectors will be denoted by $P_\perp$ and $R_\perp$. As in the last section we will let $\tilde{A} = A + E$ denote a perturbation of A. The associated projectors will be denoted by $\tilde{P}$, $\tilde{R}$, $\tilde{P}_\perp$, and $\tilde{R}_\perp$.

3.1. Projections and Acute Perturbations

Given the relation of the pseudo-inverse to the geometry of $\mathbf{C}^n$, it is natural to cast our results in terms of unitarily invariant norms. We will let $\|\cdot\|$ denote a family of unitarily invariant norms generated by the symmetric gauge function Φ.

Since we are working with unitarily invariant norms, we may freely rotate our problem to simplify it. In particular, let $V = (V_1\ V_2)$ be a unitary matrix with $\mathcal{R}(V_1) = \mathcal{R}(A^H)$, and let $U = (U_1\ U_2)$ be a unitary matrix with $\mathcal{R}(U_1) = \mathcal{R}(A)$. Then

$$U^H A V = \begin{pmatrix} U_1^H A V_1 & U_1^H A V_2 \\ U_2^H A V_1 & U_2^H A V_2 \end{pmatrix} = \begin{pmatrix} A_{11} & 0 \\ 0 & 0 \end{pmatrix},$$

where A_{11} is square and nonsingular. If we set

$$U^H E V = \begin{pmatrix} U_1^H E V_1 & U_1^H E V_2 \\ U_2^H E V_1 & U_2^H E V_2 \end{pmatrix} = \begin{pmatrix} E_{11} & E_{12} \\ E_{21} & E_{22} \end{pmatrix},$$

then

$$U^H \tilde{A} V = \begin{pmatrix} A_{11} + E_{11} & E_{12} \\ E_{21} & E_{22} \end{pmatrix} \equiv \begin{pmatrix} \tilde{A}_{11} & E_{12} \\ E_{21} & E_{22} \end{pmatrix}. \tag{3.2}$$

We will call these transformed, partitioned matrices the REDUCED FORM of the problem. Many statements about the original problem have revealing analogues in the reduced form. For example, in reduced form

$$P\tilde{A} = \begin{pmatrix} A_{11} + E_{11} & E_{12} \\ 0 & 0 \end{pmatrix}. \tag{3.3}$$

The final item in this subsection is to define the kind of perturbation under which pseudo-inverses behave well. Essentially these are perturbations under which the column and row spaces do not alter catastrophically. These perturbations are characterized in the following theorem.

Theorem 3.1. *The following statements are equivalent.*

1. $\|\tilde{P} - P\|_2 < 1$.

2. There is no vector in $\mathcal{R}(A)$ that is orthogonal to $\mathcal{R}(\tilde{A})$ and vice versa.

3. $\text{rank}(A) = \text{rank}(\tilde{A}) = \text{rank}(P\tilde{A})$.

Corresponding statements hold for the row spaces of A and $\tilde{A}$.

Proof. 1 $\Rightarrow$ 2: Suppose there were a nonzero vector $x \in \mathcal{R}(A)$ that is orthogonal to $\mathcal{R}(\tilde{A})$. This is equivalent to saying that $Px = x$ and $\tilde{P}x = 0$. It follows that $(P - \tilde{P})x = x$, which implies that $\|P - \tilde{P}\|_2 \geq 1$, a contradiction. The reciprocal relation follows by interchanging A and $\tilde{A}$ in the preceding argument.

2 $\Rightarrow$ 3: If, say, $\text{rank}(A) > \text{rank}(\tilde{A})$, then the dimension of the column space of A is greater than the dimension of the column space of $\tilde{A}$. Hence there is a vector in $\mathcal{R}(A)$ that is orthogonal to $\mathcal{R}(\tilde{A})$. Thus it remains only to show that $\text{rank}(P\tilde{A}) = \text{rank}(A)$. In reduced form this amounts to saying that the matrix $(\tilde{A}_{11}\ E_{12})$ has full row rank [cf. (3.3)]. Suppose to the contrary that $x^{\text{H}}(\tilde{A}_{11}\ E_{12}) = 0$ for some nonzero vector x. Then $(x^{\text{H}}\ 0)^{\text{H}} \in \mathcal{R}(A)$. But by (3.3), $(x^{\text{H}}\ 0)\tilde{A} = 0$, i.e., $(x^{\text{H}}\ 0)^{\text{H}} \in \mathcal{R}(\tilde{A})^{\perp}$.

3 $\Rightarrow$ 1: By Corollary I.5.4 and Theorem I.5.5 the 2-norm of $\tilde{P} - P$ is the 2-norm of $X_{\perp}^{\text{H}}Y$, where the columns of $X_{\perp}$ form an orthonormal basis for $\mathcal{R}(A)^{\perp}$, and those of Y an orthonormal basis for $\mathcal{R}(\tilde{A})$. In reduced form we may take $X = (0\ I)^{\text{H}}$. To get Y, note that since $(\tilde{A}_{11}\ E_{12})$ has full row rank, we may permute the columns of $\tilde{A}$ so that the reduced form is

$$\begin{pmatrix} B_{11} & B_{12} \\ B_{21} & B_{22} \end{pmatrix},$$

where B_{11} is nonsingular. If we set $F = B_{21}B_{11}^{-1}$, then the columns

$$Y = \begin{pmatrix} I \\ F \end{pmatrix} (I + F^{\text{H}}F)^{-\frac{1}{2}}$$

form an orthonormal basis for $\mathcal{R}(\tilde{A})$. It follows that $X^{\text{H}}Y = F(I + F^{\text{H}}F)^{-\frac{1}{2}}$. It is easy to see that if σ is a singular value of F then $\sigma(1 + \sigma^2)^{-\frac{1}{2}} < 1$ is a singular value of $X^{\text{H}}Y$, and conversely. Hence $\|\tilde{P} - P\|_2 = \|X^{\text{H}}Y\|_2 < 1$. ∎

This theorem suggests the following definition.

3. THE PSEUDO-INVERSE

Definition 3.2. *The matrix $\tilde{A}$ is an* ACUTE PERTURBATION *of A if $\|\tilde{P} - P\|_2 < 1$ and $\|\tilde{R} - R\|_2 < 1$. We also say that A and $\tilde{A}$ are acute.*

Thus the column spaces of acute perturbations are of the same dimensions and the canonical angles between them are all less than $\pi/2$, whence the name acute perturbation. The same is true of the row spaces. The following theorem characterizes acute perturbations in the reduced form.

Theorem 3.3. *In the reduced form the matrices A and $\tilde{A}$ are acute if and only if $\tilde{A}_{11}$ is nonsingular and*

$$E_{22} = E_{21}\tilde{A}_{11}^{-1}E_{12}. \tag{3.4}$$

In this case, if we set

$$F_{21} = E_{21}\tilde{A}_{11}^{-1} \qquad F_{12} = \tilde{A}_{11}^{-1}E_{12},$$

then

$$\tilde{A} = \begin{pmatrix} I \\ F_{21} \end{pmatrix} \tilde{A}_{11} \begin{pmatrix} I & F_{12} \end{pmatrix} \tag{3.5}$$

and

$$\tilde{A}^\dagger = \begin{pmatrix} I & F_{12} \end{pmatrix}^\dagger \tilde{A}_{11}^{-1} \begin{pmatrix} I \\ F_{21} \end{pmatrix}^\dagger \tag{3.6}$$

Proof. Assume that A and $\tilde{A}$ are acute, and suppose that $\tilde{A}_{11}$ is singular. Specifically, let $A_{11}x = 0$ for some $x \neq 0$, and consider the vector $y = E_{21}x$. If $y = 0$, then $(x^H\ 0)^H$ is a vector in the row space of A that is orthogonal to the row space of $\tilde{A}$. If $y \neq 0$, then $(0\ y^H)^H$ is a vector in the column space of $\tilde{A}$ that is orthogonal to the columns space of A. Either case is a contradiction.

Equation (3.4) simply says that the last columns (rows) of the partition (3.2) are linear combinations of the initial columns (rows), which is necessary since $\tilde{A}_{11}$ is nonsingular and $\text{rank}(\tilde{A}) = \text{rank}(A)$.

Conversely, if $\tilde{A}_{11}$ is nonsingular and (3.4) holds, then it is easily verified that $\text{rank}(A) = \text{rank}(\tilde{A}) = \text{rank}(P\tilde{A}) = \text{rank}(\tilde{A}R)$, which is sufficient for acuteness.

The formula (3.5) is an immediate consequence of (3.4), and (3.6) follows from Penrose's conditions. ∎

From the above characterization, it is easy to verify that

$$\lim_{\tilde{A} \to A} \operatorname{rank}(\tilde{A}) = \operatorname{rank}(A)$$

if and only if A and $\tilde{A}$ are ultimately acute. Comparing this with (3.1), we see that

> the set on which the pseudo-inverse is continuous about A is the set of acute perturbations of A.

3.2. General Results

In this subsection we shall establish results, due to Wedin, that do not require acuteness. The first result shows that nonacute perturbations are not only discontinuous but in some sense behave like poles.

Theorem 3.4. *If A and $\tilde{A}$ are not acute, then*

$$\|\tilde{A}^\dagger - A^\dagger\| \geq \frac{1}{\|E\|_2}. \tag{3.7}$$

Moreover, if $\operatorname{rank}(\tilde{A}) \geq \operatorname{rank}(A)$, *then*

$$\|\tilde{A}^\dagger\| \geq \frac{1}{\|E\|_2}. \tag{3.8}$$

Proof. Suppose that $\operatorname{rank}(\tilde{A}) \geq \operatorname{rank}(A)$. Since $\tilde{A}$ is not an acute perturbation of A, there is a nonzero vector $x \in \mathcal{R}(\tilde{A})$ that is orthogonal to $\mathcal{R}(A)$ or a nonzero vector $x \in \mathcal{R}(\tilde{A}^{\mathrm{H}})$ that is orthogonal to $\mathcal{R}(A^{\mathrm{H}})$. Assume without loss of generality that the former is true and that $\|x\|_2 = 1$. Then

$$\begin{aligned}
1 &= x^{\mathrm{H}} x \\
&= x^{\mathrm{H}} \tilde{P} x \\
&= x^{\mathrm{H}} \tilde{A} \tilde{A}^\dagger x \\
&= x^{\mathrm{H}} (A + E) \tilde{A}^\dagger x \\
&= x^{\mathrm{H}} E \tilde{A}^\dagger x \\
&\leq \|E\|_2 \|\tilde{A}^\dagger x\|_2.
\end{aligned}$$

3. THE PSEUDO-INVERSE

Hence $\|\tilde{A}^\dagger\|_2 \geq \|\tilde{A}^\dagger x\|_2 \geq 1/\|E\|_2$, which establishes (3.8). To establish (3.7), note that $A^\dagger x = 0$. Hence

$$\|\tilde{A}^\dagger - A^\dagger\|_2 \geq \|(\tilde{A}^\dagger - A^\dagger)x\|_2 = \|\tilde{A}^\dagger x\|_2 \geq \frac{1}{\|E\|_2}.$$

This inequality for $\operatorname{rank}(\tilde{A}) \leq \operatorname{rank}(A)$ is valid by symmetry. ∎

For small perturbations the case of interest is $r = \operatorname{rank}(\tilde{A}) > \operatorname{rank}(A)$, since $\operatorname{rank}(\tilde{A}) = \operatorname{rank}(A)$ implies that the perturbation is acute when E is sufficiently small. In this case it is easy to understand what is going on. The matrix $\tilde{A}$ must have r nonzero singular values. Since A has fewer than r singular values, at least one of the singular values of $\tilde{A}$ must approach zero as $\tilde{A}$ approaches A. This means that $\tilde{A}^\dagger$, whose spectral norm is equal to the reciprocal of the smallest singular value of $\tilde{A}$, must diverge.

We now turn to our general perturbation bounds. The theorems are based on two lemmas: one containing perturbation bounds on products of projections, and the other an explicit formula for $\tilde{A}^\dagger$. First the bounds.

Lemma 3.5. *The projections $P_\perp$ and $\tilde{P}$ satisfy*

$$\tilde{P}P_\perp = (\tilde{A}^\dagger)^{\mathrm{H}} E^{\mathrm{H}} P_\perp, \qquad (3.9)$$

from which it follows that

$$\|\tilde{P}P_\perp\| \leq \|\tilde{A}^\dagger\|_2 \|E\|. \qquad (3.10)$$

If $\operatorname{rank}(\tilde{A}) = \operatorname{rank}(A)$, then

$$\|\tilde{P}P_\perp\| = \|\tilde{P}_\perp P\|. \qquad (3.11)$$

If $\operatorname{rank}(\tilde{A}) \geq \operatorname{rank}(A)$, then

$$\|\tilde{P}P_\perp\| \geq \|\tilde{P}_\perp P\|. \qquad (3.12)$$

Remark 3.6. Similar results hold for the product $P\tilde{P}_\perp$ as well as the row projections R and $\tilde{R}$. We will reference this lemma for any of these results.

Proof. We have

$$\tilde{P}P_\perp = \tilde{P}^{\mathrm{H}} P_\perp = (\tilde{A}^\dagger)^{\mathrm{H}} \tilde{A}^{\mathrm{H}} P_\perp$$
$$= (\tilde{A}^\dagger)^{\mathrm{H}} (A+E)^{\mathrm{H}} P_\perp$$
$$= (\tilde{A}^\dagger)^{\mathrm{H}} E^{\mathrm{H}} P_\perp.$$

The inequality (3.10) follows on taking norms in (3.9).

The inequality (3.11) follows from Theorem I.5.5, which shows that if $\mathrm{rank}(A) = \mathrm{rank}(\tilde{A})$ then $\tilde{P}P_\perp$ and $\tilde{P}_\perp P$ have the same singular values.

To establish (3.12), let $\tilde{P} = \tilde{P}_1 + \tilde{P}_2$, where $\mathrm{rank}(\tilde{P}_1) = \mathrm{rank}(A)$ and $\mathcal{R}(\tilde{P}_2) \perp \mathcal{R}(A)$. Then

$$\|P\tilde{P}_\perp\| = \|P(I - \tilde{P}_1 - \tilde{P}_2)\| = \|P(I - \tilde{P}_1)\| = \|\tilde{P}_1 P_\perp\|,$$

the last inequality following from (3.11). Now for any x we have

$$\|\tilde{P}_1 P_\perp x\|_2 \leq \|\tilde{P} P_\perp x\|_2.$$

Hence by Theorem I.5.5, $\|\tilde{P}_1 P_\perp\| \leq \|\tilde{P} P_\perp\|$. ∎

The second ingredient in our bounds is an explicit expression for $\tilde{A}^\dagger$. Actually, we will use three closely related expressions.

Lemma 3.7. *The difference $\tilde{A}^\dagger - A^\dagger$ is given by the expressions*

$$\tilde{A}^\dagger - A^\dagger = -\tilde{A}^\dagger E A^\dagger + \tilde{A}^\dagger P_\perp - \tilde{R}_\perp A^\dagger, \qquad (3.13)$$

$$\tilde{A}^\dagger - A^\dagger = -\tilde{A}^\dagger \tilde{P} E R A^\dagger + \tilde{A}^\dagger \tilde{P} P_\perp - \tilde{R}_\perp R A^\dagger, \qquad (3.14)$$

and

$$\tilde{A}^\dagger - A^\dagger = -\tilde{A}^\dagger \tilde{P} E R A^\dagger + (\tilde{A}^{\mathrm{H}} \tilde{A})^\dagger \tilde{R} E^{\mathrm{H}} P_\perp + \tilde{R}_\perp E^{\mathrm{H}} P(AA^{\mathrm{H}})^\dagger. \qquad (3.15)$$

Proof. These expressions may be verified by replacing all quantities by their definitions in terms of A, $\tilde{A}$, $A^\dagger$, and $\tilde{A}^\dagger$ (e.g., $E = \tilde{A} - A$) and simplifying. ∎

We are now in a position to establish a general bound on $\|\tilde{A}^\dagger - A^\dagger\|$. The exact form of the bound depends on the norm $\|\cdot\|$.

3. THE PSEUDO-INVERSE

Theorem 3.8 (Wedin). *The error $\tilde{A}^\dagger - A^\dagger$ has the following bound.*

$$\|\tilde{A}^\dagger - A^\dagger\| \leq \mu \max\{\|A^\dagger\|_2^2, \|\tilde{A}^\dagger\|_2^2\}\|E\|, \qquad (3.16)$$

where μ is given in the following table:

$\|\cdot\|$	arbitrary	spectral	Frobenius
μ	3	$\frac{1+\sqrt{5}}{2}$	$\sqrt{2}$

Proof. The inequality for an arbitrary norm follows immediately on taking norms in (3.15). The results for the spectral and Frobenius norms require further argument.

For the spectral norm we have from (3.13) that for any unit vector $u \in \mathbf{C}^m$

$$\|(\tilde{A}^\dagger - A^\dagger)u\|_2^2 = \| - \tilde{A}^\dagger E A^\dagger u + \tilde{A}^\dagger P_\perp u\|_2^2 + \|\tilde{R}_\perp A^\dagger u\|_2^2$$
$$= \| - \tilde{A}^\dagger E A^\dagger P u + \tilde{A}^\dagger P_\perp P_\perp u\|_2^2 + \|\tilde{R}_\perp A^\dagger P u\|_2^2$$
$$\leq (\|\tilde{A}^\dagger E A^\dagger\|_2 \|P u\|_2 + \|\tilde{A}^\dagger P_\perp\|_2 \|P_\perp u\|_2)^2$$
$$\quad + \|\tilde{R}_\perp A^\dagger\|_2^2 \|P u\|_2^2.$$
$$(3.17)$$

Let

$$\alpha_1 = \|\tilde{A}^\dagger E A^\dagger\|_2, \quad \alpha_2 = \|\tilde{A}^\dagger P_\perp\|_2, \quad \alpha_3 = \|\tilde{R}_\perp A^\dagger\|_2$$

and

$$\cos\theta = \|P u\|_2 \geq 0, \quad \sin\theta = \|P_\perp u\|_2 \geq 0.$$

Substituting these values into (3.17), we get

$$\|(\tilde{A}^\dagger - A^\dagger)u\|_2^2 \leq (\alpha_1 \cos\theta + \alpha_2 \sin\theta)^2 + \alpha_3^2 \cos^2\theta$$
$$\leq \max_{0 \leq \theta \leq \frac{\pi}{2}}[(\alpha_1 \cos\theta + \alpha_2 \sin\theta)^2 + \alpha_3^2 \cos^2\theta]$$
$$= \tfrac{1}{2}\{\alpha_1^2 + \alpha_2^2 + \alpha_3^2 + [(\alpha_1^2 + \alpha_2^2 + \alpha_3^2)^2 - 4\alpha_2^2\alpha_3^2]^{\frac{1}{2}}\}$$
$$\leq \tfrac{3+\sqrt{5}}{2}\max\{\alpha_1^2, \alpha_2^2, \alpha_3^2\}$$
$$= \left(\tfrac{1+\sqrt{5}}{2}\right)^2 \max\{\alpha_1^2, \alpha_2^2, \alpha_3^2\}.$$
$$(3.18)$$

Now

$$\alpha_1 \leq \|A^\dagger\|_2 \|\tilde{A}^\dagger\|_2 \|E\|_2.$$

By Lemma 3.5,
$$\alpha_2 = \|\tilde{A}^\dagger \tilde{P} P_\perp\|_2$$
$$\leq \|\tilde{A}^\dagger\|_2 \|\tilde{P} P_\perp\|_2$$
$$\leq \|\tilde{A}^\dagger\|_2^2 \|E\|_2,$$

and similarly
$$\alpha_3 \leq \|A^\dagger\|_2^2 \|E\|_2.$$

Hence from (3.18) we obtain

$$\|\tilde{A}^\dagger - A^\dagger\|_2 \leq \frac{1+\sqrt{5}}{2} \max\{\|A^\dagger\|_2^2, \|\tilde{A}^\dagger\|_2^2\} \|E\|_2. \tag{3.19}$$

For the Frobenius norm, we first consider the case where $\text{rank}(\tilde{A}) \leq \text{rank}(A)$. Let

$$F_1 = -\tilde{A}^\dagger \tilde{P} E R A^\dagger, \quad F_2 = \tilde{A}^\dagger \tilde{P} P_\perp, \quad F_3 = -\tilde{R}_\perp R A^\dagger$$

be the terms in (3.14). Then

$$\|\tilde{A}^\dagger - A^\dagger\|_F^2 = \|F_1 + F_2\|_F^2 + \|F_3\|_F^2. \tag{3.20}$$

Now $F_1 + F_2 = \tilde{A}^\dagger(-\tilde{P} E A^\dagger P + \tilde{P} P_\perp)$. Hence

$$\|F_1 + F_2\|_F^2 \leq \|\tilde{A}^\dagger\|_2^2 (\|\tilde{P} E A^\dagger P\|_F^2 + \|\tilde{P} P_\perp\|_F^2).$$

It follows from Lemma 3.5 that

$$\|\tilde{P} E A^\dagger P\|_F^2 + \|\tilde{P} P_\perp\|_F^2 \leq \|\tilde{P} E A^\dagger\|_F^2 + \|\tilde{P}_\perp P\|_F^2$$
$$= \|\tilde{P} E A^\dagger\|_F^2 + \|\tilde{P}_\perp E A^\dagger\|_F^2$$
$$= \|E A^\dagger\|_F^2$$
$$\leq \|A^\dagger\|_2^2 \|E\|_F^2.$$

Consequently,
$$\|F_1 + F_2\|_F \leq \|A^\dagger\|_2 \|\tilde{A}^\dagger\|_2 \|E\|_F.$$

Moreover, we have

$$\|F_3\|_F \leq \|A^\dagger\|_2 \|\tilde{R}_\perp R\|_F = \|\tilde{A}^\dagger\|_2 \|\tilde{R}_\perp E^H A^\dagger\|_F \leq \|A^\dagger\|_2^2 \|E\|_F.$$

3. THE PSEUDO-INVERSE

Hence from (3.20) we get

$$\|\tilde{A}^\dagger - A^\dagger\|_F \leq \sqrt{2}\|A^\dagger\|_2 \max\{\|A^\dagger\|_2, \|\tilde{A}^\dagger\|_2\}\|E\|_F$$
$$\leq \sqrt{2}\max\{\|A^\dagger\|_2^2, \|\tilde{A}^\dagger\|_2^2\}\|E\|_F. \qquad (3.21)$$

Since the bound is symmetric in A and B, it also holds for the case $\operatorname{rank}(A) \leq \operatorname{rank}(\tilde{A})$. ∎

Although the perturbation bounds bear a family resemblance to the bounds for matrix inverses, they cannot by themselves insure the convergence of $\tilde{A}^\dagger$ to $A^\dagger$ as $E \to 0$, since $\tilde{A}^\dagger$ may grow unboundedly. What is needed is the additional hypothesis that $\operatorname{rank}(A) = \operatorname{rank}(\tilde{A})$, which ensures that A and $\tilde{A}$ will become acute as $E \to 0$. The following theorem gives the perturbation bounds that hold for this case. It is established by essentially the same arguments as Theorem 3.8 — with the difference that some of the terms vanish in the expressions for $\tilde{A}^\dagger$. The details are left as an exercise.

Theorem 3.9 (Wedin). *Let* $A \in \mathbf{C}^{m \times n}$, *where* $m \geq n$. *If* $\operatorname{rank}(A) = \operatorname{rank}(\tilde{A})$, *then*

$$\|\tilde{A}^\dagger - A^\dagger\| \leq \mu \|A^\dagger\|_2 \|\tilde{A}^\dagger\|_2 \|E\|, \qquad (3.22)$$

where μ is given by the following table:

	norm		
	arbitrary	spectral	Frobenius
$\operatorname{rank}(A) < m, n$	3	$\frac{1+\sqrt{5}}{2}$	$\sqrt{2}$
$\operatorname{rank}(A) = n$	2	$\sqrt{2}$	1
$\operatorname{rank}(A) = m = n$	1	1	1

A trivial rearrangement of (3.22) gives a familiar looking corollary.

Corollary 3.10. *Let*

$$\kappa_2(A) = \|A\|_2 \|A^\dagger\|_2.$$

If $\operatorname{rank}(A) = \operatorname{rank}(\tilde{A})$, *then*

$$\frac{\|\tilde{A}^\dagger - A^\dagger\|}{\|\tilde{A}^\dagger\|_2} \leq \mu \kappa_2(A) \frac{\|E\|}{\|A\|_2}. \qquad (3.23)$$

There are two points to make about this corollary. First, although the number $\kappa_2(A)$ is usually called the condition number of A, the theorem shows that the "real" condition number is $\mu\kappa_2(A)$ — at least to the extent that the bounds really describe the behavior of $\tilde{A}^\dagger$. However, the usage is sanctioned by custom; and if we take the view that a condition number is any number whose size gives a rough estimate of the sensitivity of the problem, then the usage is even correct.

The second point is that as $E \to 0$ the right-hand side of (3.23) approaches zero. This means that the relative error in $\tilde{A}^\dagger$ approaches zero; i.e., $\tilde{A}^\dagger \to A^\dagger$, all under the hypothesis that $\text{rank}(A) = \text{rank}(\tilde{A})$. On the other hand if $\text{rank}(A) \neq \text{rank}(\tilde{A})$, then by Theorem 3.4 the matrix $\tilde{A}^\dagger$ cannot converge. Thus we have established the statement (3.1) with which we opened this section:

a necessary and sufficient condition for $\tilde{A}^\dagger \to A^\dagger$ as $\tilde{A} \to A$ is that $\text{rank}(\tilde{A}) \to \text{rank}(A)$.

3.3. Acute Perturbations

It is evident from the proof of Theorem 3.8 that we have given away much in deriving the bounds. In particular, if $\tilde{A}$ is a small acute perturbation of A then P and $\tilde{P}$ are nearly equal, and the same is true of R and $\tilde{R}$. Thus it follows from (3.15) that $\tilde{A}^\dagger - A^\dagger$ can be decomposed into three terms — one essentially depending on PER, one on $PER_\perp$, and one on $P_\perp ER$. However, this does not tell the whole story; for we shall show that the dependency of $\tilde{A}^\dagger - A^\dagger$ on $PER_\perp$ and $P_\perp ER$ is bounded, no matter how large these projections may be.

In order to state our results concisely, we must introduce some additional notation. Let $\|\cdot\|$ be generated by the symmetric gauge function Φ, and for any $F \in \mathbf{C}^{k \times r}(k \geq r)$ define

$$\Psi_\Phi(F) \stackrel{\text{def}}{=} \Phi\left(\frac{\sigma_1(F)}{[1+\sigma_1^2(F)]^{\frac{1}{2}}}, \ldots, \frac{\sigma_r(F)}{[1+\sigma_r^2(F)]^{\frac{1}{2}}}\right) \leq \|F\|. \quad (3.24)$$

The function Ψ_Φ is not a norm; however, it has some useful properties. First, from Theorems I.4.5 and II.3.9 and the monotonicity of Φ,

$$\Psi_\Phi(GF) \leq \Psi_\Phi(\|G\|_2 F) \leq \Psi_\Phi(\|G\|F).$$

3. THE PSEUDO-INVERSE

Second, since for $\alpha \geq 1$
$$\frac{\alpha\sigma}{(1+\alpha^2\sigma^2)^{\frac{1}{2}}} \leq \frac{\alpha\sigma}{(1+\sigma^2)^{\frac{1}{2}}},$$
we have
$$\alpha > 1 \implies \Psi_\Phi(\alpha F) \leq \alpha\Psi_\Phi(F).$$
For small F, the function Ψ_Φ is asymptotic to $\|F\|$, and for all F it is bounded: viz.,
$$\Psi_\Phi(F) \leq \|I_r\|.$$
Finally, for the spectral norm
$$\Psi_2(F) = \frac{\|F\|_2}{(1+\|F\|_2^2)^{\frac{1}{2}}}.$$

Our first result concerns a rather special matrix.

Lemma 3.11. *The matrix*
$$\begin{pmatrix} I \\ F \end{pmatrix}$$
satisfies
$$\left\|\begin{pmatrix} I \\ F \end{pmatrix}^\dagger\right\|_2 \leq 1 \tag{3.25}$$
and
$$\left\|\begin{pmatrix} I \\ F \end{pmatrix}^\dagger - (I\ 0)\right\| = \Psi_\Phi(F). \tag{3.26}$$

Proof. Let $\sigma_i(F)$ be the singular values of F. It is easily verified that
$$\begin{pmatrix} I \\ F \end{pmatrix}^\dagger = (I + F^H F)^{-1}(I\ F^H), \tag{3.27}$$
whose singular values are
$$\frac{1}{[1+\sigma_i^2(F)]^{\frac{1}{2}}} \leq 1,$$

from which (3.25) follows. Also if

$$G = \begin{pmatrix} I \\ F \end{pmatrix}^\dagger - (I\ 0),$$

then

$$GG^H = I - (I + F^H F)^{-1}.$$

It follows that the singular values of G are given by

$$\frac{\sigma_i(F)}{[1 + \sigma_i^2(F)]^{\frac{1}{2}}},$$

which establishes (3.26). ∎

We turn now to the perturbation theorem.

Theorem 3.12. *Let $\tilde{A}$ be an acute perturbation of A, and let*

$$\hat{\kappa} = \|A\| \|\tilde{A}_{11}^{-1}\|_2.$$

Then

$$\frac{\|\tilde{A}^\dagger - A^\dagger\|}{\|A^\dagger\|} \leq \hat{\kappa} \frac{\|E_{11}\|}{\|A\|} + \Psi_\Phi\left(\hat{\kappa} \frac{E_{12}}{\|A\|}\right) + \Psi_\Phi\left(\hat{\kappa} \frac{E_{21}}{\|A\|}\right), \qquad (3.28)$$

where Ψ_Φ is defined by (3.24).

Proof. Let F_{ij} be defined as in Theorem 3.3. Let

$$I_{21} = \begin{pmatrix} I \\ 0 \end{pmatrix}, \quad I_{12} = (I\ 0),$$

$$J_{21} = \begin{pmatrix} I \\ F_{21} \end{pmatrix}, \quad J_{12} = (I\ F_{12}).$$

From (3.6), $\tilde{A}^\dagger = J_{12}^\dagger \tilde{A}_{11}^{-1} J_{21}^\dagger$; hence

$$\tilde{A}^\dagger - A^\dagger = (J_{12}^\dagger - I_{12}^\dagger)A_{11}^{-1}I_{21}^\dagger + J_{12}^\dagger A_{11}^{-1}(J_{21}^\dagger - I_{21}^\dagger) + J_{12}^\dagger(\tilde{A}_{11}^{-1} - A_{11}^{-1})J_{21}^\dagger. \tag{3.29}$$

3. THE PSEUDO-INVERSE 149

From Corollary 2.7 we have the following bound:

$$\|J_{12}^\dagger(\tilde{A}_{11}^{-1} - A_{11}^{-1})J_{21}^\dagger\| \le \|A_{11}^{-1}\|\hat{\kappa}\frac{\|E_{11}\|}{\|A_{11}\|}. \tag{3.30}$$

By Lemma 3.11

$$\begin{aligned}\|(J_{12}^\dagger - I_{12}^\dagger)A_{11}^{-1}I_{21}^\dagger\| &\le \|A_{11}^{-1}\|\|J_{12}^\dagger - I_{12}^\dagger\| \\ &= \|A_{11}^{-1}\|\Psi_\Phi(F_{12}) \\ &= \|A_{11}^{-1}\|\Psi_\Phi(\tilde{A}_{11}^{-1}E_{12}) \\ &\le \|A_{11}^{-1}\|\Psi_\Phi\left(\hat{\kappa}\frac{E_{12}}{\|A\|}\right),\end{aligned} \tag{3.31}$$

and likewise

$$\|J_{12}^\dagger A_{11}^{-1}(J_{21}^\dagger - I_{21}^\dagger)\| \le \|A_{11}^{-1}\|\Psi_\Phi\left(\hat{\kappa}\frac{E_{21}}{\|A\|}\right). \tag{3.32}$$

The bound (3.28) follows on combining (3.29), (3.30), (3.31), and (3.32) and recalling that $\|A_{11}^{-1}\| = \|A^\dagger\|$. ∎

The bound (3.28) gives a rather nice dissection of $\|\tilde{A}^\dagger - A^\dagger\|$. Asymptotically, it is better than the bound that would be obtained by taking norms in (3.15); i.e.,

$$\frac{\|\tilde{A}^\dagger - A^\dagger\|}{\|A^\dagger\|} \le \hat{\kappa}\frac{\|E_{11}\| + \|E_{12}\| + \|E_{21}\|}{\|A\|}$$

(the two are asymptotically equal for E_{12} and E_{21} small). However, the bound also shows that E_{12} and E_{21} can have at most a bounded effect on $\|\tilde{A}^\dagger - A^\dagger\|$.

When A is square and nonsingular, E_{12} and E_{21} are void, and the bound reduces to that of Corollary 2.7.

If E_{11} is sufficiently small, we can estimate $\|\tilde{A}_{11}^{-1}\|_2$ in terms of $\|A_{11}^{-1}\|_2$ and $\|E\|$. This gives the following corollary.

Corollary 3.13. *In Theorem 3.12, let*

$$\kappa = \|A\|\|A^\dagger\|_2,$$

and suppose that

$$\|A^\dagger\|_2\|E_{11}\| < 1,$$

so that
$$\gamma \equiv 1 - \kappa\|E_{11}\|/\|A\| > 0.$$
Then
$$\|\tilde{A}^\dagger\| \leq \|A^\dagger\|/\gamma, \tag{3.33}$$
and
$$\frac{\|\tilde{A}^\dagger - A^\dagger\|}{\|A^\dagger\|} \leq \frac{\kappa}{\gamma}\frac{\|E_{11}\|}{\|A\|} + \Psi_\Phi\left(\frac{\kappa}{\gamma}\frac{E_{21}}{\|A\|}\right) + \Psi_\Phi\left(\frac{\kappa}{\gamma}\frac{E_{12}}{\|A\|}\right). \tag{3.34}$$

Proof. From the equation $\tilde{A}^\dagger = J_{12}^\dagger \tilde{A}_{11}^{-1} J_{21}^\dagger$, we have
$$\|\tilde{A}^\dagger\| \leq \|J_{12}^\dagger\|\|\tilde{A}_{11}^{-1}\|\|J_{21}^\dagger\|_2 \leq \|\tilde{A}_{11}^{-1}\|.$$
By Corollary 2.7,
$$\|\tilde{A}_{11}^{-1}\| \leq \frac{\|A_{11}^{-1}\|}{\gamma} = \frac{\|A^\dagger\|}{\gamma},$$
which establishes (3.33). Also $\hat{\kappa} \leq \kappa/\gamma$, and the inequality (3.34) follows from (3.28). ∎

3.4. Asymptotic Forms and Derivatives

Asymptotic forms for $\tilde{A}$ may be obtained from (3.15). Of course for $\tilde{A}^\dagger$ to approach $A^\dagger$ we must have $\operatorname{rank}(A) = \operatorname{rank}(\tilde{A})$; and since we are assuming that E is arbitrarily small, $\tilde{A}$ may be assumed to be an acute perturbation of A. In this case
$$\tilde{A}^\dagger = A^\dagger + O(\|E\|),$$
and
$$\tilde{P} = \tilde{A}\tilde{A}^\dagger = (A + E)[A^\dagger + O(\|E\|)] = P + O(\|E\|)$$
with similar expressions for the other projections. Hence from (3.15)
$$\tilde{A}^\dagger = A^\dagger - A^\dagger PERA^\dagger + (A^H A)^\dagger RE^H P_\perp - R_\perp E^H P(AA^H)^\dagger + O(\|E\|^2). \tag{3.35}$$

We could apply this formula, as we did the corresponding formula for the inverse, to calculate $\partial A^\dagger/\partial \alpha_{ij}$; however, the results are complicated. Instead let us assume that $A(\tau)$ is a differentiable function of τ with
$$\operatorname{rank}[A(\tau)] = \operatorname{rank}[A(\tau')]$$

3. THE PSEUDO-INVERSE

for all τ and τ'. Then $A(\tau)^\dagger$ is a differentiable function of τ and

$$\frac{dA^\dagger}{d\tau} = -A^\dagger P \frac{dA}{d\tau} R A^\dagger + (A^H A)^\dagger R \frac{dA^H}{d\tau} P_\perp - R_\perp \frac{dA^H}{d\tau} P(AA^H)^\dagger. \quad (3.36)$$

The reduced form of (3.35) can be computationally useful. From the results of the last section we have

$$\tilde{A}_{11}^{-1} = A_{11}^{-1} - A_{11}^{-1} E_{11} A_{11}^{-1} + O(\|E_{11}\|^2).$$

From (3.27) in the proof of Lemma 3.11 we have

$$\begin{pmatrix} I \\ F_{21} \end{pmatrix}^\dagger = (I \ A_{11}^{-H} E_{21}^H) + O(\|E_{11}\| \|E_{21}\|)$$

and

$$(I \ F_{12})^\dagger = \begin{pmatrix} I \\ E_{12}^H A_{11}^{-H} \end{pmatrix} + O(\|E_{11}\| \|E_{12}\|).$$

Hence from (3.6)

$$\tilde{A}^\dagger =$$
$$\begin{pmatrix} A_{11}^{-1} - A_{11}^{-1} E_{11} A_{11}^{-1} + O(\|E_{11}\|^2) & (A_{11}^H A_{11})^{-1} E_{21}^H + O(\|E_{11}\| \|E_{21}\|) \\ E_{12}^H (A_{11} A_{11}^H)^{-1} + O(\|E_{11}\| \|E_{12}\|) & E_{12}^H (A_{11}^H A_{11} A_{11}^H)^{-1} E_{21}^H \\ & +O(\|E_{11}\| \|E_{12}\| \|E_{21}\|) \end{pmatrix}.$$

Notes and References

Much of the material in this and the next two sections has been taken, sometimes word for word, from a survey article by Stewart [205, 1977].

The notion of acute subspaces is due to Davis and Kahan [53, 1970], who used the second condition of Theorem 3.1 as a characterization. The notion of an acute perturbation of a matrix is due to Wedin [260, 1973].

Penrose [178, 1955] established that the pseudo-inverse is continuous only if the rank is unchanged. However, he used techniques that do not give explicit perturbation bounds. The subject was revived by Golub and Wilkinson [94, 1966], whose interest in stable algorithms for solving least squares problems (see [88]) led them to derive first-order perturbation bounds for least squares solutions. The first perturbation bounds for the pseudo-inverse itself were given by Ben-Israel [24, 1966], who restricts his class of perturbations so that (in reduced form) only E_{11} is nonzero. More general bounds for acute perturbations were established by Hanson and Lawson [102, 1969], Pereyra [180, 1969], and Stewart [199, 1969]. Theorem 3.12 extends Stewart's bound to unitarily invariant norms. An identity in terms of projections related to (3.6) is given by Wedin [260, 1973], who uses it to derive bounds for acute perturbations.

The general results in the second subsection are essentially due to Wedin [260, 1973]. Theorem 3.4 is an extension by Wedin of a theorem of Stewart [199, 1969]. In an earlier report Wedin [258, 1969] considers the sharpness of the constants μ in Theorem 3.9 and shows that for the spectral norm the constant μ cannot be made smaller.

Early differentiability results have been given by Pavel-Parvu and Korganoff [176, 1969], Hearon and Evans [107, 1968] and Decell [54, 1972]. Wedin [258, 1969] derived the formula (3.36), as we do, from (3.15). The same results for functions of several variables was derived independently by Golub and Pereyra [90, 1973] in connection with separable nonlinear least squares problems. For more, see [91].

Exercises

1. Show that if X has linearly independent columns and B is positive definite, then $\mathcal{R}(X)$ and $\mathcal{R}(BX)$ are acute.

2. Let X_1 and Y_1 have full column rank and suppose that $\mathcal{R}(X_1)$ and $\mathcal{R}(Y_1)$ are acute. Show that if the columns of X_2 span $\mathcal{R}(Y_1)^\perp$ then $(X_1 \; X_2)$ is nonsingular. [Hint: Use canonical bases.]

3. Show that $\text{rank}(A) = \text{rank}(\tilde{A})$ is not sufficient for $\tilde{A}$ to be an acute perturbation of A.

4. Give an example of matrices A and $\tilde{A}$ such that $\|P_{\tilde{A}} - P_A\|_2 < 1$, while $\|R_{\tilde{A}} - R_A\|_2 = 1$.

5. Let $\kappa_2(A) = \|A\|_2 \|A^\dagger\|_2$. Show that if $\text{rank}(\tilde{A}) < \text{rank}(A)$ then

$$\frac{\|\tilde{A} - A\|_2}{\|A\|_2} \geq \kappa(A),$$

and the bound can be attained.

6. (Wedin [258]). Show that the constants for the spectral norm in Theorem 3.9 are optimal. [Note: this is a hard problem, Wedin solves it, not by exhibiting matrices for which the bounds are attained, but by exhibiting matrices for which the bounds are asymptotically sharp.]

7. Let Ψ_Φ be defined by (3.24). Show that

$$|\alpha| \leq 1 \implies |\alpha|\Psi_\Phi(F) \leq \Psi_\Phi(\alpha F)$$

and

$$\frac{\|F\|}{[1 + \sigma_{\max}^2(F)]^{\frac{1}{2}}} \leq \Psi_\Phi(F) \leq \|F\|.$$

8. Let A be of full rank. Calculate $\partial A^\dagger / \partial \alpha_{ij}$.

4. Projections

In this section we shall consider how the projection P varies with A. Since $P = AA^\dagger$, we can obtain perturbation bounds for P from the theory developed in the last section. However we can derive sharper bounds by working directly with one of the decompositions of $\tilde{A}^\dagger$. In particular we shall work with the decomposition (3.6) based on the reduced forms of A and $\tilde{A}$. The use of this form presupposes that A and $\tilde{A}$ are acute. This is no loss, since, as with the pseudo-inverse, we must require $\operatorname{rank}(A) = \operatorname{rank}(\tilde{A})$ to ensure the continuity of P_A, which in turn implies that the perturbation is ultimately acute.

Theorem 4.1. Let $\tilde{A}$ be an acute perturbation of A, and let $\hat{\kappa}$ be defined as in Theorem 3.12. Then

$$\|\tilde{P} - P\|_2 \leq \frac{\hat{\kappa}\|E_{21}\|_2/\|A\|}{[1 + (\hat{\kappa}\|E_{21}\|_2/\|A\|)^2]^{\frac{1}{2}}} < 1. \quad (4.1)$$

Proof. With F_{21} defined as in Theorem 3.3, we have

$$\mathcal{R}(\tilde{A}) = \mathcal{R}\begin{pmatrix} I \\ F_{21} \end{pmatrix}.$$

The matrix

$$\begin{pmatrix} I \\ F_{21} \end{pmatrix} (I + F_{21}^{\mathrm{H}} F_{21})^{-1} (I \ F_{21}^{\mathrm{H}})$$

is a Hermitian, idempotent matrix whose column space is $\mathcal{R}(\tilde{A})$; hence it is $\tilde{P}$. It follows that

$$\tilde{P} - P = \begin{pmatrix} (I + F_{21}^{\text{H}} F_{21})^{-1} - I & (I + F_{21}^{\text{H}} F_{21})^{-1} F_{21}^{\text{H}} \\ F_{21}(I + F_{21}^{\text{H}} F_{21})^{-1} & F_{21}(I + F_{21}^{\text{H}} F_{21})^{-1} F_{21}^{\text{H}} \end{pmatrix}, \quad (4.2)$$

from which it is easily verified that

$$(\tilde{P} - P)^2 = \begin{pmatrix} F_{21}^{\text{H}} F_{21}(I + F_{21}^{\text{H}} F_{21})^{-1} & 0 \\ 0 & F_{21}(I + F_{21}^{\text{H}} F_{21})^{-1} F_{21}^{\text{H}} \end{pmatrix}. \quad (4.3)$$

Now the nonzero singular values of the diagonal blocks in (4.3) are

$$\sigma_i^2(F_{21})/[1 + \sigma_i^2(F_{21})],$$

where the $\sigma_i(F_{21})$ are the nonzero singular values of F_{21}. The bound follows from the fact that the largest singular value σ_1 of F_{21} satisfies

$$\sigma_1(F_{21}) = \|F_{21}\|_2 \leq \hat{\kappa} \frac{\|E_{21}\|_2}{\|A\|}. \quad \blacksquare$$

In terms of projections, the bound (4.1) can be written in the form

$$\|\tilde{P} - P\|_2 \leq \frac{\hat{\kappa} \|P_\perp E R\|_2 / \|A\|}{[1 + (\hat{\kappa} \|P_\perp E R\|_2 / \|A\|)^2]^{\frac{1}{2}}} < 1.$$

The bound is interesting in several ways. First, it is independent of E_{12} and E_{22}. Second, its dependence on E_{11} is only through the constant $\hat{\kappa}$. Third, the bound is always less than unity. Finally, it goes to zero along with E_{21}.

If the hypotheses of Corollary 3.13 are satisfied (that is, when $\|A_{11}^{-1}\|_2 \|E_{11}\| < 1$), then we may replace $\hat{\kappa}$ by κ/γ in (4.1). Thus, κ serves as a condition number for P_A.

Asymptotic forms may be obtained in the usual way from (4.2). Indeed,

$$\tilde{P} - P = \begin{pmatrix} O(\|E_{21}\|^2) & F_{21}^{\text{H}} + O(\|E_{21}\|^3) \\ F_{21} + O(\|E_{21}\|^3) & O(\|E_{21}\|^2) \end{pmatrix}.$$

In terms of projections

$$\tilde{P} = P + P_\perp ERA^\dagger + A^{\dagger H}RE^H P_\perp + O(\|P_\perp ER\|^2).$$

It follows that if $A(\tau)$ is differentiable and varies without changing rank, then $P(\tau)$ is differentiable and

$$\frac{dP}{d\tau} = P_\perp \frac{dA}{d\tau} RA^\dagger + A^{\dagger H} R \frac{dA^H}{d\tau} P_\perp. \qquad (4.4)$$

Notes and References

Theorem 4.1 is due to Stewart [205, 1977]. The expression (4.4) for the derivative is due to Golub and Pereyra [90, 1973].

5. The Linear Least Squares Problem

In this section we will derive perturbation bounds for the solution of the least squares problem of minimizing $\|b - Ax\|_2$ and bounds for the resulting residual vector. Although the solution of minimum norm is given by $x = A^\dagger b$, the perturbation theory of Section 3 again does not give the best possible results.

We shall assume throughout this section that $\tilde{A}$ is an acute perturbation of A, and we shall work with the reduced form of the problem. In this form x is replaced by $V^H x$ and b is replaced by $U^H b$. If x and b are partitioned in the forms

$$x = \begin{pmatrix} x_1 \\ x_2 \end{pmatrix}, \quad b = \begin{pmatrix} b_1 \\ b_2 \end{pmatrix},$$

where $x_1, b_1 \in \mathbf{C}^r$, then

$$x_1 = A_{11}^{-1} b_1$$

and

$$x_2 = 0.$$

Moreover, the norm of the residual vector

$$r = b - Ax$$

is given by
$$\|r\|_2 = \|b_2\|_2.$$

In the theorems to follow we shall freely use the definitions made in the previous sections (e.g., $\hat{\kappa}$, κ, and γ). As in Sections 3 and 4, the number $\hat{\kappa}$ may be replaced by κ/γ whenever $\|A^\dagger\|_2\|E_{11}\| < 1$.

One additional piece of notation will be needed. In analogy with (2.13), define
$$\eta = \frac{\|A\|\|x\|_2}{\|b_1\|_2}.$$

Since $b_1 = A_{11}x_1$, we have $\eta \geq 1$. Also $\|x\| \leq \|A^\dagger\|\|b_1\|$, which shows that $\eta \leq \kappa$. When A is ill conditioned, that is, when $A^\dagger$ is large, the vector x may be either large or small. In the first case η is near κ, and we shall say that x reflects the ill-conditioning of A.

5.1. Perturbation of the Coefficients

We begin by bounding the effects of perturbations in b.

Theorem 5.1. Let $\tilde{b} = b + e$, $x = A^\dagger b$, and $\tilde{x} = A^\dagger \tilde{b}$. Then
$$\frac{\|\tilde{x} - x\|_2}{\|x\|_2} \leq \frac{\kappa}{\eta} \frac{\|Pe\|_2}{\|Pb\|_2}. \tag{5.1}$$

Proof. With the obvious partitioning of e we have $\tilde{x} - x = A_{11}^{-1}e_1$, so that
$$\|\tilde{x} - x\|_2 \leq \|A_{11}^{-1}\|_2\|e_1\|_2. \tag{5.2}$$
But $\|x\|_2 = \eta\|b_1\|_2/\|A\|$, which combined with (5.2) yields (5.1). ∎

Theorem 5.1 shows that the perturbation in x is determined by the projection of e onto $\mathcal{R}(A)$. However, Pe is normalized by $\|Pb\|_2$, and if this latter quantity is small, the perturbation may be large. Since
$$\|b\|_2^2 = \|Pb\|_2^2 + \|r\|_2^2,$$
this observation may be summarized by saying that large residuals are troublesome, a statment that will be amply confirmed later.

Since η can be as large as κ, the number κ cannot be taken as a condition number for perturbations in b without further qualification.

5. LEAST SQUARES PROBLEMS

If x does not reflect the ill-conditioning of A, then η is near unity and κ is a condition number. As η grows the solution becomes increasingly insensitive to perturbations in b.

We next turn the effects of a perturbation in A on x.

Theorem 5.2. Let $x = A^\dagger b$ and $\tilde{x} = \tilde{A}^\dagger b$, where $\tilde{A} = A + E$ is an acute perturbation of A. Then

$$\frac{\|\tilde{x} - x\|_2}{\|x\|_2} \leq \hat{\kappa}\frac{\|E_{11}\|_2}{\|A\|} + \Psi\left(\hat{\kappa}\frac{E_{12}}{\|A\|}\right) \\ + \hat{\kappa}^2 \frac{\|E_{12}\|_2}{\|A\|}\left(\eta^{-1}\frac{\|b_2\|_2}{\|b_1\|_2} + \frac{\|E_{21}\|_2}{\|A\|}\right). \quad (5.3)$$

Proof. Write

$$\tilde{x} - x = J_{12}^\dagger(\tilde{A}_{11}^{-1} - A_{11}^{-1})b_1 + (J_{12}^\dagger - I_{12}^\dagger)A_{11}^{-1}b_1 + J_{12}^\dagger\tilde{A}_{11}^{-1}(J_{21}^\dagger - I_{21}^\dagger)b. \quad (5.4)$$

Then

$$\|J_{12}^\dagger(\tilde{A}_{11}^{-1} - A_{11}^{-1})b_1\|_2 \leq \hat{\kappa}\frac{\|E_{11}\|_2}{\|A\|}\|x\|_2, \quad (5.5)$$

and

$$\|(J_{12}^\dagger - I_{12}^\dagger)A_{11}^{-1}b_1\|_2 \leq \Psi_2\left(\hat{\kappa}\frac{E_{12}}{\|A\|}\right)\|x\|_2. \quad (5.6)$$

Now

$$J_{12}^\dagger\tilde{A}_{11}^{-1}(J_{21}^\dagger - I_{21}^\dagger)b = J_{12}^\dagger\tilde{A}_{11}^{-1}[(I + F_{21}^\mathrm{H}F_{21})^{-1} - I]b_1 \\ + J_{12}^\dagger\tilde{A}_{11}^{-1}(I + F_{21}^\mathrm{H}F_{21})^{-1}F_{21}^\mathrm{H}b_2. \quad (5.7)$$

To bound the first term in (5.7), note that

$$(I + F_{21}^\mathrm{H}F_{21})^{-1} - I = -(I + F_{21}^\mathrm{H}F_{21})^{-1}F_{21}^\mathrm{H}F_{21}.$$

Hence

$$\|J_{12}^\dagger\tilde{A}_{11}^{-1}[(I + F_{21}^\mathrm{H}F_{21}) - I]b_1\|_2 \\ \leq \|\tilde{A}_{11}^{-1}\|_2\|(I + F_{21}^\mathrm{H}F_{21})^{-1}\|_2\|F_{21}^\mathrm{H}\|_2\|F_{21}b_1\|_2 \\ \leq \|\tilde{A}_{11}^{-1}\|_2^2\|E_{21}\tilde{A}_{11}^{-1}b_1\|_2 \\ \leq \|\tilde{A}_{11}^{-1}\|_2^2\|E_{21}\|_2^2\|x\|_2 \\ = \left[\hat{\kappa}\frac{\|E_{21}\|_2}{\|A\|}\right]^2\|x\|_2. \quad (5.8)$$

For the second term in (5.7) we have

$$\begin{aligned}
\|J_{21}^\dagger \tilde{A}_{11}^{-1}(I + F_{21}^H F_{21})^{-1} F_{21} b_2\|_2 \\
\leq \|\tilde{A}_{11}^{-1}\|_2^2 \|E_{21}\| \|b_2\| \\
= \|\tilde{A}_{11}^{-1}\|_2^2 \|E_{21}\|_2 \frac{\|b_2\|_2}{\|b_1\|_2} \eta^{-1} \|x\|_2 \|A\| \\
\leq \eta^{-1} \hat{\kappa}^2 \frac{\|E_{21}\|_2}{\|A\|} \frac{\|b_2\|}{\|b_1\|} \|x\|_2.
\end{aligned} \qquad (5.9)$$

The bound (5.3) follows on combining (5.4)–(5.9). ∎

The first two terms in (5.3) are unexceptionable. The first term corresponds to the classical result for linear systems and is the only nonzero term when A is square and nonsingular. The second term depends on $PER_\perp$ and vanishes when A is of full column rank, as it is in many applications.

The third term requires more explanation. If terms of second order in $\|E_{21}\|$ are ignored, it is essentially

$$\hat{\kappa}^2 \frac{\|b_2\|_2}{\|b_1\|_2} \frac{\|E_{21}\|_2}{\|A\|} \equiv \tan\theta \frac{\hat{\kappa}^2}{\eta} \frac{\|E_{21}\|_2}{\|A\|},$$

where θ is the angle subtended by b and $\mathcal{R}(A)$. The number $\hat{\kappa}\eta^{-1}\tan\theta$ can vary from 0 to ∞. It is small when θ is small (i.e., when the residual vector is small). It is also reduced in size when $\|E_{11}\|_2$ is small and x reflects the ill-conditioning of A so that $\eta \cong \kappa \cong \hat{\kappa}$. When x does not reflect the ill-conditioning of A and θ is significant, it is of order $\hat{\kappa}^2$, thus making the third term in (5.3) the dominant one.

We have bounded the third term in the decomposition (5.4) in such a way as to reflect its behavior when E_{21} is small. In fact it is bounded for all values of E_{21}, and the third term in (5.3) may be replaced by

$$\hat{\kappa}\eta \frac{\|b\|_2}{\|b_1\|_2} \Psi_2\left(\hat{\kappa}\frac{E_{21}}{\|A\|}\right).$$

For the full rank case there is a structured perturbation theorem for least squares solutions.

Theorem 5.3 (Björck). *Let A be of full column rank and let $x = A\tilde{b}$. Let S and s be nonnegative, and assume that*

$$|E| \leq \epsilon S \quad \text{and} \quad |e| \leq \epsilon s. \qquad (5.10)$$

5. LEAST SQUARES PROBLEMS

If $\tilde{x}$ is any solution of the least squares problem of minimizing $\|\tilde{b} - \tilde{A}\tilde{x}\|_2 \equiv \|\tilde{r}\|_2$ (n.b., $\tilde{A}$ is not assumed to be an acute perturbation of A), then for any absolute norm $\|\cdot\|$

$$\|\tilde{x} - x\| \leq \epsilon[\||A^\dagger|(s + S|\tilde{x}|)\| + \||(A^H A)^{-1}|S^T|\tilde{r}|\|]. \tag{5.11}$$

Proof. By Theorem 1.5, we have

$$\begin{pmatrix} I & A+E \\ (A+E)^H & 0 \end{pmatrix} \begin{pmatrix} \tilde{r} \\ \tilde{x} \end{pmatrix} = \begin{pmatrix} \tilde{b} \\ 0 \end{pmatrix}. \tag{5.12}$$

Hence

$$\begin{pmatrix} I & A \\ A^H & 0 \end{pmatrix} \begin{pmatrix} \tilde{r} \\ \tilde{x} \end{pmatrix} = \begin{pmatrix} b + e - E\tilde{x} \\ -E^H \tilde{r} \end{pmatrix}.$$

Since

$$\begin{pmatrix} I & A \\ A^H & 0 \end{pmatrix}^{-1} = \begin{pmatrix} P_\perp & A^{\dagger H} \\ A^\dagger & -(A^H A)^{-1} \end{pmatrix},$$

we have

$$\tilde{x} - A^\dagger b = A^\dagger e - A^\dagger E\tilde{x} + (A^H A)^{-1} E^T \tilde{r}.$$

Since $x = A^\dagger b$, we have on taking absolute values and using (5.10)

$$|\tilde{x} - x| \leq \epsilon[|A^\dagger|(s + S|\tilde{x}|) + |(A^H A)^{-1}|S^T|\tilde{r}|]. \tag{5.13}$$

The inequality (5.11) now follows on taking norms in (5.13). ∎

Remark 5.4. The proof of the theorem works if the matrix $(A + E)^H$ in (5.12) is replaced by $(A + F)^H$, where $|F| \leq \epsilon S$.

As it stands, the theorem is unsatisfactory for applications in which E is not known explicitly, for in that case it is impossible to compute $\tilde{r}$. However, if we compute $\hat{r} = b - A\tilde{x}$, then $\tilde{r} = \hat{r} - E\tilde{x}$. It follows that

$$|\tilde{r}| \leq |\hat{r}| + \epsilon S|\tilde{x}|,$$

and we may use this upper bound in place of $|\tilde{r}|$ in (5.11). Note that the adjustment is only of order ϵ^2 and will usually be negligible.

5.2. The Residual

Since the residual vector is given by $r = Pb$, the theory of Section 4 may be applied to give perturbation bounds for the residual. Specifically, if

$$\tilde{x} = \tilde{A}^\dagger b$$

and

$$\tilde{r} = b - \tilde{A}\tilde{x} = \tilde{P}b,$$

then

$$\|\tilde{r} - r\|_2 \leq \|\tilde{P} - P\|_2 \|b\|_2,$$

and $\|\tilde{P} - P\|_2$ can be bounded by (4.1).

5.3. Backward Perturbations

The problem of backward perturbations for least squares problems is far more difficult than the corresponding problem for linear systems. To see why, let $\tilde{x}$ be a purported solution of the problem of minimizing $\|b - Ax\|_2$, and let $\hat{r} = b - A\tilde{x}$. What we would like is to find a perturbation E such that $\tilde{x}$ is an exact solution of the problem of minimizing $\|b - (A+E)\tilde{x}\| \equiv \|\tilde{r}\|_2$. For linear systems all we need do is produce an E for which $\tilde{r}$ is zero. For the least squares problem, however, we must choose E so that the residual is orthogonal to the column space of $A + E$; i.e., so that $(A + E)^{\mathrm{H}}\tilde{r} = 0$. Since $\tilde{r}$ is defined in terms of E, this equation is nonlinear in E, and at present we know only special solutions.

Theorem 5.5. *Let $\tilde{x}$ be given. Let $x = A^\dagger b, r = b - Ax$, and $\hat{r} = b - A\tilde{x}$. Then $\tilde{x} = (A + E_i)^\dagger b$ ($i = 1, 2$) for*

$$E_1 = -\frac{\hat{r}\hat{r}^{\mathrm{H}} X}{\|\hat{r}\|_2^2},$$

in which case

$$\|E_1\|_2 = \frac{\|X^{\mathrm{H}}\hat{r}\|_2}{\|\hat{r}\|_2},$$

and for

$$E_2 = \frac{(\hat{r} - r)\tilde{x}^{\mathrm{H}}}{\|\tilde{x}\|_2^2},$$

5. LEAST SQUARES PROBLEMS

in which case

$$\|E_2\|_2 = \frac{\|\hat{r} - r\|_2}{\|\tilde{x}\|_2} = \frac{(\|\hat{r}\|_2^2 - \|r\|_2^2)^{\frac{1}{2}}}{\|\tilde{x}\|_2}. \tag{5.14}$$

Proof. The proof for E_1 is a straightforward, if tedious, verification that $(A + E)^{\text{H}}\tilde{r} = 0$.

For $\|E_2\|$, note that $\hat{r} - r = A(x - \tilde{x}) \in \mathcal{R}(A)$. Hence $\mathcal{R}(A + E_2) \subset \mathcal{R}(A)$. But it is easy to verify that $b - (A + E_2)\tilde{x} = r$. Hence

$$b - (A + E_2)\tilde{x} \in \mathcal{R}(A)_\perp \subset \mathcal{R}(A + E_2)_\perp,$$

which is sufficient for $\tilde{x}$ to be a solution of the perturbed least squares problem.

The first equality in (5.14) follows on taking norms. The second follows from the Pythagorean equality and the observation that since $\hat{r} \in \mathcal{R}(A)$, we have $\hat{r} \perp r$. ∎

The perturbation E_1 and its norm can be computed if we are given $\tilde{x}$. It is small when the residual $\hat{r}$ is nearly orthogonal to the column space of A. The perturbation E_2 cannot be computed, since it involves the true residual r, which is not known. However, it has the theoretical consequence that there is little use hunting for the exact minimizing x. Provided the residual is nearly minimal, the approximate solution $\tilde{x}$, however inaccurate, is the exact solution of a slightly perturbed problem.

The matrix E_2 is only one of a class of backward perturbation theorems. For example, if it is desirable that some of the columns of A not be altered by the perturbation, we can proceed as follows. Let $\check{x}$ be the vector obtained from $\tilde{x}$ by setting to zero the components corresponding to the columns that are not to be disturbed. Then

$$\check{E} = \frac{(\hat{r} - r)\check{x}^{\text{H}}}{\|\check{x}\|_2^2}$$

is the required matrix. Of course $\|\check{x}\|_2 \leq \|\tilde{x}\|_2$, so that $\|\check{E}\|_2 \geq \|E_2\|_2$; however $\|\check{E}\|_2$ may still be small enough for practical purposes.

The attempt to formulate a structured backward perturbation theorem for the least squares problem leads to an intractable optimization problem. However we may apply the Oettli–Prager theorem (Theorem 2.17) to the expanded equations to get the following useful result. The proof is left as an exercise.

Theorem 5.6 (Björck). Let $S \in \mathbf{R}^{m \times n}$ and $s \in \mathbf{R}^m$ be nonnegative. For $b, \hat{r} \in \mathbf{C}^m$ and $\hat{x} \in \mathbf{C}^n$ let

$$\epsilon = \max\left\{\max_i \frac{|\hat{r} + A\hat{x} - b|_i}{(S|\hat{x}| + s)_i}, \max_i \frac{|A^{\mathrm{H}}\hat{r}|_i}{(S^{\mathrm{T}}|\hat{r}|)_i}\right\}.$$

(here $0/0 = 0$ and otherwise $\rho/0 = \infty$). If $\epsilon \neq \infty$, then there are matrices E and F satisfying $|E|, |F| \leq \epsilon S$ and a vector e satisfying $|e| \leq \epsilon s$ such that

$$\begin{pmatrix} I & A+E \\ (A+F)^{\mathrm{H}} & 0 \end{pmatrix} \begin{pmatrix} \hat{r} \\ \hat{x} \end{pmatrix} = \begin{pmatrix} b+e \\ 0 \end{pmatrix}.$$

This theorem does not say that $\hat{r}$ and $\hat{x}$ are solutions of a slightly perturbed least square problem, since the perturbation E in A is different from the perturbation F^{H} in A^{H}. Nonetheless, by Remark 5.4, the common bound on E and F can be used in Theorem 5.3 to assess the accuracy of $\tilde{x}$.

5.4. Asymptotic Forms and Derivatives

An asymptotic form of the perturbed least squares solution $\tilde{x}$ can be obtained from (3.15):

$$\tilde{x} = x - A^{\dagger}PERx + R_{\perp}E^{\mathrm{H}}P(A^{\mathrm{H}})^{\dagger}x + (A^{\mathrm{H}}A)^{\dagger}RE^{\mathrm{H}}P_{\perp}b + O(\|E\|^2).$$

The corresponding derivative formula is

$$\frac{dx}{dt} = -A^{\dagger}P\frac{dA}{d\tau}Rx + R_{\perp}\frac{dA^{\mathrm{H}}}{d\tau}P(A^{\mathrm{H}})^{\dagger}x + (A^{\mathrm{H}}A)^{\dagger}R\frac{dA^{\mathrm{H}}}{d\tau}P_{\perp}b + O(\|E\|^2).$$

In reduced form

$$\begin{pmatrix} \tilde{x}_1 \\ \tilde{x}_2 \end{pmatrix} = \begin{pmatrix} x_1 \\ x_2 \end{pmatrix} + \begin{pmatrix} -A_{11}^{-1}E_{11}x_1 + (A_{11}^{\mathrm{H}}A_{11})^{-1}E_{21}b_2 \\ E_{12}^{\mathrm{H}}A_{11}^{-\mathrm{H}}x_1 \end{pmatrix} + O(\|E\|^2).$$

5. Least Squares Problems

Notes and References

The first perturbation analysis of the least squares problem is due to Golub and Wilkinson [94, 1966], who gave first order bounds. They were the first to note the dependence of the solution on κ^2. Rigorous upper bounds were derived by Hanson and Lawson [102, 1969], Pereyra [180, 1969], Stewart [199, 1969], and Wedin [258, 1969]. More recent treatments have been given by Lawson and Hanson [142, 1974] and Adbelmalek [1, 1974]. Van der Sluis [245, 1975] gives an especially detailed treatment. He was the first to point out the mitigating effect of η in (5.3).

Strictly speaking, κ is not a condition number for the least squares problem — at least not in the simple sense we have been using the term. Nonetheless, it is called the condition number of A everywhere in the literature.

Statisticians are concerned with the effects of errors in the matrix A, a problem they treat under the names "errors in the variables" or "measurement error models." One approach to the problem is to pose a probabilistic model of the error and investigate its effects [116, 49, 80]. Another approach is to compute "regression diagnostics" to tell when the error is having harmful effects [22, 213]. Yet another approach is given in Exercise 5.8.

The structured perturbation theorem (Theorem 5.3) is due to Björck [36, 1989], who also noted that it remains valid when the perturbations of A in the augmented equations are different (Remark 5.4). Arioli, Duff, and de Rijk [6, 1989] have used this fact to analyze the errors in algorithms based on the expanded equations (1.7).

Theorem 5.5 is due to Stewart [206, 1977], and Theorem 5.6 to Björck [36, 1989]. The problem of obtaining optimal backward perturbation bounds, structured or not, is unsolved. See [112] for further details.

Exercises

1. Show that $\kappa_2(A^H A) = \kappa_2^2(A)$.

THE FOLLOWING EXERCISES USE FIRST ORDER PERTURBATION THEORY TO EXPLORE THE SENSITIVITY OF LEAST SQUARES SOLUTIONS TO ERRORS IN INDIVIDUAL COLUMNS.

2. Let A be of full column rank and $x = A^\dagger b$. Let $E = e \mathbf{1}_i$. Show that

$$\tilde{x} - x = -\xi_i A^\dagger e + c_i^{(-1)} e^H r + O(\|e\|_2^2),$$

where $c_i^{(-1)}$ is the ith column of the cross-product matrix $C = A^{\mathrm{H}}A$ and $r = b - Ax$ is the residual vector.

3. Show that
$$\|\tilde{x} - x\|_2 \leq \|e\|_2 \sqrt{(|\xi_i|\|A^\dagger\|_2)^2 + (\|c_i^{(-1)}\|_2\|r\|_2)^2} + O(\|e\|_2^2).$$

4. Show that
$$|\tilde{\xi}_i - \xi_i| \leq \|e\|_2 \sqrt{\|a_j^{(\dagger)}\|_2^2 + (|\gamma_{ij}^{(-1)}|\|r\|_2)^2} + O(\|e\|_2^2),$$
where $a_j^{(\dagger)}$ is the jth row of $A^\dagger$ and $\gamma_{ij}^{(-1)}$ is the (i,j)-element of C^{-1}.

———◇———

5. (A quick and dirty bound). Let A and $\tilde{A}$ have full rank. Starting from the normal equations
$$\tilde{A}^{\mathrm{H}}\tilde{A}\tilde{x} = \tilde{A}^{\mathrm{H}}b,$$
use the perturbation theory for linear systems to show that for any consistent norm $\|\cdot\|$
$$\frac{\|\tilde{x} - x\|}{\|\tilde{x}\|} \leq \kappa(A)\frac{\|E\|}{\|A\|}\left(1 + \frac{\|b\|}{\|\tilde{x}\|}\right) + \kappa^2(A)\frac{\|E\|}{\|A\|}\left(1 + \frac{\|E\|}{\|A\|}\right).$$

6. (Higham and Stewart [114, 1987]) . Let A be of full column rank and let $C = A^{\mathrm{H}}A$. Show that if F is sufficiently small then $C + F$ can be written in the form
$$C + F = (A + E)^{\mathrm{H}}(A + E),$$
where
$$\|E\|_{\mathrm{F}} \lesssim \frac{1}{2}\|A^\dagger\|_2\|F\|_{\mathrm{F}}.$$
Show that this bound is realistic.

7. The vector $\hat{r}$ in Theorem 5.6 can be regarded as an arbitrary parameter. Write down the bounds obtained when $\hat{r} = b - A\hat{x}$ and $\hat{r} = 0$. [Note: the problem of determining the optimal $\hat{r}$ is open.]

8. (Stewart [217, 1989]). Let A have full rank. Show that
$$\tilde{x} = A^\dagger(b + Ex) + O(\|E\|^2).$$
Give an expression and a bound for the $O(\|E\|^2)$ term. [Note: To the statistician, this result says that if E is small enough, the least squares solution $\tilde{x}$ behaves as if it came from an unperturbed problem in which the error in the right-hand side has been inflated.]

Chapter IV
The Perturbation of Eigenvalues

Of all the problems in matrix perturbation theory the perturbation of the eigenvalues of a matrix presents the most varied technical difficulties. The problem itself is simply stated. Given a matrix $A \in \mathbf{C}^{n \times n}$ and a perturbation E of A, how are the spectra $\mathcal{L}(A)$ and $\mathcal{L}(A+E)$ related? But this simplicity is elusive. In the first place, the term "related" has more than one natural sense, as we shall see in the first section of this chapter. More important, different classes of matrices, even matrices having the same eigenvalues, behave differently under perturbation.

Example 1. *Let $A_0 = 0$. Then the eigenvalues of A_0 are all zero. Let E be a perturbation of A_0. Since $A_0 + E = E$, it follows from Theorem II.2.6 that*

$$\lambda \in \mathcal{L}(A_1 + E) \implies |\lambda| \leq \|E\|.$$

for any consistent matrix norm $\|\cdot\|$. On the other hand, let

$$A_1 = \begin{pmatrix} 0 & 1 & 0 & 0 \\ 0 & 0 & 1 & 0 \\ 0 & 0 & 0 & 1 \\ 0 & 0 & 0 & 0 \end{pmatrix}.$$

Then the eigenvalues of A_1 are also zero. However, if $\epsilon > 0$, the eigenvalues of

$$\tilde{A}_1 = \begin{pmatrix} 0 & 1 & 0 & 0 \\ 0 & 0 & 1 & 0 \\ 0 & 0 & 0 & 1 \\ \epsilon & 0 & 0 & 0 \end{pmatrix}$$

are $\epsilon^{\frac{1}{4}}\omega_i$, where the ω_i are the primitive 4th roots of unity (i.e., 1, i, -1, $-i$). Thus while a perturbation of order, say, 10^{-8} will induce a perturbation of order only 10^{-8} in the eigenvalues of A_0, it can induce a perturbation of as large as 10^{-2} in the eigenvalues of A_1.

This example shows that a general perturbation theory for eigenvalues has to be pessimistic, since it must account for the ill-conditioned behavior of the eigenvalues of matrices like A_1. The cure for this problem is to develop individual perturbation theories for different classes of matrices, which is what we will do in this chapter. We begin in Section 1 with the general case. In Section 2 we introduce the very useful Gerschgorin theorem and use it to compute the derivative of a simple eigenvalue. In Section 3 we treat normal and diagonalizable matrices, and in Section 4 Hermitian matrices. The chapter concludes with a section on special topics.

> As in the last chapter we will use the tilde conventions to denote perturbations. Specifically, A will denote a (complex) matrix of order n, and $\tilde{A} = A + E$ will denote a perturbation of A. The eigenvalues of A are $\mathcal{L}(A) = \{\lambda_1, \ldots, \lambda_n\}$ and those of $\tilde{A}$ are $\mathcal{L}(\tilde{A}) = \{\tilde{\lambda}_1, \ldots, \tilde{\lambda}_n\}$. As usual the characteristic polynomials of A and $\tilde{A}$ will be written $\phi_A(\lambda)$ and $\phi_{\tilde{A}}(\lambda)$.

1. General Perturbation Theorems

1.1. Continuity: Ostrowski–Elsner Theorems

The first thing that we will establish about eigenvalues is that they are continuous, which follows from a fact and a theorem. The fact is that the characteristic polynomial of a matrix, being itself a polynomial in

1. General Perturbation Theorems

the elements of the matrix, is a continuous function of the matrix. The theorem is Rouché's theorem. In the form we will use here, it states that if ϕ and η are analytic in a simply connected region Ω and $\mathcal{D} \subset \Omega$ is a disk for which

$$|\eta(\zeta)| < |\phi(\zeta)|, \quad \zeta \in \partial\mathcal{D},$$

where $\partial\mathcal{D}$ is the boundary of $\mathcal{D}$, then $\phi(\zeta)$ and $\phi(\zeta) + \eta(\zeta)$ have the same number of zeros in $\mathcal{D}$.

Theorem 1.1. *Let λ be an eigenvalue of A of algebraic multiplicity m. Then for any norm $\|\cdot\|$ and all sufficiently small $\epsilon > 0$ there is a $\delta > 0$ such that if $\|E\| < \delta$, the disk $\mathcal{D}(\lambda, \epsilon) = \{\zeta \in \mathbf{C} : |\zeta - \lambda| \leq \epsilon\}$ contains exactly m eigenvalues of $\tilde{A}$.*

Proof. Let ϵ be so small that $\mathcal{D}(\lambda, \epsilon)$ contains only the eigenvalue λ of A. Let $\eta(\zeta) = \phi_{\tilde{A}}(\zeta) - \phi_A(\zeta)$. By the continuity of the characteristic polynomial, as $\tilde{A} \to A$ the function $\eta(\zeta)$ converges to zero on the compact set $\partial\mathcal{D}$. Since $\phi_A(\zeta)$ is nonzero on $\partial\mathcal{D}$, there is a $\delta > 0$ such that $|\eta(\zeta)| < |\phi_A(\zeta)|$ on $\partial\mathcal{D}$ whenever $\|E\| < \delta$. By Rouché's theorem ϕ_A and $\phi_{\tilde{A}} = \phi_A + \eta$ have the same number of zeros in $\mathcal{D}$. ∎

Theorem 1.1 is an example of a qualitative perturbation theorem: it states that a perturbation must be small without providing a bound on the size of the perturbation. We now turn to a theorem of Elsner, which provides explicit bounds. However, we first need to introduce some notation to describe how the the eigenvalues of two matrices are situated with respect to one another.

Definition 1.2. *Let A have eigenvalues $\lambda_1, \ldots, \lambda_n$ and $\tilde{A}$ have eigenvalues $\tilde{\lambda}_1, \ldots, \tilde{\lambda}_n$. Then the SPECTRAL VARIATION OF $\tilde{A}$ WITH RESPECT TO A is*

$$\operatorname{sv}_A(\tilde{A}) \stackrel{\text{def}}{=} \max_i \min_j |\tilde{\lambda}_i - \lambda_j|. \tag{1.1}$$

The HAUSDORFF DISTANCE between the eigenvalues of A and $\tilde{A}$ is

$$\operatorname{hd}(A, \tilde{A}) \stackrel{\text{def}}{=} \max\{\operatorname{sv}_A(\tilde{A}), \operatorname{sv}_{\tilde{A}}(A)\}. \tag{1.2}$$

The (OPTIMAL) MATCHING DISTANCE between the eigenvalues of A and $\tilde{A}$ is

$$\operatorname{md}(A, \tilde{A}) \stackrel{\text{def}}{=} \min_\pi \{\max_i |\tilde{\lambda}_{\pi(i)} - \lambda_i|\}, \tag{1.3}$$

where π is taken over all permutations of $\{1, 2, \ldots, n\}$.

The function $\mathrm{sv}_A(\tilde{A})$ is not a metric: it may be zero, even when the eigenvalues of A and $\tilde{A}$ are different (e.g., when $n = 2$ and $\lambda_1 = \tilde{\lambda}_1 = \tilde{\lambda}_2 = 0$ while $\lambda_2 = 1$). Geometrically, the function sv has the following interpretation.

If
$$\mathcal{D}_i = \{\zeta : |\zeta - \lambda_i| \leq \mathrm{sv}_A(\tilde{A})\}, \quad i = 1, \ldots, n,$$
then
$$\mathcal{L}(\tilde{A}) \subset \bigcup_{i=1}^n \mathcal{D}_i.$$

In other words, the eigenvalues of $\tilde{A}$ lie in the union of disks of radius $\mathrm{sv}_A(\tilde{A})$ centered at the eigenvalues of A.

The Hausdorff distance bounds the spectral variation and is actually a metric. The matching distance bounds the Hausdorff distance and is also a metric. To say that the matching distance is small is one of the nicest things that can be said of the eigenvalues of a matrix and its perturbation. It means that they can be grouped into nearby pairs.

We are now in a position to bound the Hausdorff distance between two matrices.

Theorem 1.3 (Elsner). *For any A and $\tilde{A}$,*

$$\mathrm{hd}(A, \tilde{A}) \leq (\|A\|_2 + \|\tilde{A}\|_2)^{1-\frac{1}{n}} \|E\|_2^{\frac{1}{n}}. \tag{1.4}$$

Proof. Since the right-hand side of (1.4) is symmetric in A and $\tilde{A}$, it is sufficient to prove that it bounds $\mathrm{sv}_A(\tilde{A})$. Assume the maximum in (1.1) is attained for the eigenvalue $\tilde{\lambda}$ of $\tilde{A}$, and let $\tilde{x}_1, \ldots, \tilde{x}_n$ be orthonormal vectors with $\tilde{A}\tilde{x}_1 = \tilde{\lambda}\tilde{x}_1$. Then

$$\begin{aligned}
\mathrm{sv}_A(\tilde{A})^n &\leq \prod_i |\lambda_i - \tilde{\lambda}| \\
&= \det(A - \tilde{\lambda}I) \\
&\leq \prod_i \|(A - \tilde{\lambda}I)\tilde{x}_i\|_2 \quad \text{[Hadamard inequality (I.2.2)]} \\
&= \|(A - \tilde{A})\tilde{x}_1\|_2 \prod_{i>1} \|(A - \tilde{\lambda}I)\tilde{x}_i\|_2 \\
&\leq \|E\|_2 (\|A\|_2 + \|\tilde{A}\|_2)^{n-1}.
\end{aligned}$$

The result follows on taking nth roots in the above inequality and from the symmetry of the resulting bound. ∎

1. GENERAL PERTURBATION THEOREMS

As we have mentioned above, the most desirable bound is one on the matching distance. In some cases bounds on the spectral variation or the Hausdorff distance can be converted into such a bound. Since the technique, with appropriate variations, can be applied to other problems, we will develop it informally and then summarize the results.

Let us begin by relaxing our bound a little and writing

$$\mathrm{sv}_A(\tilde{A}) \leq \mu \|E\|_2^{\frac{1}{n}} \equiv \epsilon,$$

where

$$\mu = (\max\{2\|A + \tau E\|_2 : 0 \leq \tau \leq 1\})^{1-\frac{1}{n}}.$$

As above, set

$$\mathcal{D}_i = \{\zeta : |\zeta - \lambda_i| \leq \epsilon\}, \quad i = 1, \ldots, n.$$

The purpose of this adjustment is to make the bound monotone in τE. We claim that

if any m of the disks $\mathcal{D}_i$ are isolated from the others, then their union contains exactly m eigenvalues of $\tilde{A}$.

To see this, assume without loss of generality that the m disks isolated from the others are $\mathcal{D}_1, \mathcal{D}_2, \ldots, \mathcal{D}_m$. For $0 \leq \tau \leq 1$, let $\tilde{A}_\tau = \tau \tilde{A} + (1-\tau)A = A + \tau E$, and let

$$\mathcal{D}_i^{(\tau)} = \{\zeta : |\zeta - \lambda_i| \leq \mu \|\tau E\|_2^{\frac{1}{n}}\}.$$

Since

$$\|A\|_2 + \|\tilde{A}_t au\|_2 \leq \mu,$$

by Theorem 1.3, we have $\mathrm{sv}_A(\tilde{A}_\tau) \leq \mu \|\tau E\|_2^{\frac{1}{n}} = \tau^{\frac{1}{n}}\epsilon$, and the eigenvalues of $\tilde{A}_\tau$ lie in the union of the disks $\mathcal{D}_i^{(\tau)}$.

Now $\cup_{i=1}^m \mathcal{D}_i^{(0)}$ contains exactly m eigenvalues of $\tilde{A}_0 = A$, namely $\lambda_1(A), \lambda_2(A), \ldots, \lambda_m(A)$. Since $\tau^{\frac{1}{n}}\epsilon$ is an increasing function of τ, as τ varies from zero to one the region $\cup_{i=1}^m \mathcal{D}_i^{(\tau)}$ remains disjoint from the other disks. Since by Theorem 1.1 the eigenvalues of $\tilde{A}_\tau$ are continuous in τ, they cannot jump from one disjoint region to another. Hence $\cup_{i=1}^m \mathcal{D}_i^{(1)}$ must contain exactly m eigenvalues of $\tilde{A}_1 = \tilde{A}$.

It is now easy to obtain a bound on md($A, \tilde{A}$). Let $\mathcal{C}_1, \mathcal{C}_2, \ldots, \mathcal{C}_k$ be the connected components of $\cup_{i=1}^{n} \mathcal{D}_i$. If $\mathcal{C}_l$ is the union of m_l of the disks $\mathcal{D}_i$, then it contains exactly m_l eigenvalues of A and m_l eigenvalues of $\tilde{A}$. Choose the permutation π to associate the eigenvalues of A in each $\mathcal{C}_l$ with the corresponding eigenvalues of $\tilde{A}$. Since each eigenvalue of A in $\mathcal{C}_l$ is within $(2m_l - 1)\delta(A, \tilde{A})$ of any point in $\mathcal{C}_l$, each eigenvalue of $\tilde{A}$ in $\mathcal{C}_l$ is within $(2m_l - 1)\delta(A, \tilde{A}) \leq (2n - 1)\delta(A, \tilde{A})$ of the corresponding eigenvalues of A.

We have just established the following theorem.

Theorem 1.4 (Ostrowski, Elsner).

$$\mathrm{md}(A, \tilde{A}) \leq (2n - 1)(\|A\|_2 + \|\tilde{A}\|_2)^{1-\frac{1}{n}} \|E\|_2^{\frac{1}{n}}.$$

Actually, we have only used the fact that Theorem 1.3 gives a bound on $\mathrm{sv}_A(\tilde{A})$. Elsner, by an application of Hall's theorem (Theorem II.3.14), has shown that the factor $2n - 1$ can be replaced by $2\lfloor n/2 \rfloor$. To summarize:

Theorem 1.5. *Let $\tau \geq 0$. If $\beta(\tau)$ is a nondecreasing bound on $\mathrm{sv}_A(A + \tau E)$, then*

$$\mathrm{md}(A, \tilde{A}) \leq (2n - 1)\beta(1).$$

If $\beta(\tau)$ is a nondecreasing bound on $\mathrm{hd}(A, A + \tau E)$, then

$$\mathrm{md}(A, \tilde{A}) \leq 2\lfloor n/2 \rfloor \beta(1). \tag{1.5}$$

1.2. The Bauer–Fike and Henrici Theorems

As was pointed out in the introduction to this chapter, any general perturbation bound on the eigenvalues of a matrix will have to be pessimistic. In Theorems 1.3 and 1.4, this shows itself by the fact that $\delta(A, \tilde{A})$ is proportional to the nth root of the error $\nu(A, \tilde{A})$. Example 1—or rather a trivial extension of it—shows that this nth root is necessary. However, in most cases it is unrealistic.

To see one way in which this can come about, let us return to Example 1 and set $A_\eta = \eta A_1$, where η is presumed small. Let E be given with $\|E\|_2 = \epsilon$. If ϵ is much smaller than η, Elsner's bound will generally be of the right order, unless E has special structure. On the other hand, if $\epsilon \geq \eta$, then $\|A_\eta + E\|_2 \leq \eta + \epsilon \leq 2\epsilon$, and no eigenvalue of

1. GENERAL PERTURBATION THEOREMS

A_η can change by more than 2ϵ. The reason that the fourth root of the error is unrealistic in the second case is that $A_\eta + E$ can be regarded as a perturbation of the zero matrix, which is well behaved.

In this subsection we will derive a bound, due to Henrici, that takes this phenomenon into account. It is based on a general theorem of Bauer and Fike.

Theorem 1.6 (Bauer-Fike). *Let Q be nonsingular, and let $\|\cdot\|$ be consistent. If $\tilde\lambda \in \mathcal{L}(\tilde A)$ is not an eigenvalue of A, then*

$$\|Q^{-1}(A - \tilde\lambda I)^{-1} Q\|^{-1} \leq \|Q^{-1} E Q\|. \qquad (1.6)$$

Proof. We have

$$Q^{-1}(\tilde A - \tilde\lambda I) Q = Q^{-1}[(A - \tilde\lambda I) + E] Q$$
$$= Q^{-1}(A - \tilde\lambda I) Q \{ I + [Q^{-1}(A - \tilde\lambda I)^{-1} Q][Q^{-1} E Q] \}.$$

Since $\tilde A - \tilde\lambda I$ is singular,

$$\begin{aligned} 1 &\leq \|[Q^{-1}(A - \tilde\lambda I)^{-1} Q][Q^{-1} E Q]\| \\ &\leq \|Q^{-1}(A - \tilde\lambda I)^{-1} Q\| \|Q^{-1} E Q\|, \end{aligned} \qquad (1.7)$$

and this last inequality is equivalent to (1.6). ∎

Note that if the left-hand side of (1.6) is regarded as zero when $\tilde\lambda \in \mathcal{L}(A)$, then the inequality holds for all eigenvalues of $\tilde A$. In the sequel we will not be overfussy in dealing with trivial singularities of this kind.

Our first application of the Bauer–Fike theorem is to prove Henrici's perturbation theorem. It is phrased in terms of a deviation from normality. Recall that if a matrix is normal, its Schur form is diagonal. Consequently the size of the off-diagonal terms in the Schur form can be used to measure the departure of a matrix from normality.

Definition 1.7. *Let ν be a norm on $\mathbf{C}^{n\times n}$. Let $\mathcal{U}$ be the set of unitary U such that $U^H A U$ is upper triangular. For each $U \in \mathcal{U}$ write $U^H A U = \Lambda_U + R_U$, where R_U is strictly upper triangular. Then the ν-DEPARTURE FROM NORMALITY of A is the number*

$$\delta_\nu(A) \stackrel{\mathrm{def}}{=} \min_{U \in \mathcal{U}} \nu(R_U).$$

The departure from normality is not easy to calculate, since the Schur form is not unique. However, if A has eigenvalues λ_i, then by the unitary invariance of the Frobenius norm we have for any Schur form

$$\|A\|_F^2 = \sum |\lambda_i|^2 + \|R\|_F^2.$$

Thus we have the following theorem.

Theorem 1.8. *For any matrix A with eigenvalues λ_i,*

$$\delta_F(A) = \sqrt{\|A\|_F^2 - \sum |\lambda_i|^2}.$$

We are now in a position to state and prove Henrici's theorem.

Theorem 1.9 (Henrici). *Let ν be a norm on $\mathbf{C}^{n \times n}$ such that $\nu(C) \geq \|C\|_2$ for all $C \in \mathbf{C}^{n \times n}$. Then for every eigenvalue $\tilde{\lambda}$ of $\tilde{A}$ there is an eigenvalue λ of A such that*

$$\frac{\left(\frac{|\tilde{\lambda}-\lambda|}{\delta_\nu(A)}\right)^n}{1 + \left(\frac{|\tilde{\lambda}-\lambda|}{\delta_\nu(A)}\right) + \cdots + \left(\frac{|\tilde{\lambda}-\lambda|}{\delta_\nu(A)}\right)^{n-1}} \leq \frac{\|E\|_2}{\delta_\nu(A)}. \tag{1.8}$$

Proof. Let $\tilde{\lambda}$ be an eigenvalue of $\tilde{A}$, and let $U^H A U = \Lambda + R$ be a Schur form of A. Then by (1.6),

$$\|(\Lambda - \tilde{\lambda} I + R)^{-1}\|_2^{-1} \leq \|E\|_2. \tag{1.9}$$

Since R is strictly upper triangular,

$$(\Lambda - \tilde{\lambda} I + R)^{-1} = \{I - (\Lambda - \tilde{\lambda} I)^{-1} R + \cdots + (-1)^{n-1} [(\Lambda - \tilde{\lambda} I)^{-1} R]^{n-1}\}(\Lambda - \tilde{\lambda} I)^{-1}.$$

Thus if $\delta = \min\{\lambda \in \mathcal{L}(A) : |\tilde{\lambda} - \lambda|\}$,

$$\|(\Lambda - \tilde{\lambda} I + R)^{-1}\|_2 \leq \delta^{-1}\{1 + \delta^{-1} \delta_\nu(A) + \cdots + [\delta^{-1} \delta_\nu(A)]^{n-1}\}.$$

Hence

$$\|(\Lambda - \tilde{\lambda} I + R)^{-1}\|_2^{-1} \geq \frac{\delta}{1 + \delta_\nu(A)/\delta + \cdots + [\delta_\nu(A)/\delta]^{n-1}}. \tag{1.10}$$

1. GENERAL PERTURBATION THEOREMS 173

The theorem follows on combining (1.9) and (1.10) and dividing by $\delta_\nu(A)$. ∎

The remarkable thing about Henrici's theorem is that it provides a continuous transition between the two cases mentioned at the beginning of this subsection: namely, the case in which the perturbation bound is proportional to the nth root of the error and the case in which it is proportional to the error itself. To see this, let

$$\psi(\eta) = \eta^n/(1 + \eta + \cdots + \eta^{n-1}), \qquad (1.11)$$

so that the left-hand side of the bound (1.8) has the form $\psi[|\tilde\lambda - \lambda|/\delta_\nu(A)]$. For η small, $\psi(\eta) \cong \eta^n$, and the bound takes the asymptotic form

$$\frac{\operatorname{sv}_A(\tilde A)}{\delta_\nu(A)} \lesssim \left(\frac{\|E\|_2}{\delta_\nu(A)}\right)^{\frac{1}{n}}.$$

When η is large, $\psi(\eta) \cong \eta$, and the bound takes the asymptotic form

$$\operatorname{sv}_A(\tilde A) \lesssim \|E\|_2.$$

Specifically, we have the following corollary.

Corollary 1.10. *If* $\|E\|_2/\delta_\nu(A) < n^{-1}$, *then*

$$\frac{\operatorname{sv}_A(\tilde A)}{\delta_\nu(A)} \leq n^{\frac{1}{n}}\left(\frac{\|E\|_2}{\delta_\nu(A)}\right)^{\frac{1}{n}}. \qquad (1.12)$$

If $\|E\|_2/\delta_\nu(A) > 1$, *then*

$$\operatorname{sv}_A(\tilde A) \leq \|E\|_2 + \delta_\nu(A). \qquad (1.13)$$

Proof. If $\psi(\eta) < 1/n$, then $\eta < 1$. Hence if $\|E\|_2/\delta_\nu(A) < n^{-1}$,

$$n^{-1}\left(\frac{\operatorname{sv}_A(\tilde A)}{\delta_\nu(A)}\right)^n \leq \psi\left(\frac{\operatorname{sv}_A(\tilde A)}{\delta_\nu(A)}\right) \leq \frac{\|E\|_2}{\delta_\nu(A)},$$

from which (1.12) follows. On the other hand, if $\psi(\eta) > 1$, then $\eta > 1$ and

$$\psi(\eta) = \frac{\eta}{1 + \eta^{-1} + \cdots + \eta^{-(n-1)}} \geq \eta(1 - \eta^{-1}) = \eta - 1.$$

Hence if $\|E\|_2/\delta_\nu(A) > 1$,

$$\frac{\mathrm{sv}_A(\tilde{A})}{\delta_\nu(A)} \le \psi^{-1}\left(\frac{\|E\|_2}{\delta_\nu(A)}\right) \le \frac{\|E\|_2}{\delta_\nu(A)} + 1,$$

which is equivalent to (1.13). ∎

Using Theorem 1.5 and the monotonicity of ψ, we have the following bound on $\mathrm{md}(A, \tilde{A})$.

Corollary 1.11. *Let ψ be defined by (1.11). Then*

$$\mathrm{md}(A, \tilde{A}) \le (2n - 1)\delta_\nu(A)\psi^{-1}\left(\frac{\|E\|_2}{\delta_\nu(A)}\right).$$

An unsavory aspect of Henrici's theorem, one that it shares with Theorems 1.3 and 1.4, is the nth root of the error in its bounds. The examples that show that it's presence is necessary all depend on the matrix having a Jordan block equal to its order. The following theorem shows that for a matrix with smaller Jordan blocks the root is smaller. Its proof is similar to Henrici's theorem and is left as an exercise.

Theorem 1.12. *Let $Q^{-1}AQ = J$ be the Jordan canonical form of A. Let m be the size of the largest Jordan block in J. Then for any eigenvalue $\tilde{\lambda} \in \lambda(\tilde{A})$ there is an eigenvalue λ of A such that*

$$\frac{|\tilde{\lambda} - \lambda|^m}{1 + |\tilde{\lambda} - \lambda| + \cdots + |\tilde{\lambda} - \lambda|^{m-1}} \le \|Q^{-1}EQ\|_2. \tag{1.14}$$

1.3. Residual Bounds

Let the columns of X form a basis for an invariant subspace of A. From Theorem 3.9, we know that there is a unique matrix M (which is now easily seen to be $X^\dagger AX$) such that

$$AX - XM = 0.$$

The matrix M is the representation of A on $\mathcal{R}(X)$ with respect to the basis X, and hence the eigenstructure of M is a substructure of the eigenstructure of A.

1. GENERAL PERTURBATION THEOREMS

Now suppose that the columns of X span a subspace that is only approximately invariant. For example, X may come from a numerical algorithm for approximating invariant subspaces. Then for any M the residual

$$R = AX - XM \qquad (1.15)$$

is nonzero, although presumably with a proper choice of M it can be made small. An important problem in perturbation theory is: Given some norm of R, determine how near $\mathcal{R}(X)$ is to an invariant subspace of A and how the eigenvalues of M relate to those of A. We will consider the invariant subspace problem in the next chapter. Here we will focus on the eigenvalue problem.

The key tool in our investigation is the following backward perturbation theorem. Its proof, which is purely computational, is left as an exercise.

Theorem 1.13. Let $A \in \mathbf{C}^{n \times n}$, $X \in \mathbf{C}^{n \times p}$, and $M \in \mathbf{C}^{p \times p}$. Let R be defined by (1.15). If Y^H is any matrix satisfying $Y^H X = I$ and

$$\tilde{A} = A - RY^H, \qquad (1.16)$$

then

$$\tilde{A}X - XM = 0.$$

The theorem says that if R is small then $\mathcal{R}(X)$ is an exact invariant subspace of a matrix $\tilde{A}$ that is near A — in fact within $\|RY^H\|$ of A in any norm $\|\cdot\|$. Moreover, M is the representation of $\tilde{A}$ on $\mathcal{R}(X)$, and its eigenvalues are therefore eigenvalues of $\tilde{A}$. Since we know $\|E\|$, we may use any appropriate perturbation theorem for eigenvalues to assess the accuracy of the eigenvalues of M. For example, from Corollary 1.11 we have the following corollary.

Corollary 1.14. Let $\mu_1, \ldots, \mu_p$ be the eigenvalues of M. Then there are eigenvalues $\lambda_{j_1}, \ldots, \lambda_{j_p}$ of A such that

$$|\mu_i - \lambda_{j_i}| \leq (2n-1)\delta_\nu(A)\psi^{-1}\left(\frac{\|RY^H\|_2}{\delta_\nu(A)}\right).$$

The problem of choosing M and Y to minimize $\|RY^H\|$ still remains. In general, the problem is intractable; however, for unitarily invariant norms it has an elegant solution.

Theorem 1.15. *In the notation of Theorem 1.13, assume that* $X^H X = I$. *Let* $\|\cdot\|$ *be a unitarily invariant norm. Then* $\|R\|$ *is minimized for* $M = X^H A X$, *and* RY^H *is minimized for* $M = X^H A X$ *and* $Y = X$.

Proof. Let $(X \ X_\perp)$ be unitary. Then from (1.15),

$$\|R\| = \|(X \ X_\perp)^H R\| = \left\| \begin{pmatrix} X^H A X - M \\ X_\perp A X \end{pmatrix} \right\|.$$

It follows from Corollary 3.8 that $\|R\|$ is minimized when $X^H A X - M = 0$.

To minimize $\|RY^H\|$, note that $Y^H X = I$ implies that $Y = X + X_\perp S$, for some S. Then

$$\|RY^H\| = \|(X \ X_\perp)^H RY^H (X \ X_\perp)\|$$

$$= \left\| \begin{pmatrix} X^H A X - M & (X^H A X - M)S^H \\ X_\perp A X & X_\perp A X S^H \end{pmatrix} \right\|.$$

Again by Corollary 3.8, the norm of RY^H is minimized when $X^H A X - M = 0$ and $S = 0$. ∎

Notes and References

Perturbation theory for eigenvalues comes in two flavors. In this book we consider comparatively unstructured errors and attempt to bound the perturbations in terms of some norm of the errors. Other approaches impose some structure on the errors; for example, they may be analytic functions of a complex variable. The problem is then to determine how this structure affects the perturbed eigenvalues: e.g., when are they analytic functions of the variable, what kind of paths do they follow in the complex plane? For more results of this kind see the books by Kato [135] and Baumgärtel [17]. The approach taken here is the one generally followed by numerical analysts; for example, Householder [121, 1964] and Wilkinson [269, 1965]. Particular mention should be made of the elegant little book by Bhatia [28, 1987], which is a required supplement to this chapter.

Rouché's theorem may be found in most texts on complex analysis. As Example 1 shows, we cannot expect much more than continuity in the eigenvalues, at least for defective eigenvalues. However, the reader should not

1. GENERAL PERTURBATION THEOREMS

conclude from this example that all perturbations of defective eigenvalues are multiples of primitive roots of unity. A counterexample is given in the exercises.

The term "spectral variation" is found in Henrici [109, 1962] but may be of earlier vintage. The Hausdorff distance between two sets may be found in the second edition of Hausdorff's famous book on set theory [104, 1914]. In general, the Hausdorff distance is a metric only over the class of closed bounded sets, which is just what the set of eigenvalues of a matrix is. The term "optimal matching distance" seems to be due to Bhatia [28], although the concept has been around for some time (e.g., Henrici calls it the eigenvalue distance).

The first general perturbation bounds for eigenvalues were given by Ostrowski [169, 1957]. Theorem 1.3 is due to Elsner [66, 1985], who also shows that the bounds are in some sense the best possible (Exercise 1.4). It is perhaps significant both Ostrowski and Elsner use Hadamard's inequality in deriving their bounds.

The fact that one can count the eigenvalues in the connected components of inclusion regions provided by the bound on the spectral variation was first noted by Gerschgorin [85, 1931] and also by Ostrowski [169, 1957], who used it to establish the "$2n-1$" bound on the matching distance. The "$2\lfloor n/2 \rfloor$" bound is due to Elsner [65, 1982].

Bauer and Fike [15, 1960] have not been treated fairly in the literature. Their names have become associated with a weak corollary of Theorem 1.6 (Theorem 3.3), which is frequently trotted out as a straw man by people who have not read the original paper. The generality of their technique makes it applicable in a variety of situations (e.g., Henrici's theorem, Gerschgorin's theorem in the next section, and also a useful theorem of Demmel [55]).

Henrici's theorem [109, 1962] is but one of many results that Henrici casts in terms of the departure from normality. The observation (Corollary 1.10) that the theorem provides a smooth transition from nonlinear to linear behavior in the bounds is new.

For a single eigenvalue, Theorem 1.13 may be found in Wilkinson's book [269, 1965]. The optimality conditions of Theorem 1.15 are part of the folklore, at least for the Frobenius norm. The observations that the conditions are optimal for all unitarily invariant norms appears to be new.

In some applications we may have, in addition to a residual for an approximate invariant subspace, a residual for the corresponding left invariant subspace. Kahan, Parlett, and Jiang [134, 1982] show how to use this information to derive a backward perturbation theorem (See Exercise 1.12).

Exercises

1. Show that the eigenvalues of the matrix $J_n(0) + \epsilon \mathbf{1}_i \mathbf{1}_1^T$ ($i = 1, \ldots, n$) are zero with multiplicity $n - i$ and $\epsilon^{\frac{1}{i}}$ times the primitive ith roots of unity.

2. (Wilkinson [269, p.80]). Let $A = \text{diag}[J_3(0), J_2(0)]$. Show that there is a perturbation of A of order ϵ for which all the eigenvalues of the perturbed matrix are of order $\epsilon^{\frac{5}{2}}$.

3. Let $[A]$ be the equivalence class of matrices having the same eigenvalues as A. Show that the Hausdorff distance and the matching distance are metrics over the space of such equivalence classes.

4. (Elsner [66]). Show that equality holds in the bound (1.4) if and only if $A = \omega \|A\|_2 I$ ($|\omega| = 1$) and $\tilde{A}$ has $-\omega \|\tilde{A}\|_2$ as an eigenvalue.

5. (Kato [135, p.109]). Let $\mathcal{M}(\tau)$ be an unordered n-tuple of n numbers that depend continuously on the parameter τ in an interval $\mathcal{I}$. Show that there are functions $\mu_i(\tau)$ ($i = 1, \ldots, n$), continuous on $\mathcal{I}$, such that $\mathcal{M}(\tau)$ consists of $\mu_1(\tau), \ldots, \mu_n(\tau)$.

6. Establish the bound (1.5). [Note: This is a difficult problem. The idea is to declare eigenvalues λ and $\tilde{\lambda}$ related if they can be connected by a suitably short chain of disks. One then applies Hall's theorem (Theorem 3.14) to the 0-1 matrix of this relation. See [65, 28] for details.]

7. Show that if $\|(\tilde{\lambda} I - A)^{-1}\|_2 \geq \eta$, then there is an eigenvalue λ of A satisfying
$$|\tilde{\lambda} - \lambda| \leq 2(\|A\|_2 + \eta^{-1})\eta^{-\frac{1}{n}}.$$

8. (Henrici [109]). For any A
$$\rho_F(A) \leq \sqrt[4]{\frac{n^3 - n}{12}} \sqrt{\|A^H A - A A^H\|_F}.$$

Characterize the matrices for which equality holds. [Note: Recall that A is normal if and only if $\|A^H A - A A^H\|_F = 0$. It is therefore not surprising that the size of $\|A^H A - A A^H\|_F$ is related to the departure from normality. This problem is not an easy.]

1. GENERAL PERTURBATION THEOREMS

9. (A structured backward perturbation theorem for eigenvectors). Let $r = (\tilde{\lambda} I - A\tilde{x})$ and let S be nonnegative. Set

$$\epsilon = \max_i \frac{|\rho_i|}{(S|\tilde{x}|)_i}$$

(Here $0/0 = 0$ and otherwise $\rho/0 = \infty$). If $\epsilon \neq \infty$, there is a matrix E satisfying $|E| \leq \epsilon S$ such that $(\tilde{\lambda}, \tilde{x})$ is an eigenpair of $A + E$.

THE FOLLOWING EXERCISES CONCERN BACKWARD PERTURBATIONS WHEN RESIDUALS FOR LEFT AND RIGHT INVARIANT SUBSPACES ARE KNOWN. THEY ARE BASED ON A GENERAL THEOREM OF DAVIS, KAHAN, AND WEINBERGER ON DILATIONS [52, 1982], WHICH IS ESTABLISHED IN THE NEXT TWO EXERCISES. THE DILATION PROBLEM MAY BE STATED AS FOLLOWS. GIVEN THE PARTITIONED MATRIX

$$A = \begin{pmatrix} A_{11} & A_{12} \\ A_{21} & A_{22} \end{pmatrix},$$

DETERMINE A_{22} SO THAT $\|A\|_2$ IS MINIMIZED. FOR A HISTORY OF THE PROBLEM AND APPLICATIONS, SEE THE ARTICLE JUST CITED.

10. Let $\|A_{11}\|_2 \leq \nu$. Show that

$$\left\| \begin{pmatrix} A_{11} \\ A_{21} \end{pmatrix} \right\|_2 \leq \nu$$

if and only if $A_{21} = K_{21}(\nu^2 I - A_{11}^H A_{11})^{\frac{1}{2}}$, where $\|K_{21}\|_2 \leq 1$. In particular we may take $K_{21} = A_{21}(\nu^2 I - A_{11}^H A_{11})^{\dagger/2}$.

11. Let A be as above and let

$$\max \left\{ \left\| \begin{pmatrix} A_{11} \\ A_{21} \end{pmatrix} \right\|_2, \|(A_{11} \; A_{12})\|_2 \right\} \leq \nu.$$

Then A_{22} may be chosen so that $\|A\|_2 \leq \nu$. In particular if $K_{21} = A_{21}(\nu^2 I - A_{11}^H A_{11})^{\dagger/2}$ and $K_{12} = (\nu^2 I - A_{11} A_{11}^H)^{\dagger/2} A_{12}$, then the most general form of A_{22} is

$$A_{22} = -K_{21} A_{11}^H K_{12} + \nu(I - K_{21} K_{21}^H)^{\frac{1}{2}} C (I - K_{12}^H K_{12})^{\frac{1}{2}},$$

where C is an arbitrary matrix satisfying $\|C\|_2 \leq 1$. [Hint: Apply the previous exercise three times: twice to define K_{21} and K_{12} and once to the partition

$$A = \begin{pmatrix} A_1 \\ A_2 \end{pmatrix}.$$

N.b., the last step is nontrivial.]

12. (Kahan, Parlett, and Jiang [134]). Let $A \in \mathbf{C}^{n \times n}$. Let $X, Y \in \mathbf{C}^{n \times p}$ have orthonormal columns, and assume that $Y^H X$ is nonsingular. For any $M \in \mathbf{C}^{p \times p}$ let $N = (Y^H X)^{-1} M (Y^H X)$. Set

$$R = AX - XM \quad \text{and} \quad S^H = Y^H A - N Y^H.$$

Then there is at least one matrix E such that

$$(A + E)X = XM \quad \text{and} \quad Y^H(A + E) = N Y^H.$$

Moreover the smallest solution in the Frobenius norm satisfies

$$\|E\|_F = \sqrt{\|R\|_F^2 + \|S\|_F^2 - \|F_{11}\|},$$

where $F_{11} = Y^H R = S^H X$. The smallest solution in the spectral norm satisfies

$$\|E\|_2 = \max\{\|R\|_2, \|S\|_2\}.$$

[Hint: Let $(X\ X_\perp)$ and $(Y\ Y_\perp)$ be orthogonal and set

$$\begin{pmatrix} Y^H \\ Y_\perp^H \end{pmatrix} E (X\ X_\perp) = \begin{pmatrix} F_{11} & F_{12} \\ F_{21} & F_{22} \end{pmatrix}.$$

Then show that only F_{22} is free.]

——◇——

2. Gerschgorin Theory: Differentiability

The results of the last section do not suggest a way to assign a condition number to an eigenvalue. The problem is that eigenvalues associated with a nontrivial Jordan block are not differentiable functions of the elements of the matrix. However, this does not mean that individual eigenvalues cannot behave in a locally linear fashion and hence have condition numbers. This section is devoted to one of the most powerful tools for probing the sensitivity of a single eigenvalue — the Gerschgorin theorem.

2. Gerschgorin Theory

2.1. Gerschgorin's Theorem

Strictly speaking, Gerschgorin's theorem is not a perturbation theorem; it states that the eigenvalues of a matrix lie in the union of certain disks in the complex plane. However, as we shall see in the next subsection, it can be used to establish extremely accurate perturbation bounds.

There are several ways of establishing Gerschgorin's theorem. Here we will approach it through the Bauer–Fike theorem of the last section.

Theorem 2.1 (Gerschgorin). *For $A \in \mathbf{C}^{n \times n}$ let*

$$\alpha_i = \sum_{j \neq i} |\alpha_{ij}|$$

and

$$\mathcal{G}_i(A) = \{z \in \mathbf{C} : |z - \alpha_{ii}| \leq \alpha_i\}. \tag{2.1}$$

Then

$$\mathcal{L}(A) \subset \bigcup_{i=1}^{n} \mathcal{G}_i(A). \tag{2.2}$$

Moreover, if m of the GERSCHGORIN DISKS $\mathcal{G}_i(A)$ *are isolated from the other $n - m$ disks, then there are precisely m eigenvalues of A in their union.*

Proof. Let $D = \mathrm{diag}(\alpha_{11}, \alpha_{22}, \ldots, \alpha_{nn})$. In the Bauer–Fike theorem make the following substitutions:

$$\begin{aligned} Q &\leftarrow I, \\ A &\leftarrow D, \\ \tilde{A} &\leftarrow A, \\ \|\cdot\| &\leftarrow \|\cdot\|_\infty. \end{aligned}$$

Then it is easy to verify that the first inequality in (1.7) is equivalent to saying that each eigenvalue of A lies in a Gerschgorin disk.

The proof of the second part of the theorem uses the techniques developed in the previous subsection and is left as an exercise. ∎

The following illustrates how much an improvement Gerschgorin's theorem can be over Elsner's theorem. It also illustrates a deficiency in the straightforward use of Gerschgorin theorem.

Example 2.2. *Consider the matrix*

$$A = \begin{pmatrix} 1 & 10^{-4} \\ 10^{-4} & 2 \end{pmatrix}. \tag{2.3}$$

Regarding A as a perturbation of the matrix $\mathrm{diag}(1,2)$, *we find from Theorem 1.3 that one eigenvalue must lie in the interval* $[1 - 0.021, 1 + 0.021]$ *and the other in the interval* $[2 - 0.021, 2 + 0.021]$ *(actually the theorem yields intervals that are just barely greater than .04 in length). On the other hand, by Gerschgorin's theorem each of the intervals* $[1 - 10^{-4}, 1 + 10^{-4}]$ *and* $[2 - 10^{-4}, 2 + 10^{-4}]$ *must contain an eigenvalue of A. Thus Gerschgorin's theorem is better than Elsner's by more than two orders of magnitude.*

However, the eigenvalues of A are approximately $1 - 10^{-8}$ *and* $2 + 10^{-8}$. *Thus Gerschgorin's theorem is still off by four orders of magnitude.*

It is worth noting that in the above example we have replaced disks in the complex plane with intervals on the real axis. The rationale for this is the following. The two disks — either the ones provided by Elsner's's theorem or by Gerschgorin's theorem — contain only one eigenvalue each. Since A is real, its complex eigenvalues must occur in complex conjugate pairs. Hence the eigenvalues in the disks must be real and are contained in the intersection of the disks with the real line.

2.2. Diagonal Similarities

Example 2.2 shows that the bounds provided by Gerschgorin's theorem need not be very sharp. Now in principle there is no reason why Gerschgorin's theorem should provide sharp bounds. However, the matrix A of (2.3) has a special structure; it is almost diagonal, and it turns out that we can exploit this structure to obtain sharper bounds.

The general technique is seen at its simplest with the matrix of Example 2.2. Let $D_\alpha = \mathrm{diag}(\alpha, 1)$, and let

$$A_\alpha = D_\alpha A D_\alpha^{-1} = \begin{pmatrix} 1 & 10^{-4}\alpha \\ 10^{-4}\alpha^{-1} & 2 \end{pmatrix}.$$

2. GERSCHGORIN THEORY

Since A_α is similar to A it has the same eigenvalues; however, A and A_α have different Gerschgorin disks. As α becomes small the first disk shrinks, while the other grows. Eventually, the second disk expands to engulf the first, but until it does, the first provides an ever-improving bound on the eigenvalue. In particular, as long as

$$10^{-4}\alpha + 10^{-4}\alpha^{-1} < 1,$$

the two Gerschgorin disks will remain isolated. It is easy to see that this will be true as long as α is just a little greater than 10^{-4}, say $\alpha = 1.01 \cdot 10^{-4}$. This isolates an eigenvalue of A in the interval $[1 - 1.01 \cdot 10^{-8}, 1 + 1.01 \cdot 10^{-8}]$, which is a very sharp bound.

Of course this is a trivial example. However, the technique it illustrates — that of reducing one Gerschgorin disk until the others overwhelm it — is widely applicable. We will give another example in the proof of the following theorem.

Theorem 2.3. *Let λ be a simple eigenvalue of the matrix A, with right and left eigenvectors x and y, and let $\tilde{A} = A + E$ be a perturbation of A. Then there is a unique eigenvalue $\tilde{\lambda}$ of $\tilde{A}$ such that*

$$\tilde{\lambda} = \lambda + \frac{y^{\mathrm{H}} E x}{y^{\mathrm{H}} x} + O(\|E\|^2). \tag{2.4}$$

Proof. Let $\delta > 0$ be the distance between λ and the other eigenvalues of A. Let $J = Y^{\mathrm{H}} A X$ be the Jordan canonical form of A, in which the superdiagonals are equal to $\delta/3$ or zero [see (I.3.3)]. Note that the first columns x and y of X and Y are the right and left eigenvectors corresponding to λ, and since $Y^{\mathrm{H}} = X^{-1}$ we have $y^{\mathrm{H}} x = 1$; i.e., the denominator in (2.4) is nonzero.

Now consider the matrix $\tilde{J} = Y^{\mathrm{H}}(A + E)X$. This matrix has the form illustrated below for $n = 5$:

$$\tilde{J} = \begin{pmatrix} \lambda + y^{\mathrm{H}} E x & \epsilon & \epsilon & \epsilon & \epsilon \\ \epsilon & \mu & \tau & \epsilon & \epsilon \\ \epsilon & \epsilon & \mu & \tau & \epsilon \\ \epsilon & \epsilon & \epsilon & \mu & \tau \\ \epsilon & \epsilon & \epsilon & \epsilon & \mu \end{pmatrix}.$$

Here we have used ϵ to stand generically for a quantity bounded by $\|Y\|\|E\|\|X\|$; μ for an eigenvalue of A other than λ plus ϵ; and τ for a quantity bounded by $\epsilon + \delta/3$. By a diagonal similarity transformation, we may replace $\tilde{J}$ by a matrix of the form

$$\tilde{J}_\alpha = \begin{pmatrix} \lambda + y^H Ex & \alpha\epsilon & \alpha\epsilon & \alpha\epsilon & \alpha\epsilon \\ \alpha^{-1}\epsilon & \mu & \tau & \epsilon & \epsilon \\ \alpha^{-1}\epsilon & \epsilon & \mu & \tau & \epsilon \\ \alpha^{-1}\epsilon & \epsilon & \epsilon & \mu & \tau \\ \alpha^{-1}\epsilon & \epsilon & \epsilon & \epsilon & \mu \end{pmatrix}.$$

Now the first Gerschgorin disk of $\tilde{J}_\alpha$ has center $\lambda + y^H Ex$ and radius bounded by $(n-1)\alpha\epsilon$. The other disks have center μ and radii bounded by $\alpha^{-1}\epsilon + \tau + (n-3)\epsilon$. Hence if

$$\alpha^{-1}\epsilon + \delta/3 + n\epsilon + (n-1)\alpha\epsilon < \delta, \qquad (2.5)$$

the first Gerschgorin disk will be disjoint from the others.

Now let E be small enough so that

$$\frac{2}{3}\delta - n\epsilon > \frac{\delta}{2}.$$

Then if

$$n\epsilon\alpha^2 - \frac{\delta}{2}\alpha + \epsilon < 0,$$

the inequality (2.5) is satisfied. This latter condition will be satisfied if

$$\alpha = \frac{4\epsilon}{\delta},$$

provided we require E to be so small that

$$\frac{16n\epsilon^2}{\delta^2} < 1.$$

In this case, the radius of the first Gerschgorin disk is bounded by $4n\epsilon^2/\delta = O(\epsilon^2)$. Since this disk is centered at $\lambda + y^H Ex$, the unique eigenvalue it contains is our $\tilde{\lambda}$. ∎

An immediate consequence of Theorem 2.3 is that the simple eigenvalues of a matrix are differentiable functions of the elements of the matrix.

2. GERSCHGORIN THEORY

Corollary 2.4. *Under the hypotheses of Theorem 2.3 the eigenvalue λ is a differentiable function of A. Moreover,*

$$\frac{\partial \lambda}{\partial \alpha_{ij}} = \frac{\bar{\eta}_i \xi_j}{y^{\text{H}} x}. \tag{2.6}$$

Proof. By definition a function $f(A)$ is differentiable if there is a linear operator f'_A such that $f(A + E) = f(A) + f'_A(E) + o(\|E\|)$. Equation (2.4) exhibits such an operator for the eigenvalue λ: namely, $E \mapsto \frac{y^{\text{H}} E x}{y^{\text{H}} x}$. To establish (2.6), note that

$$\frac{\partial \lambda}{\partial \alpha_{ij}} = \lim_{\tau \to 0} \frac{\lambda(A + \tau \mathbf{1}_i \mathbf{1}_j^{\text{T}}) - \lambda(A)}{\tau}.$$

But by (2.4)

$$\lambda(A + \tau \mathbf{1}_i \mathbf{1}_j^{\text{T}}) - \lambda(A) = \tau \frac{y^{\text{H}} \mathbf{1}_i \mathbf{1}_j^{\text{T}} x}{y^{\text{H}} x} + O(\tau^2),$$

from which the result follows immediately. ∎

The proof of Theorem 2.3 is almost as interesting as the theorem itself, since it gives us insight into the factors that make the higher order terms important. Specifically, we require that terms involving ϵ/δ be sufficiently small. The denominator δ shows that if a simple eigenvalue is near its neighbors, the range of perturbations for which the derivative provides an adequate approximation will be restricted. The size of the numerator depends not only on E, but on the sizes of the reducing transformations X and Y. If these are large, we again can expect higher order terms to become significant. It is worth noting that according to the remark following (I.3.3), a small value of δ will tend to aggravate this effect.

Equation (2.4) can be written

$$\tilde{\lambda} = \frac{y^{\text{H}}(A + E)x}{y^{\text{H}} x} + O(\|E\|^2). \tag{2.7}$$

The quantity $y^{\text{H}}(A + E)x / y^{\text{H}} x$ is called a RAYLEIGH QUOTIENT, and one way of stating the theorem is to say that the Rayleigh quotient provides a first-order approximation to the perturbed eigenvalue. We will generalize the notion of a Rayleigh quotient in Section V.2, where we will give explicit bounds for the second order terms.

The theorem also provides us with a condition number for a simple eigenvalue. We see from (2.4) that

$$|\tilde{\lambda} - \lambda| \lesssim \frac{\|y\| \cdot \|x\|}{|y^H x|} \|E\|,$$

for any consistent pair of matrix and vector norms. Thus the quantity

$$\nu = \frac{\|y\| \cdot \|x\|}{|y^H x|} \tag{2.8}$$

is a condition number for λ.

When $\|\cdot\| = \|\cdot\|_2$, the number ν is the secant of the angle between x and y. It is one when x and y lie in the same direction, and grows unboundedly as x and y approach orthogonality. Note that if λ is simple its left and right eigenvectors cannot be orthogonal, although it is easy to construct examples where they are as close to orthogonality as we like. Also note that the left and right eigenvectors corresponding to a nontrivial Jordan block have to be orthogonal.

Notes and References

Gerschgorin [85, 1931] established his theorem as a corollary to the theorem that a diagonally dominant matrix is nonsingular. In particular, the union of the Gerschgorin disks of A is the complement of the set of all ζ for which $\zeta I - A$ is diagonally dominant. In a restricted form the diagonal dominance theorem is due to Lévy [146, 1881], and in a general form to Desplanques [58, 1887]. The theorem kept getting itself rediscovered until Olga Taussky put a stop to it with a paper appropriately entitled *A Recurring Theorem on Determinants* [239, 1949]. Rohrbach [187, 1931] used the technique to establish eigenvalue bounds but did not define the regions now called Gerschgorin disks.

Actually, the theorem stated by Gerschgorin is not true unless the matrix is irreducible (see Exercise 2.7).

More generally, if π is any proposition such that $\pi(A)$ is true if and only if A is nonsingular, then the complement of the set $\{\zeta : \pi(\zeta I - A) \text{ is true}\}$ contains all the eigenvalues of A. By varying π one can get different regions, some of which are treated in the exercises.

2. GERSCHGORIN THEORY

In his paper, Gerschgorin noted that the union of k isolated disks contains exactly k Eigenvalues. Although the idea of using diagonal similarity to reduce the radius of an isolated disk is due to Gerschgorin (and in a different sense to Rohrbach), it was Wilkinson [269, 1965] who refined the technique and applied it to a variety of problems. Although ad hoc techniques for reducing the diameter of an isolated disk suffice for most applications, there are algorithms for determining the optimal disk [253, 154].

Exercises

1. A matrix A is strictly diagonally dominant if

$$|\alpha_{ii}| > \sum_{j \neq i} |\alpha_{ij}|, \quad i = 1, \ldots, n.$$

Show that a strictly diagonally dominant matrix is nonsingular and use this fact to prove Gerschgorin's theorem.

2. Let $Ax = \lambda x$, and suppose $|\xi_j| \geq |\xi_i|$ ($i = 1, \ldots, n$). Show that λ lies in the Gerschgorin disk centered at α_{jj}.

3. (Ostrowski [167]). Let $\rho_i = \sum_{j \neq i} |\alpha_{ij}|$ and $\gamma_i = \sum_{j \neq i} |\alpha_{ji}|$. Show that if for some $\tau \in [0, 1]$

$$|\alpha_{ii}| > \gamma_i^\tau \rho_i^{1-\tau}, \quad i = 1, \ldots, n,$$

then A is nonsingular.

4. (Ostrowski [167]). In the notation of the last exercise suppose that

$$|\alpha_{ii}||\alpha_{jj}| > \gamma_i^\tau \rho_i^{1-\tau} \gamma_j^\tau \rho_j^{1-\tau}, \quad i,j = 1, \ldots, n, \ i \neq j.$$

Show that A is nonsingular.

5. (Qi [182]). Let A be of order n. Let

$$\delta_i = \max\left\{\sum_{j \neq i} |\alpha_{ij}|, \sum_{j \neq i} |\alpha_{ji}|\right\}, \quad i = 1, \ldots, n.$$

Show that the singular values of A lie in the union of the intervals $[|\alpha_{ii}| - \delta_i, |\alpha_{ii}| + \delta_i]$ ($i = 1, \ldots, n$).

6. (Feingold and Varga [72]). Let A be partitioned in the form

$$A = \begin{pmatrix} A_{11} & A_{12} & \cdots & A_{1k} \\ A_{21} & A_{22} & \cdots & A_{2k} \\ \vdots & \vdots & & \vdots \\ A_{k1} & A_{k2} & \cdots & A_{kk} \end{pmatrix},$$

and let $\|\cdot\|$ be a consistent norm. Show that if λ is an eigenvalue of A, then for some i

$$\|(\lambda I - A_{ii})^{-1}\|^{-1} \leq \sum_{j \neq i} \|A_{ij}\|.$$

7. A square matrix A is REDUCIBLE if there is a permutation matrix P such that

$$P^{\mathrm{T}} A P = \begin{pmatrix} A_{11} & A_{12} \\ 0 & A_{22} \end{pmatrix},$$

where A_{11} and A_{22} are square. Show that an irreducible diagonally dominant matrix for which at least one of the diagonals is strictly dominant is nonsingular.

8. (Taussky [238]). Let A be irreducible. Show that $\lambda \in \mathcal{L}(A)$ lies on the boundary of one Gerschgorin disk, then it lies on the boundaries of all the Gerschgorin disks.

9. Let $A = X_1 J_{m_1}(\lambda_1) Y_1^{\mathrm{H}} + \cdots + X_k J_{m_k}(\lambda_m) Y_m^{\mathrm{H}}$ be the Jordan decomposition of A, and assume that λ_1 has multiplicity m_1. Show that if E is sufficiently small there are exactly m eigenvalues of $\tilde{A}$ that are in $\mathcal{L}[J_{m_1}(\lambda_1) + Y_1^{\mathrm{H}} E X_1 + O(\|E\|^2)]$.

10. (Wilkinson [271]). Let x and y be left and right eigenvectors corresponding to the simple eigenvalue λ. Let $\theta = \angle(x,y)$. Show that there is a matrix E satisfying

$$\frac{\|E\|_2}{\|A\|_2} \leq \cot \theta$$

such that λ is a multiple eigenvalue of $A + E$. Otherwise put, if a simple eigenvalue of a matrix has a large condition number, then the matrix is near one with a multiple eigenvalue.

3. Normal and Diagonalizable Matrices

A normal matrix is any matrix satisfying $A^H A = AA^H$. From this it follows that Hermitian matrices, skew Hermitian matrices, and unitary matrices are all normal. Given the importance of this class of matrices, it is natural to seek a special perturbation theory for its eigenvalues. The main complicating factor here is that normal matrices, unlike Hermitian matrices, can have complex eigenvalues which cannot be ordered by size. Nonetheless, normal matrices have enough structure to enable us to prove the striking Hoffman–Wielandt theorem.

Since any normal matrix can be diagonalized by a unitary transformation, the normal matrices are special cases of diagonalizable matrices; that is, matrices that can be diagonalized by similarity transformations (these matrices are sometimes called normalizable). In the second subsection we will treat the perturbation of eigenvalues of diagonalizable matrices.

3.1. The Hoffman–Wielandt Theorem

In Section 1 we saw that it is relatively easy to obtain bounds on the spectral variation $\mathrm{sv}_A(\tilde{A})$ of a matrix $\tilde{A}$ with respect to A. Although it is usually possible to escalate such a bound into a bound on $\mathrm{md}(A, \tilde{A})$, we pay the price of a factor of $2n - 1$ in the bound. The essence of the Hoffman–Wielandt theorem is that when A and $\tilde{A}$ are normal we do not have to pay such a price to get a bound on

$$\mathrm{md}_2(A, \tilde{A}) \stackrel{\mathrm{def}}{=} \min_\pi \sum_i \sqrt{|\tilde{\lambda}_{\pi(i)} - \lambda_i|^2}, \qquad (3.1)$$

where π ranges over all permutations of the integers $1, 2, \ldots, n$. (The subscript 2 refers to the 2-norm. In this notation the usual matching distance is md_∞.)

Theorem 3.1 (Hoffman–Wielandt). *Let A and $\tilde{A}$ be normal. Then*

$$\mathrm{md}_2(A, \tilde{A}) \leq \|\tilde{A} - A\|_F, \qquad (3.2)$$

where $\mathrm{md}_2(A, \tilde{A})$ is defined by (3.1).

Proof. Since $\|\cdot\|_F$ is unitarily invariant, we may assume that $A = \Lambda = \text{diag}(\lambda_1, \ldots, \lambda_n)$. Let $\tilde{A} = W\tilde{\Lambda}W^H$, where W is unitary and $\tilde{\Lambda} = \text{diag}(\tilde{\lambda}_1, \ldots, \tilde{\lambda}_n)$. We will have established the theorem if we can show that $\|\Lambda - V\tilde{\Lambda}V^H\|_F$, regarded as a function of the unitary matrix V, is minimized when $V = P_\pi$ is a permutation matrix corresponding to some permutation π. For in that case,

$$\text{md}_2^2(A, \tilde{A}) = \text{md}_2^2(\Lambda, \tilde{\Lambda}) \leq \sum_i |\lambda_i - \tilde{\lambda}_{\pi(i)}|^2$$
$$\leq \|\Lambda - W\tilde{\Lambda}W^H\|_F^2 = \|A - \tilde{A}\|_F^2.$$

Denoting the elements of V by ν_{ij}, we have by direct calculation

$$\|\Lambda - V\tilde{\Lambda}V^H\|_F^2 = \sum_i |\lambda_i|^2 + \sum_i |\tilde{\lambda}_i|^2 - \varphi(V),$$

where

$$\varphi(V) = \sum_{i,j} (\lambda_i \overline{\tilde{\lambda}_j} + \overline{\lambda}_i \tilde{\lambda}_j)|\nu_{ij}|^2.$$

Thus our problem reduces to showing that $\varphi(V)$ is maximized when V is some permutation matrix.

Since V is unitary, the matrix whose elements are $|\nu_{ij}|^2$ is doubly stochastic. For any doubly stochastic matrix S define

$$\psi(S) = \sum_{i,j} (\lambda_i \overline{\tilde{\lambda}_j} + \overline{\lambda}_i \tilde{\lambda}_j)\sigma_{ij}.$$

It is clear that $\max_V \varphi(V) \leq \max_S \psi(S)$, since not every doubly stochastic matrix has elements of the form $|\nu_{ij}|^2$, where $V = (\nu_{ij})$ is unitary. Therefore, if we can show that ψ is maximized when S is a permutation matrix P_π, then since P_π is unitary, it also maximizes φ.

By Birkhoff's theorem (Theorem II.3.16), any doubly stochastic matrix S can be written as a convex combination of the permutation matrices P_π: namely,

$$S = \sum_\pi \alpha_\pi P_\pi,$$

where the α_π are nonnegative and sum to one. Since ψ is linear in S,

$$\psi(S) = \sum_\pi \alpha_\pi \psi(P_\pi).$$

3. NORMAL AND DIAGONALIZABLE MATRICES

It follows that if π is the permutation for which $\psi(P_\pi)$ is maximal, then $\psi(S) \le \psi(P_\pi)$. Hence $\varphi(P_\pi)$ is also maximal, and π is the permutation required by the theorem. ∎

The hypothesis that *both* A and $\tilde{A}$ be normal is necessary. For example, let

$$A = \begin{pmatrix} 0 & 0 \\ 0 & 4 \end{pmatrix}$$

and

$$\tilde{A} = \begin{pmatrix} -1 & -1 \\ 1 & 1 \end{pmatrix},$$

so that A is normal but $\tilde{A}$ is not. The eigenvalues of A are 0 and 4 while those of $\tilde{A}$ are both zero. Hence

$$\mathrm{md}_2^2(A, \tilde{A}) = 16 > 12 = \|\tilde{A} - A\|_\mathrm{F}^2.$$

This fact complicates the practical application of the Hoffman–Wielandt theorem, since the sum of two normal matrices may not be normal. Even the sum of a normal matrix and a Hermitian matrix may fail to be normal. Thus the class of perturbations that the theorem can handle is strictly limited.

A case in point is the attempt to derive residual bounds from a backward perturbation result like Theorem 1.13. The difficulty is that the matrix $\tilde{A}$, defined by (1.16), need not be normal. However, the following result gives a residual bound for a single eigenvalue.

Theorem 3.2. *Let A be normal. If $\|x\|_2 = 1$, then*

$$\min_{1 \le i \le n} |\lambda_i - x^\mathrm{H} A x| \le \|Ax - (x^\mathrm{H} A x)x\|_2. \qquad (3.3)$$

Proof. Since A is normal, there is a unitary matrix U such that $A = U\Lambda U^\mathrm{H}$, where $\Lambda = \mathrm{diag}(\lambda_1, \ldots, \lambda_n)$. Hence

$$\|[A - (x^\mathrm{H} A x)I]x\|_2 = \|U(\Lambda - (x^\mathrm{H} A x)I)U^\mathrm{H} x\|_2$$
$$\ge \min_{1 \le i \le n} |\lambda_i - x^\mathrm{H} A x|,$$

from which (3.3) follows. ∎

3.2. Diagonalizable Matrices

The chief general result for diagonalizable matrices follows from the Bauer–Fike theorem (Theorem 1.6).

Theorem 3.3. *Suppose that A is diagonalizable; i.e., $X^{-1}AX = \Lambda$, where Λ is diagonal. Let $\|\cdot\|$ be a consistent matrix norm such that $\|\mathrm{diag}(\delta_1,\ldots,\delta_n)\| = \max_i |\delta_i|$. Then*

$$\mathrm{sv}_A(\tilde{A}) \leq \|X^{-1}EX\| \tag{3.4}$$

and

$$\mathrm{sv}_A(\tilde{A}) \leq \kappa(X)\|E\|, \tag{3.5}$$

where as usual $\kappa(X) = \|X\|\|X^{-1}\|$. Moreover,

$$\mathrm{md}(A,\tilde{A}) \leq (2n-1)\|X^{-1}EX\| \leq (2n-1)\kappa(X)\|E\|. \tag{3.6}$$

Proof. Let $\tilde{\lambda}$ be an eigenvalue of $\tilde{A}$. Under the hypotheses of the theorem, the inequality (1.6) in the Bauer–Fike theorem assumes the form

$$\|(\Lambda - \tilde{\lambda}I)^{-1}\|^{-1} \leq \|X^{-1}EX\|,$$

from which (3.4) follows immediately. The inequality (3.5) follows from consistency. Finally, (3.6) follows from Theorem 1.5. ∎

The bounds (3.4) and (3.5) hold for the widely used norms $\|\cdot\|_p$ ($p = 1, 2, \infty$) (and in fact for all the Hölder norms). They hold trivially for all normalized unitarily invariant norms, since these norms dominate the spectral norm.

Corollary 3.4. *If A is normal, then*

$$\mathrm{sv}_A(\tilde{A}) \leq \|E\|_2.$$

Although (3.4) is stronger than (3.5), we will usually have no more than an estimate of $\|E\|$, in which case we are forced to use the weaker bound. Here the condition number of the matrix of eigenvectors serves as an overall condition number for the eigenvalue problem of A. Unfortunately, if we replace X by XD, where D is diagonal, $\kappa(X)$ changes, even though X continues to diagonalize A. Moreover, by making one column of X very large or very small we can make $\kappa(X)$ arbitrarily

3. NORMAL AND DIAGONALIZABLE MATRICES

large—a situation we called artificial ill-conditioning in the last chapter.

These considerations lead us to ask: What is the optimal scaling of X? In general this is a very difficult question; however, for the Frobenius norm we can give an answer.

Theorem 3.5. *Let $X \in \mathbb{C}^{n \times n}$ be nonsingular, and let $Y^H X = I$. Then*

$$\kappa_F(X) \geq \sum \|y_i\|_2 \|x_i\|_2,$$

with equality if and only if there is an $\alpha \neq 0$ such that

$$\|y_i\|_2 = \alpha \|x_i\|_2, \quad i = 1, \ldots, n. \tag{3.7}$$

Proof. By the Cauchy inequality

$$\kappa_F^2(X) = (\|x_1\|_2^2 + \cdots + \|x_n\|_2^2)(\|y_1\|_2^2 + \cdots + \|y_n\|_2^2)$$
$$\geq (\|x_1\|_2 \|y_1\|_2 + \cdots + \|x_n\|_2 \|y_n\|_2)^2.$$

Equality holds if and only if $(\|x_1\|_2, \ldots, \|x_n\|_2)$ and $(\|y_1\|_2, \ldots, \|y_n\|_2)$ are proportional, which is equivalent to (3.7). ∎

There are two observations to be made about this theorem. First, the proportional scaling (3.7) is probably not a bad strategy for other balanced norms like $\|\cdot\|_p$ ($p = 1, 2, \infty$). Second, if the eigenvalues of Λ are simple, the optimal $\kappa_F(X)$ is the sum of the individual condition numbers of the eigenvalues [cf. (2.8)]. This shows that the bounds in Theorem 3.3 are realistic in the sense that if the optimal $\kappa_F(X)$ is large, then there must exist at least one ill-conditioned eigenvalue.

Notes and References

For the Hoffman–Wielandt theorem, see [117, 1953]. Wilkinson [269, 1965] gives an elementary proof that does not use Birkhoff's theorem.

The Hoffman–Wielandt theorem can be rewritten in a suggestive manner. Let Φ be a symmetric gauge function and let $\|\cdot\|_\Phi$ be the associated unitarily invariant norm. Set

$$\mathrm{md}_\Phi(A, \tilde{A}) = \min_\pi \Phi(|\tilde{\lambda}_{\pi(1)} - \lambda_1|^2, \ldots, |\tilde{\lambda}_{\pi(n)} - \lambda_n|^2),$$

where as usual π ranges over all permutations of the integers $1, \ldots, n$. Then for $\Phi(x) \equiv \|x\|_2$, the Hoffman-Wielandt theorem states that

$$\mathrm{md}_\Phi(A, \tilde{A}) \leq \|\tilde{A} - A\|_\Phi. \tag{3.8}$$

It is natural to conjecture that (3.8) remains true for normal matrices and arbitrary unitarily invariant norms. The conjecture is untrue, even for orthogonal matrices; however, many partial results are known. The following survey is largely based on the book by Bhatia [28], which contains proofs and further references.

Mirsky [158, 1960] showed that the conjecture is true for Hermitian matrices. See Section 3 for a proof and applications.

Wittmeyer [274, 1936], claims that the theorem is true for normal matrices and the 2-norm, but he refers the reader to his Ph.D. thesis for the proof. Since others have tried and failed to establish this result, it must remain open until Wittmeyer's proof can be examined.

Bhatia and Davis [29, 1984] have shown that the conjecture is true for orthogonal matrices and the 2-norm. Another proof was given by Bhatia and Holbrook [32, 1985].

Other partial results are obtained by relaxing the bound. Bhatia, Davis, and McIntosh [31, 1983] have shown that for unitary matrices

$$\mathrm{md}_\Phi(A, \tilde{A}) \leq \frac{\pi}{2}\|\tilde{A} - A\|_\Phi,$$

and they give an example to show that $\frac{\pi}{2}$ is the best possible constant (Exercise 3.4). They also show that for normal matrices

$$\mathrm{md}(A, \tilde{A}) \leq \gamma\|\tilde{A} - A\|_2,$$

where $\gamma \leq 2.91$ [30, 1987]; i.e., the conjecture is true for normal matrices and the 2-norm, provided we multiply the right-hand side by a factor of about three. For most practical applications this is good enough.

The inequality (3.5) is due to Bauer and Fike [15], but as we have pointed out it is a weak corollary of their more general results.

Exercises

1. Let A and $\tilde{A}$ be normal of order n. Show that

$$\|\tilde{A} - A\|_\mathrm{F} \geq \max_\pi \sum_i \sqrt{|\tilde{\lambda}_{\pi(i)} - \lambda_i|^2},$$

3. NORMAL AND DIAGONALIZABLE MATRICES

where π ranges over all permutations of the integers $1, 2, \ldots, n$.

2. Let A and $\tilde{A}$ be normal. If there are convex sets $\mathcal{A}$ and $\tilde{\mathcal{A}}$ such that

   1. $\mathcal{A}$ contains k eigenvalues of A,
   2. $\tilde{\mathcal{A}}$ contains at least $n - k + 1$ eigenvalues of $\tilde{A}$,
   3. the distance from $\mathcal{A}$ to $\tilde{\mathcal{A}}$ is δ,

   then
   $$\delta \leq \|\tilde{A} - A\|_2.$$

3. (Bhatia and Davis [29]). Let A and $\tilde{A}$ be orthogonal matrices with their eigenvalues lying in a semicircle of the unit circle. Order the eigenvalues by the order in which they appear on the semicircle, say counterclockwise. Show that
   $$\max_i |\tilde{\lambda}_i - \lambda_i| \leq \|\tilde{A} - A\|_2.$$

4. Let Φ be the symmetric gauge function defined by $\Phi(x) = \|x\|_1$. Let
   $$A_\pm = \begin{pmatrix} 0 & 1 & 0 & \cdots & 0 \\ 0 & 0 & 1 & \cdots & 0 \\ \vdots & \vdots & \vdots & & \vdots \\ 0 & 0 & 0 & \cdots & 1 \\ \pm 1 & 0 & 0 & \cdots & 0 \end{pmatrix}.$$
   Show that $\|A_+ - A_-\|_\Phi = 2$, whereas $\lim_{n \to \infty} \mathrm{md}_\Phi(A_+, A_-) = \pi$.

5. Give an example of a doubly stochastic matrix S whose elements are not of the form $|u_{ij}|$, where U is unitary.

6. (Bauer–Householder [16]). Let $\Lambda = X^{-1}AX$ be diagonal. Let α and β be polynomials and w a vector with $\beta(A)w \neq 0$. Show that there is an eigenvalue of A in the region
   $$\left\{ \left|\frac{\alpha(\zeta)}{\beta(\zeta)}\right| \leq \kappa(X) \frac{\|\alpha(A)w\|_2}{\|\beta(A)w\|_2} \right\}.$$

7. Let $X \in \mathbf{C}^{n \times n}$ and let $\kappa_{\mathrm{opt}}(X)$ be the smallest value of $\kappa_F(XD)$, where D is nonsingular and diagonal (see Theorem 3.5). Show that if $\|x_i\|_2 = 1$ ($i = 1, \ldots, n$), then
   $$\kappa_F(X) \leq \sqrt{n}\, \kappa_{\mathrm{opt}}(X).$$

4. Hermitian Matrices

In this section we will treat the perturbation of eigenvalues of Hermitian matrices. This is an area rich in results, and we will only be able to sample some of the more important.

We will begin with two classical results: Sylvester's inertia theorem and Cauchy's interlacing theorem. We will then establish Wielandt's elegant generalization of Fischer's characterization of the eigenvalues of a Hermitian matrix. This result in turn yields a host of powerful perturbation bounds.

Throughout this section A will denote a Hermitian matrix with eigenvalues

$$\lambda_1 \geq \lambda_1 \geq \cdots \geq \lambda_n,$$

and $\tilde{A} = A + E$ will denote a Hermitian perturbation of A with eigenvalues

$$\tilde{\lambda}_1 \geq \tilde{\lambda}_1 \geq \cdots \geq \tilde{\lambda}_n.$$

4.1. Inertia and Interlacing

A fundamental problem of matrix theory is to determine what remains invariant under some class of transformations. For example, the eigenvalues and Jordan structure of a matrix are not altered by similarity transformations. For Hermitian matrices it is natural to consider transformations that leave the matrix Hermitian, which leads us to the class of CONGRUENCE TRANSFORMATIONS; that is, transformations of the form

$$X^{\mathrm{H}} A X,$$

where X is nonsingular. Unless X is unitary, the eigenvalues of A need not remain invariant under this transformation. However, the number of positive, negative, and zero eigenvalues does not change.

Theorem 4.1 (Sylvester, Jacobi). *Let A be Hermitian, and define the* INERTIA *of A to be the ordered triplet*

$$\mathrm{inertia}(A) = [\pi(A), \nu(A), \zeta(A)],$$

4. HERMITIAN MATRICES

where $\nu(A)$, $\zeta(A)$, and $\pi(A)$ are respectively the number of negative, zero, and positive eigenvalues of A. Then for any nonsingular X,

$$\text{inertia}(X^H A X) = \text{inertia}(A).$$

Proof. The proof is by contradiction. Suppose, for example, that A has more positive eigenvalues than $X^H A X$. Let $\mathcal{Y}$ be the space spanned by the eigenvectors corresponding to positive eigenvalues of A. Then

$$y \in \mathcal{Y} \implies y^H A y > 0.$$

Let $\mathcal{Z}$ be the space spanned by all vectors of the form Xz, where z is an eigenvector corresponding to a negative or zero eigenvalue of $X^H A X$. Then

$$z \in \mathcal{Z} \implies z^H A z \le 0.$$

Hence $\mathcal{Y} \cap \mathcal{Z} = \{0\}$. But by hypothesis, $\dim(\mathcal{Y}) + \dim(\mathcal{Z}) > n$, where n is the order of A. Hence $\mathcal{X}$ and $\mathcal{Y}$ have a vector in common — a contradiction. ∎

An important consequence of the inertia theorem is Cauchy's beautiful theorem relating the eigenvalues of a principal submatrix to the eigenvalues of the original matrix.

Theorem 4.2 (Cauchy). *Let B be a principal submatrix of A of order $n-1$ with eigenvalues $\mu_1 \ge \mu_2 \ge \cdots \ge \mu_{n-1}$. Then*

$$\lambda_1 \ge \mu_1 \ge \lambda_2 \ge \mu_2 \ge \cdots \ge \mu_{n-1} \ge \lambda_n.$$

Proof. Without loss of generality assume that B is the leading principle submatrix of A, so that we may write

$$A = \begin{pmatrix} B & a \\ a^H & \alpha \end{pmatrix}.$$

Assume that the theorem is false. Then for some i either $\mu_i > \lambda_i$ or $\lambda_{i+1} > \mu_i$. Let i be the first such index.

We will treat the case $\mu_i > \lambda_i$, the other case being similar. Let $\mu_i > \tau > \lambda_i$. Then $B - \tau I$ is nonsingular, and the matrix

$$H = \begin{pmatrix} B - \tau I & 0 \\ 0 & \alpha - \tau - a^H(B - \tau I)^{-1} a \end{pmatrix} =$$

$$\begin{pmatrix} I & 0 \\ -a^H(B-\tau I)^{-1} & 1 \end{pmatrix} \begin{pmatrix} B - \tau I & a \\ a^H & \alpha - \tau I \end{pmatrix} \begin{pmatrix} I & -(B - \tau I)^{-1} a \\ 0 & 1 \end{pmatrix}$$

is congruent to $A - \tau I$. Hence by the inertia theorem, H has the same number of positive eigenvalues as $A - \tau I$, namely $i - 1$. But H has at least as many positive eigenvalues as $B - \tau I$, namely i. The contradiction establishes the theorem. ∎

If, in the theorem, C is a principal submatrix of A of order $n - 2$, then the eigenvalues $\nu_1 \geq \nu_2 \geq \cdots \geq \nu_{n-2}$ of C satisfy $\mu_1 \geq \nu_1 \geq \mu_2 \geq \nu_2 \geq \cdots \geq \nu_{n-1} \geq \mu_{n-1}$. Hence

$$\lambda_i \geq \nu_i \geq \lambda_{i+2}, \qquad i = 1, 2, \ldots, n - 2.$$

Continuing through submatrices in this manner, we have the following corollary.

Corollary 4.3. *Let B be a principle submatrix of order $n - k$ of A with eigenvalues $\mu_1 \geq \mu_2 \geq \cdots \geq \mu_{n-k}$. Then*

$$\lambda_i \geq \mu_i \geq \lambda_{i+k}, \qquad i = 1, 2, \ldots, n - k.$$

Finally, we observe that the interlacing theorem holds for more than just principal submatrices. Let $U \in \mathbf{C}^{n \times (n-k)}$ have orthonormal columns and let V be chosen so that $(U\ V)$ is unitary. Then applying Corollary 4.3 to the matrix $(U\ V)^{\mathrm{H}} A (U\ V)$, we have the following corollary.

Corollary 4.4. *Let $U \in \mathbf{C}^{n \times (n-k)}$ have orthonormal columns. Let the eigenvalues of $U^{\mathrm{H}} A U$ be $\mu_1 \geq \mu_2 \geq \cdots \geq \mu_{n-k}$. Then*

$$\lambda_i \geq \mu_i \geq \lambda_{i+k}, \qquad i = 1, 2, \ldots, n - k.$$

4.2. Wielandt's Theorem and Its Consequences

It is a consequence of Theorem I.3.13 that

$$\lambda_1 = \max_{x^{\mathrm{H}} x = 1} x^{\mathrm{H}} A x.$$

An important generalization of this fact is Fischer's theorem, which states that

$$\lambda_i = \max_{\dim(\mathcal{X}) = i} \min_{\substack{x \in \mathcal{X} \\ x^{\mathrm{H}} x = 1}} x^{\mathrm{H}} A x.$$

4. HERMITIAN MATRICES 199

In this subsection we will establish a further generalization, due to Wielandt, which has far-ranging implications. The proof, which has been adapted directly from Wielandt's paper, is complicated and may be omitted without loss of continuity.

Theorem 4.5 (Wielandt). Let $1 \leq i_1 < i_2 < \cdots < i_k \leq n$. Then

$$\lambda_{i_1} + \lambda_{i_2} + \cdots + \lambda_{i_k} = \max_{\substack{\mathcal{X}_{i_1} \subset \mathcal{X}_{i_2} \subset \cdots \subset \mathcal{X}_{i_k} \\ \dim(\mathcal{X}_{i_j}) = i_j}} \min_{\substack{X = (x_{i_1}\, x_{i_2}\, \cdots\, x_{i_k}), x_{i_j} \in \mathcal{X}_{i_j} \\ X^H X = I}} \operatorname{trace}(X^H A X),$$

(4.1)

and

$$\lambda_{i_1} + \lambda_{i_2} + \cdots + \lambda_{i_k} = \min_{\substack{\mathcal{X}_{i_1} \supset \mathcal{X}_{i_2} \supset \cdots \supset \mathcal{X}_{i_k} \\ \dim(\mathcal{X}_{i_j}) = n - i_j + 1}} \max_{\substack{X = (x_{i_1}\, x_{i_2}\, \cdots\, x_{i_k}), x_{i_j} \in \mathcal{X}_{i_j} \\ X^H X = I}} \operatorname{trace}(X^H A X).$$

(4.2)

Remark 4.6. Note that the words max and min (instead of sup and inf) imply that the maximizing or minimizing objects actually exist.

Proof. We will establish (4.1), from which it is an easy exercise to establish (4.2). We begin by showing that there is a particular sequence $\mathcal{X}_{i_1} \subset \mathcal{X}_{i_2} \subset \cdots \subset \mathcal{X}_{i_k}$ of subspaces with $\dim(\mathcal{X}_{i_j}) = i_j$ such that if $X = (x_{i_1}\, x_{i_2}\, \ldots\, x_{i_k})$ $(x_{i_j} \in \mathcal{X}_{i_j})$ has orthonormal columns, then $\operatorname{trace}(X^H A X) \geq \sum_{i_j} \lambda_{i_j}$. In fact, let $\mathcal{X}_{i_j}$ be the space spanned by the eigenvectors of A corresponding to $\lambda_1, \lambda_2, \ldots, \lambda_{i_j}$. Then x_{i_j} is a linear combination of these eigenvectors, and since $x_{i_j}^H x_{i_j} = 1$, we have $x_{i_j}^H A x_{i_j} \geq \lambda_{i_j}$. Hence

$$\operatorname{trace}(X^H A X) = \sum_{i_j} x_{i_j}^H A x_{i_j} \geq \sum_{i_j} \lambda_{i_j}.$$

In view of the result of the last paragraph, it will be sufficient to establish that

$$\max_{\substack{\mathcal{X}_{i_1} \subset \mathcal{X}_{i_2} \subset \cdots \subset \mathcal{X}_{i_k} \\ \dim(\mathcal{X}_{i_j}) = i_j}} \min_{\substack{X = (x_{i_1}\, x_{i_2}\, \cdots\, x_{i_k}), x_{i_j} \in \mathcal{X}_{i_j} \\ X^H X = I}} \operatorname{trace}(X^H A X) \leq \lambda_{i_1} + \lambda_{i_2} + \cdots + \lambda_{i_k}.$$

The proof will be by induction on n. Note that the theorem is trivially true when $k = n$, since in this case $X^H A X$ is similar to A. Hence the theorem is true for $n = 1$.

Let us therefore assume that $n > 1$ and $k < n$. Let $\mathcal{X}_{i_1} \subset \mathcal{X}_{i_2} \subset \cdots \subset \mathcal{X}_{i_k}$ with $\dim(\mathcal{X}_{i_j}) = i_j$ be given. We must show that there is a matrix $X = (x_{i_1}\ x_{i_2}\ \ldots\ x_{i_k})$, $x_{i_j} \in \mathcal{X}_{i_j}$ with orthonormal columns such that $\operatorname{trace}(X^{\mathrm{H}}AX) \leq \lambda_{i_1} + \lambda_{i_2} + \cdots + \lambda_{i_k}$.

First assume that $i_k < n$. Let $\hat{\mathcal{X}}_{n-1}$ be an $(n-1)$-dimensional subspace containing $\mathcal{X}_{i_k}$. Let $Z = (z_1\ z_2\ \ldots\ z_{n-1})$ be a matrix with orthonormal columns such that $\mathcal{R}[(z_1\ \ldots\ z_{i_j})] = \mathcal{X}_{i_j}$ and $\mathcal{R}(Z) = \hat{\mathcal{X}}_{n-1}$. Let $B = Z^{\mathrm{H}}AZ$. Then by Corollary 4.4 the eigenvalues μ_i of B satisfy

$$\mu_i \leq \lambda_i, \qquad i = 1,\ldots,n-1. \tag{4.3}$$

Now let
$$\mathcal{Y}_{i_j} = \{Z^{\mathrm{H}}x : x \in \mathcal{X}_{i_j}\}.$$

Observe that since $\mathcal{X}_{i_j} \subset \mathcal{R}(Z)$, if $y \in \mathcal{Y}_{i_j}$ then $x = Zy \in \mathcal{X}_{i_j}$. Moreover, $y^{\mathrm{H}}By = x^{\mathrm{H}}Ax$. By the induction hypotheses there are orthonormal vectors $y_{i_j} \in \mathcal{Y}_{i_j}$ such that

$$\sum_j y_{i_j}^{\mathrm{H}} B y_{i_j} \leq \sum_j \mu_{i_j}. \tag{4.4}$$

Hence if $x_{i_j} = Zy_{i_j}$, then $x_{i_j} \in \mathcal{X}_{i_j}$ and $\sum_j y_{i_j}^{\mathrm{H}} B y_{i_j} = \sum_j x_{i_j}^{\mathrm{H}} A x_{i_j}$. Hence by (4.3) and (4.4)

$$\operatorname{trace}(X^{\mathrm{H}}AX) = \sum_j x_{i_j}^{\mathrm{H}} A x_{i_j} \leq \sum_j \lambda_{i_j},$$

which is what we were to establish.

Now assume that $i_k = n$. Let l be the largest index such that $i_l + 1 < i_{l+1}$. For notational convenience let $i_l = p$ and $i_{l+1} = q$. Let $\hat{\mathcal{X}}_{n-1}$ be an $(n-1)$-dimensional subspace that contains $\mathcal{X}_p$ and the eigenvectors corresponding to $\lambda_q, \ldots, \lambda_n$. Since $q, q+1, \ldots, n-1$ are among the indices i_j, we have

$$\mathcal{X}_p \subset \mathcal{X}_q \cap \hat{\mathcal{X}}_{n-1} \subset \cdots \subset \mathcal{X}_{n-1} \cap \hat{\mathcal{X}}_{n-1} \subset \hat{\mathcal{X}}_{n-1}.$$

Since for $i = q, \ldots, n-1$, $\dim(\mathcal{X}_i \cap \hat{\mathcal{X}}_{n-1}) \geq i-1$, we can find subspaces $\hat{\mathcal{X}}_{q-1}, \ldots, \hat{\mathcal{X}}_{n-2}$ such that

$$\hat{\mathcal{X}}_{q-1} \subset \mathcal{X}_q, \ldots, \hat{\mathcal{X}}_{n-2} \subset \mathcal{X}_{n-1}$$

4. HERMITIAN MATRICES

and
$$\mathcal{X}_{i_1} \subset \cdots \subset \mathcal{X}_p \subset \hat{\mathcal{X}}_{q-1} \subset \cdots \subset \hat{\mathcal{X}}_{n-1}.$$

Now apply the construction of the previous case to give a matrix B with eigenvalues $\mu_1 \geq \cdots \geq \mu_{n-1}$ satisfying (4.3) and a unitary matrix

$$X = (x_{i_1} \ \ldots \ x_p \ x_q \ \ldots \ x_{n-1})$$

such that

$$x_{i_j} \in \mathcal{X}_{i_j}, \quad j = 1, \ldots, l,$$
$$x_i \in \hat{\mathcal{X}}_i \subset \mathcal{X}_{i+1}, \quad i = q-1, \ldots, n-1,$$

and

$$\begin{aligned} \operatorname{trace}(X^{\mathrm{H}} A X) &\leq \sum_{j=1}^{l} \mu_{i_j} + \sum_{i=q-1}^{n-1} \mu_i \\ &\leq \sum_{j=1}^{l} \lambda_{i_j} + \sum_{i=q-1}^{n-1} \mu_i. \end{aligned} \quad (4.5)$$

By construction $\hat{\mathcal{X}}_{n-1}$ contains the eigenvectors of of A corresponding to $\lambda_q, \ldots, \lambda_n$. Hence these are also eigenvalues of B. Since $\mu_{q-1}, \ldots, \mu_{n-1}$ are the smallest eigenvalues of B, we have

$$\sum_{i=q-1}^{n-1} \mu_i \leq \sum_{i=q}^{n} \lambda_i,$$

and the result follows upon substituting this inequality in (4.5). ∎

When $k = 1$, Wielandt's theorem gives Fischer's characterization of the eigenvalues of a Hermitian matrix.

Corollary 4.7 (Fischer). *The eigenvalues of A are given by*

$$\lambda_i = \max_{\dim(\mathcal{X})=i} \min_{\substack{x \in \mathcal{X} \\ x^{\mathrm{H}} x=1}} x^{\mathrm{H}} A x$$

and

$$\lambda_i = \min_{\dim(\mathcal{X})=n-i+1} \max_{\substack{x \in \mathcal{X} \\ x^{\mathrm{H}} x=1}} x^{\mathrm{H}} A x.$$

For $i = 1$ the second of the above characterizations reduces to

$$\lambda_1 = \max_{x^{\mathrm{H}} x=1} x^{\mathrm{H}} A x,$$

as was pointed out at the beginning of this section. This latter characterization has important implications for perturbation theory. For suppose, as usual, that $\tilde{A} = A + E$, where E is also Hermitian. Then denoting the largest eigenvalues of $\tilde{A}$ and E by $\tilde{\lambda}_1$ and ϵ_1, we have

$$\tilde{\lambda}_1 = \max_{x^H x=1} x^H \tilde{A} x \leq \max_{x^H x=1} x^H A x + \max_{x^H x=1} x^H E x \leq \lambda_1 + \epsilon_1.$$

In other words, since $|\epsilon_1| \leq \|E\|_2$, the perturbation E can increase the largest eigenvalue of A by no more than $\|E\|_2$.

We will now proceed to generalize this result. As we did earlier, we first establish a result for sums of eigenvalues and then specialize it to a single eigenvalue.

Theorem 4.8. *Let the eigenvalues of E be*

$$\epsilon_1 \geq \epsilon_2 \geq \cdots \geq \epsilon_n,$$

and let $i_1, \ldots, i_k$ be distinct integers between one and n inclusive. Then

$$\lambda_{i_1} + \cdots + \lambda_{i_k} + \epsilon_{n-k+1} + \cdots + \epsilon_n \leq \tilde{\lambda}_{i_1} + \cdots + \tilde{\lambda}_{i_k}$$
$$\leq \lambda_{i_1} + \cdots + \lambda_{i_k} + \epsilon_1 + \cdots + \epsilon_k.$$

Proof. Without loss of generality, we may assume that $i_1 < \cdots < i_n$. We will first establish the second inequality. By Remark 4.6 following Theorem 4.5, there are subspaces $\mathcal{X}_{i_1} \subset \mathcal{X}_{i_2} \subset \cdots \subset \mathcal{X}_{i_k}$ such that

$$\tilde{\lambda}_{i_1} + \cdots + \tilde{\lambda}_{i_k} = \min_{\substack{X=(x_{i_1}\, x_{i_2}\, \cdots\, x_{i_k}), x_{i_j} \in \mathcal{X}_{i_j} \\ X^H X = I}} \operatorname{trace}(X^H \tilde{A} X).$$

Moreover, there are vectors $x_{i_j} \in \mathcal{X}_{i_j}$ such that $X = (x_{i_1}\, \ldots\, x_{i_k})$ is unitary and

$$\lambda_{i_1} + \cdots + \lambda_{i_k} \geq \operatorname{trace}(X^H A X).$$

It follows that

$$\tilde{\lambda}_{i_1} + \cdots + \tilde{\lambda}_{i_k} \leq \operatorname{trace}[X^H(A+E)X] \leq \lambda_{i_1} + \cdots + \lambda_{i_k} + \operatorname{trace}(X^H E X).$$

But by Corollary 4.4,

$$\operatorname{trace}(X^H E X) \leq \epsilon_1 + \cdots + \epsilon_k,$$

4. HERMITIAN MATRICES

which establishs the second inequality.
The first inequality may be obtained from the second by writing $A = \tilde{A} - E$, from which it follows that

$$\lambda_{i_1} + \cdots + \lambda_{i_k} \leq \tilde{\lambda}_{i_1} + \cdots + \tilde{\lambda}_{i_k} - \epsilon_{n-k+1} - \cdots - \epsilon_n. \quad \blacksquare$$

When $k = 1$, the theorem provides a perturbation bound.

Corollary 4.9 (Weyl). *For $i = 1, \ldots, n$*

$$\tilde{\lambda}_i \in [\lambda_i + \epsilon_n, \lambda_i + \epsilon_1].$$

There are three things to note about this corollary.

First, the corollary is similar to the Gerschgorin theorem in that it provides a set of n intervals (disks) whose union includes the eigenvalues of $\tilde{A}$. However, we know just which eigenvalue to look for in each interval. Moreover, it is impossible for an eigenvalue corresponding to one of a cluster of overlapping intervals to migrate outside its own interval.

Second, the intervals are not symmetric about the eigenvalues λ_i. In fact if ϵ_n is positive, the ith interval will not contain λ_i. This occurs when E is positive definite. In other words,

if a Hermitian matrix is perturbed by a positive definite matrix, its eigenvalues must increase.

Third, there is a weaker, more conventional form of the theorem which is stated in the following corollary.

Corollary 4.10.
$$\max\{|\tilde{\lambda}_i - \lambda_i|\} \leq \|E\|_2. \qquad (4.6)$$

This result follows directly from the preceding corollary and the observation that $\|E\|_2 = \max\{|\epsilon_1|, |\epsilon_n|\}$. In the next subsection we will generalize this corollary.

4.3. Mirsky's Theorem

Equation (4.6) can be rewritten in a more symmetric form: namely,

$$\|\text{diag}(\tilde{\lambda}_i - \lambda_i)\|_2 \leq \|E\|_2. \qquad (4.7)$$

This suggests that we attempt to replace $\|\cdot\|_2$ with other norms to obtain new perturbation bounds. In fact, (4.7) is valid for any unitarily invariant norm; however, to prove it we first establish an analogous result for singular values.

Theorem 4.11 (Mirsky). *Let X and $\tilde{X}$ be matrices of the same dimensions with singular values*

$$\sigma_1 \geq \sigma_2 \geq \cdots \geq \sigma_p,$$
$$\tilde{\sigma}_1 \geq \tilde{\sigma}_2 \geq \cdots \geq \tilde{\sigma}_p.$$

Then for any unitarily invariant norm $\|\cdot\|$,

$$\|\mathrm{diag}(\tilde{\sigma}_i - \sigma_i)\| \leq \|\tilde{X} - X\|.$$

Proof. Without loss of generality we may assume that X and $\tilde{X}$ are square (otherwise pad them out with zero rows or columns to make them so). Now by Theorem I.4.2 the eigenvalues of the matrix

$$\begin{pmatrix} 0 & X \\ X^{\mathrm{H}} & 0 \end{pmatrix}$$

are $\pm\sigma_1, \ldots, \pm\sigma_p$, and similarly for $\tilde{X}$. Finally if $\epsilon_1 \geq \cdots \geq \epsilon_n$ are the singular values of $\tilde{X} - X$, then the eigenvalues of

$$\begin{pmatrix} 0 & \tilde{X} - X \\ (\tilde{X} - X)^{\mathrm{H}} & 0 \end{pmatrix}$$

are $\pm\epsilon_1, \ldots, \pm\epsilon_p$.

In Theorem 4.8 let

$$i_k = \begin{cases} k & \text{if } \tilde{\sigma}_k \geq \sigma_k, \\ n+k & \text{if } \tilde{\sigma}_k < \sigma_k. \end{cases}$$

It then follows that

$$|\tilde{\sigma}_1 - \sigma_1| + \cdots + |\tilde{\sigma}_k - \sigma_k| \leq \epsilon_1 + \cdots + \epsilon_k, \qquad k = 1, \ldots, n.$$

Therefore by Theorem II.3.17 the inequality

$$\Phi(\tilde{\sigma}_1 - \sigma_1, \ldots, \tilde{\sigma}_p - \sigma_p) \leq \Phi(\epsilon_1, \ldots, \epsilon_p)$$

holds for any symmetric gauge function Φ. The result now follows from von Neumann's characterization of unitarily invariant norms (Theorem II.3.6). ∎

An immediate consequence of Mirsky's theorem is the generalization of (4.7). Specifically, we have the following corollary.

Corollary 4.12. *Let Φ be a symmetric gauge function and $\|\cdot\|_\Phi$ its corresponding unitarily invariant norm. Then*

$$\|\mathrm{diag}(\tilde{\lambda}_i - \lambda_i)\|_\Phi \leq \|E\|_\Phi. \tag{4.8}$$

Proof. Let $\rho = \min\{\lambda_n, \tilde{\lambda}_n\}$. Then the eigenvalues of the matrices $A - \rho I$ and $\tilde{A} - \rho I$ are nonnegative; i.e., their singular values and their eigenvalues are the same. Mirsky's theorem now applies to give (4.8). ∎

When Φ generates the Frobenius norm, we obtain a Hermitian analogue of the Hoffman–Wielandt theorem.

Corollary 4.13.

$$\sqrt{\sum_{i=1}^n (\tilde{\lambda}_i - \lambda_i)^2} \leq \|E\|_F.$$

Note that this result is stronger than the Hoffman–Wielandt theorem, since it specifies an ordering of the eigenvalues that satisfy the inequality, whereas the Hoffman–Wielandt theorem merely asserts that such an ordering exists.

4.4. Residual Bounds

In this and the next subsections we will consider applications of the Mirsky theorem. The subject of this subsection is residual bounds.

As in Section 1 we are given a matrix A (now Hermitian) of order n and a matrix X whose column space approximates an invariant subspace of A. This means that for some choice of M, the residual

$$R = AX - XM$$

will be small. In particular, if X has orthonormal columns, then by Theorem 1.15 any unitarily invariant norm of R is minimized when

$M = X^H A X$. Moreover, since A is Hermitian, we can use Mirsky's theorem to get a bound on the eigenvalues of M as an approximation to those of A.

Theorem 4.14. *Let $X \in \mathcal{R}^{n \times k}$ have orthonormal columns. Let $M = X^H A X$ and let $R = AX - XM$. Let Φ be a symmetric gauge function on $\mathcal{R}^n$, and let $\|\cdot\|_\Phi$ denote the corresponding family of unitarily invariant norms. If the eigenvalues of A are $\lambda_1 \geq \cdots \geq \lambda_n$, and the eigenvalues of M are $\mu_1 \geq \cdots \geq \mu_k$, then there are integers $i_1 < i_2 < \cdots < i_k$ such that*

$$\|\mathrm{diag}(\mu_j - \lambda_{i_j})\|_\Phi \leq \|X R^H + R X^H\|_\Phi = \Phi(\rho_1, \rho_1, \rho_2, \rho_2, \ldots), \quad (4.9)$$

where $\rho_1 \geq \rho_2 \geq \cdots$ are the singular values of R.

Proof. We will establish (4.9) for the case $2k \leq n$, leaving the other case as an exercise. For $M = X^H A X$, let

$$E = -(X R^H + R X^H).$$

Then E is Hermitian, and it is readily verified that

$$(A + E) X = X M. \quad (4.10)$$

Thus $\mathcal{R}(X)$ is an invariant subspace of $A + E$, and to each eigenvalue μ_j of M there corresponds an eigenvalue $\tilde{\lambda}_{i_j}$ of $A + E$.

By Mirsky's theorem $\|\mathrm{diag}(\tilde{\lambda}_i - \lambda_i)\|_\Phi \leq \|E\|_\Phi$. Hence $\|\mathrm{diag}(\mu_j - \lambda_{i_j})\|_\Phi \leq \|E\|_\Phi = \|X R^H + R X^H\|_\Phi$, and it remains only to establish the equality in (4.9) or equivalently that the singular values of E are $\rho_1, \rho_1, \rho_2, \rho_2, \ldots$. But if $X_\perp$ is chosen so that $(X \ X_\perp)$ is unitary, then

$$(X \ X_\perp)^H E (X \ X_\perp) = \begin{pmatrix} 0 & R^H X_\perp \\ X_\perp^H R & 0 \end{pmatrix}.$$

Since $\mathcal{R}(R) \subset \mathcal{R}(X_\perp)$ and the columns of $X_\perp$ are orthonormal, the singular values of $X_\perp^H R$ are the same as those of R, and hence those of E are those of R repeated. ∎

It is worthwhile to list the bounds for the spectral and Frobenius norms.

4. HERMITIAN MATRICES

Corollary 4.15. *For the spectral norm we have*

$$\max_j \{|\mu_j - \lambda_{i_j}|\} \leq \|R\|_2,$$

and for the Frobenius norm

$$\sqrt{\sum_j (\mu_j - \lambda_{i_j})^2} \leq \sqrt{2}\|R\|_F. \tag{4.11}$$

Remark 4.16. By an application of the argument leading to the Hoffman–Wielandt theorem, Kahan has been able to remove the factor $\sqrt{2}$ in (4.11). See Exercise 4.8.

The residual bounds derived above can be very good or very bad, and the following example shows.

Example 4.17. *If*

$$A = \begin{pmatrix} 0 & \epsilon \\ \epsilon & 0 \end{pmatrix}$$

and $X = \mathbf{1}_1$, *then* $M = 0$ *and* $\|R\|_2 = \epsilon$, *so that the bound is attained (the eigenvalues of A are $\pm\epsilon$). On the other hand, if*

$$A = \begin{pmatrix} 0 & \epsilon \\ \epsilon & 1 \end{pmatrix},$$

then the residual bound for $M = 0$ is the same, but the smallest eigenvalue of A is approximately $-\epsilon^2$!

The distinction between the two examples is that in the second the unwanted part of the spectrum is well removed from the part we are attempting to bound. In Section 3 we will show how to use such information to get a better bound.

Two more comments. First, the eigenvalues of $M = X^H A X$ are sometimes called the RAYLEIGH–RITZ APPROXIMATIONS to the eigenvalues of A. Second, although we motivated this subsection by taking X to be a matrix of approximate eigenvectors, all that is required to get accurate eigenvalues is that R be small. Indeed, the part in proof where we show that the eigenvalues of M are the same as those of $A + E$ can be turned into an algorithm for getting approximate eigenvectors from X, a procedure that is sometimes called Rayleigh–Ritz improvement.

4.5. Approximation by a Low-Rank Matrix

The second application of Mirsky's theorem is to the determination of low-rank approximations to a fixed matrix. As above, let Φ be a symmetric gauge function and $\|\cdot\|_\Phi$ be the corresponding unitarily invariant norm. Let $X \in \mathbf{C}^{m \times n}$ have the singular value decomposition

$$X = U\Sigma V^{\mathrm{H}}, \qquad (4.12)$$

where $\sigma_1 \geq \cdots \geq \sigma_m \geq 0$. We wish to find a matrix Y of rank not greater than k that is as near as possible to X in the Φ-norm.

First let Y be any matrix of rank not greater than k. Then the singular values of Y are $\tau_1 \geq \cdots \geq \tau_k \geq 0 = \cdots = 0$; i.e., the last $m-k$ singular values are zero. It follows from Mirsky's theorem that

$$\|Y - X\|_\Phi \geq \Phi(\tau_1 - \sigma_1, \ldots, \tau_k - \sigma_k, \sigma_{k+1}, \ldots, \sigma_m)$$
$$\geq \Phi(0, \ldots, 0, \sigma_{k+1}, \ldots, \sigma_m).$$

In other words, any approximation of rank not greater than k must be at least $\Phi(0, \ldots, 0, \sigma_{k+1}, \ldots, \sigma_m)$ removed from X in the Φ-norm.

Now let

$$\Sigma_k = \mathrm{diag}(\sigma_1, \ldots, \sigma_k, 0, \ldots, 0) \qquad (4.13)$$

and

$$X_k = U\Sigma_k V^{\mathrm{H}}. \qquad (4.14)$$

Then it is easily verified that X_k has rank not greater than k and $\|X_k - X\|_\Phi = \Phi(0, \ldots, 0, \sigma_{k+1}, \ldots, \sigma_m)$. Thus we have proved the following approximation theorem.

Theorem 4.18 (Schmidt, Mirsky). *Let X have the singular value decomposition (4.12), where $\sigma_1 \geq \cdots \sigma_m \geq 0$. Let Φ be a symmetric gauge function and $\|\cdot\|_\Phi$ be the corresponding unitarily invariant norm. If Y is a matrix of rank less than or equal to k, then*

$$\|Y - X\|_\Phi \geq \Phi(0, \ldots, 0, \sigma_{k+1}, \ldots, \sigma_m).$$

Moreover, equality is attained for the matrix X_k defined by (4.13) and (4.14).

4. HERMITIAN MATRICES

Notes and References

Although Sylvester published the inertia theorem in 1852 [233] (also see [234, 1853]), the theorem was found in Jacobi's papers and published posthumously [123, 1857] by Borchart, who gives 1847 as the date of discovery [39]. Hermite, who published his own proof [110, 1857], also names Jacobi as having discovered the principle. According to one biographer [195], at the time Jacobi was suffering from diabetes and from personal reverses stemming from the revolutions of 1848, which probably accounts for his failure to publish.

The interlacing theorem (Theorem 4.2) is due to Cauchy [42, 1829].

Wielandt [267, 1955] proved his theorem because he was unable "to succeed in completing the interesting sketch of a proof given by Lidskii [147, 1950]" of Theorem 4.8 (see [28, p.50] for more details and further references). Amir-Moéz [4, 1956] generalized Wielandt's characterization by replacing the sums and traces by any function of the eigenvalues in question that is nondecreasing in its arguments. The special, but very important case in Corollary 4.7 is due to Fischer [74, 1905], who actually established it for matrix pencils (see Corollary VI.1.16). Courant [46, 1920] extended the result to differential operators, and the theorem is frequently called the Courant–Fischer theorem.

Weyl [265, 1912] proved more than is stated in Corollary 4.9 (see Exercises 4.3–4.4). He also claims the analogous results for singular values à la Schmidt [192].

For Mirsky's theorem see [159, 1963], which in addition contains an admirable survey of unitarily invariant norms and related topics.

For the spectral norm and arbitrary M, the residual bound of Corollary 4.15 is due to Kahan [130, 1967] (finally published as a part of [52, 1982]), who uses the dilation theorem (Exercise 1.10) specialized to Hermitian matrices. The generalization in Theorem 4.14 to unitarily invariant norms is new. The proof given here is closely related to a proof for the spectral norm given by Parlett [175, pp. 219–220]. It should be noted that this result is but one — and one of the simplest — of a host of useful residual bounds. See [144, 145, 264] and especially the book by Parlett [175], which contains a unified treatment of many of these topics.

As was noted in the text, the eigenvalues of M in Theorem 4.14 are frequently called Rayleigh–Ritz approximations to the eigenvalues of A. Both Rayleigh and Ritz were concerned with approximating the eigenvalues of an

infinite operator by replacing it with a matrix eigenvalue problem. Rayleigh
[183, 1899] found the natural frequencies vibrating systems by restricting
its degrees of freedom to a finite number of modes, which were to be chosen to accentuate the fundamental frequency. Ritz [186, 1909] approximated
the eigenvalues of the vibrating string by minimizing the variational equation
over a finite dimensional subspace. Neither gives a formal justification for his
method. A curious custom has grown up of calling eigenvector approximations obtained from M and X "Ritz vectors," although Ritz himself merely
said that he was unable to establish their convergence using the techniques
he had developed earlier in his paper.

The Schmidt–Mirsky theorem (Theorem 4.18) is commonly attributed to
Eckart and Young [63, 1936], who established it for the Frobenius norm.
But Schmidt [192, 1907] proved it for integral operators and the Hilbert–
Schmidt norm — the natural extension of the Frobenius norm. Mirsky [159,
1963] generalized it to unitarily invariant norms.

When a Hermitian matrix is perturbed at random, a multiple eigenvalue
will tend to break up into simple eigenvalues, and the perturbation in these
eigenvalues will all be of a size. When the perturbation is not random,
however, the perturbations can be quite disparate. Sun [231, 1989] has
investigated the case where the elements of A depend analytically on several
parameters.

Exercises

1. Let
$$A = \begin{pmatrix} B & c \\ c^H & \delta \end{pmatrix}.$$
Show that there is an eigenvalue λ of A satisfying $|\lambda - \delta| \leq \|c\|_2$.

2. (Lidskii [147]). In the notation of Theorem 4.8, let $e = (\epsilon_1, \ldots, \epsilon_n)^T$.
Show that $(\tilde\lambda_1 - \lambda_1, \ldots, \tilde\lambda_n - \lambda_n)$ lies in the convex hull of the set $\{Pe : P \text{ a permutation}\}$.

THE FOLLOWING TWO EXERCISES SHOW IN MODERN NOTATION
WHAT WEYL [265] ACTUALLY PROVED.

3. Let A and B be Hermitian with B of rank k. Then the largest eigenvalue
of $A - B$ is not less than the $(k+1)$th largest eigenvalue of A.

4. Let A, B, and C have eigenvalues $\alpha_1 \geq \cdots \geq \alpha_n$, $\beta_1 \geq \cdots \geq \beta_n$, and $\gamma_1 \geq \cdots \geq \gamma_n$. Then
$$\gamma_{i+j+1} \leq \alpha_{i+1} + \beta_{j+1}.$$

———◇———

THE FOLLOWING EXERCISES DEVELOP THE KATO–TEMPLE RESIDUAL BOUND [43, SECTION 6.5] FOR AN ISOLATED EIGENVALUE AND ITS EIGENVECTOR. IN WHAT FOLLOWS $\|x\|_2 = 1$, $\mu = x^H A x$, AND $r = Ax - \mu x$.

5. Let $\mu \in (\alpha, \beta)$, where (α, β) contains no eigenvalues of A. Then
$$(\beta - \mu)(\mu - \alpha) \leq \|r\|_2.$$

6. Let $\underline{\mu} < \mu < \overline{\mu}$, where $(\underline{\mu}, \overline{\mu})$ contains exactly one eigenvalue λ of A. Then
$$\lambda \in \left[\mu - \frac{\|r\|_2^2}{\overline{\mu} - \mu}, \mu + \frac{\|r\|_2^2}{\mu - \underline{\mu}}\right].$$

———◇———

7. (Kahan [129]). Show that for the 2-norm, the hypothesis $M = X^H A X$ can be removed from Theorem 4.14. Specifically, for arbitrary Hermitian M, the inequality (4.9) can be replaced by
$$\|\text{diag}(\mu_j - \lambda_{i_j})\|_2 \leq \|R\|_2.$$
[Hint: Use the dilation theorem (Exercise 1.11).]

8. (Kahan [129]). Show that the factor $\sqrt{2}$ can be removed from (4.11). [Hint: Assume without loss of generality that A and M are diagonal, and regard R as a function of X, or more generally of $U = (X \; \hat{X})$, where U is unitary. Let $W = |U|$, and let $\delta_{ij} = (\lambda_i - \mu_j)$ when $j \leq k$ and otherwise be zero. Show that $\|R\|_F^2 = \sum_i \sum_j \omega_{ij} \delta_{ij}$. Conclude from Birkhoff's theorem that $\|R\|_2$ is minimized when U is a permutation matrix.]

5. Some Further Results

This subsection is devoted to some useful results that could not be made to fit comfortably into the preceding subsections. In the first subsection we treat the problem of non-Hermitian perturbations of Hermitian matrices; and in the second, the perturbation of eigenvalues of matrices that are similar to Hermitian or normal matrices.

5.1. Non-Hermitian Perturbations

The results of this subsection concern non-Hermitian perturbations of Hermitian matrices and except as noted are due to Kahan. Throughout we will assume that A is a Hermitian matrix with eigenvalues $\lambda_1 \geq \cdots \geq \lambda_n$. We will further assume that $\tilde{A}$ is a non-Hermitian perturbation of A; that is, $E = \tilde{A} - A$ is not Hermitian. The eigenvalues of $\tilde{A}$, which may be complex, will be written $\mu_k + i\nu_k$, where $\mu_1 \geq \cdots \geq \mu_n$. Finally we will write

$$E_\Re = \frac{E + E^{\mathrm{H}}}{2}$$

and

$$E_\Im = \frac{E - E^{\mathrm{H}}}{2i} = \frac{\tilde{A} - \tilde{A}^{\mathrm{H}}}{2i}$$

for the "real" and "imaginary" parts of E. It can be verified by direct computation that

$$\|E\|_{\mathrm{F}}^2 = \|E_\Re\|_{\mathrm{F}}^2 + \|E_\Im\|_{\mathrm{F}}^2. \tag{5.1}$$

Theorem 5.1. *Let*

$$\mathcal{D}_k = \{\mu + i\nu : |\mu + i\nu - \lambda_k| \leq \|E\|_2 \text{ and } |\nu| \leq \|E_\Im\|_2\}.$$

Then

$$\lambda(\tilde{A}) \subset \bigcup_{k=1}^n \mathcal{D}_k.$$

Proof. By Corollary 3.4, for any $\mu + i\nu \in \lambda(\tilde{A})$ there is an eigenvalue λ_k of A such that

$$|\mu + i\nu - \lambda_k| \leq \|E\|_2.$$

It remains only to show that $|\nu| \leq \|E_\Im\|_2$.

Let x be a normalized eigenvector of $\tilde{A}$ corresponding to $\mu + i\nu$; i.e.,

$$\tilde{A}x = (\mu + i\nu)x, \qquad \|x\|_2 = 1.$$

Then

$$x^{\mathrm{H}}\tilde{A}x = \mu + i\nu$$

and

$$x^{\mathrm{H}}\tilde{A}^{\mathrm{H}}x = \mu - i\nu.$$

5. Some Further Results

Hence
$$x^H E_\Im x = \frac{x^H(\tilde{A} - \tilde{A}^H)x}{2i} = \nu,$$
from which it follows that $|\nu| \leq \|E_\Im\|_2$. ∎

If one of the regions $\mathcal{D}_k$ is isolated from the others, it contains only one eigenvalue, namely $\mu_k + i\nu_k$, which is perforce real. Thus the theorem says something new only for clusters of eigenvalues whose regions overlap. Specifically, if the m regions $\mathcal{D}_k, \ldots, \mathcal{D}_{k+m-1}$ overlap, then they contain precisely m eigenvalues of $\tilde{A}$, namely $\mu_k + i\nu_k, \ldots, \mu_{k+m-1} + i\nu_{k+m-1}$. The regions themselves are disks trimmed at the top and bottom by horizontal lines $\Im(z) = \pm\|E_\Im\|_2$. As the perturbation becomes increasingly Hermitian, these lines approach one another, restricting the sizes of the imaginary parts of the eigenvalues of $\tilde{A}$.

There is another version of the theorem that is reminiscent of the Hoffman–Wielandt theorem.

Theorem 5.2. *In the notation above*

$$\sqrt{\sum_{k=1}^n \nu_k^2} \leq \|E_\Im\|_F \tag{5.2}$$

and

$$\sqrt{\sum_{k=1}^n (\mu_k - \lambda_k)^2} \leq \|E_\Re\|_F + \sqrt{\|E_\Im\|_F^2 - \sum_{k=1}^n \nu_k^2}. \tag{5.3}$$

From this it follows that

$$\sqrt{\sum_{k=1}^n |(\mu_k + i\nu_k) - \lambda_k|^2} \leq \sqrt{2}\|E\|_F. \tag{5.4}$$

Proof. By passing to the Schur form of $\tilde{A}$, we may assume that

$$\tilde{A} = M + iN + R, \tag{5.5}$$

in which $M = \text{diag}(\mu_1, \ldots, \mu_n)$, $N = \text{diag}(\nu_1, \ldots, \nu_n)$, and R is strictly upper triangular. Thus

$$A + E_\Re = M + \frac{R + R^H}{2}$$

and
$$E_\Im = N + \frac{R - R^H}{2i}.$$

Now since N and $(R-R^H)/2i$ have disjoint sets of nonzero elements,

$$\begin{aligned}\|E_\Im\|_F^2 &= \|N\|_F^2 + \left\|\frac{R-R^H}{2i}\right\|_F^2 \\ &= \|N\|_F^2 + \tfrac{1}{\sqrt{2}}\|R\|_F^2 \\ &\geq \|N\|_F^2 = \sum_{k=1}^n \nu_k^2,\end{aligned}$$

which establishes (5.2). On the other hand, since A and M are Hermitian

$$\begin{aligned}\sqrt{\sum_{k=1}^n (\mu_k - \lambda_k)^2} &\leq \|M - A\|_F \\ &= \left\|E_\Re - \frac{R+R^H}{2}\right\|_F \\ &\leq \|E_\Re\|_F + \left\|\frac{R+R^H}{2}\right\|_F \\ &= \|E_\Re\|_F + \tfrac{1}{\sqrt{2}}\|R\|_F \\ &= \|E_\Re\|_F + \sqrt{\|E_\Im\|_F^2 - \|N\|_F^2} \\ &= \|E_\Re\|_F + \sqrt{\|E_\Im\|_F^2 - \sum_{k=1}^n \nu_k^2},\end{aligned}$$

which establishes (5.2).

To establish the combined bound (5.4), write

$$\begin{aligned}\sum_{k=1}^n |(\mu_k + i\nu_k) - \lambda|^2 &= \sum_{k=1}^n (\mu_k - \lambda_k)^2 + \sum_{k=1}^n \nu_k^2 \\ &\leq \left(\|E_\Re\|_F + \sqrt{\|E_\Im\|_F^2 - \sum_{k=1}^n \nu_k^2}\right)^2 + \sum_{k=1}^n \nu_k^2 \\ &= \|E_\Re\|_F^2 + 2\|E_\Re\|_F \sqrt{\|E_\Im\|_F^2 - \sum_{k=1}^n \nu_k^2} + \|E_\Im\|_F^2 \\ &\leq (\|E_\Re\|_F + \|E_\Im\|_F)^2 \\ &\leq 2(\|E_\Re\|_F^2 + \|E_\Re\|_F^2) \\ &= 2\|E\|_F^2. \blacksquare\end{aligned}$$

5.2. Similarity Bounds

In Theorem 3.3 we assumed that A was diagonalizable and derived a bound on $\operatorname{sv}_A(\tilde{A})$ that depended on the condition number of the diagonalizing transformation. In this subsection we will assume that both A and $\tilde{A}$ can be reduced to either Hermitian or normal matrices by similarity transformations, and obtain perturbation bounds on their eigenvalues.

We begin with the Hermitian case. The principal result is based on the following lemma.

Lemma 5.3. *Let H and K be $n \times n$ Hermitian matrices, and let $\Sigma = \operatorname{diag}(\sigma_1, \ldots, \sigma_n)$ with $\sigma_1 \geq \cdots \geq \sigma_n \geq 0$. Then*

$$\|H\Sigma - \Sigma K\|_2 \geq \sigma_n \|H - K\|_2.$$

Proof. Let λ be the eigenvalue of $H - K$ of largest absolute value, so that $\|H - K\|_2 = |\lambda|$. Let x be the corresponding normalized eigenvector. Then

$$\begin{aligned} x^{\mathrm{H}}(H\Sigma - \Sigma K)x &= x^{\mathrm{H}}(H - K)\Sigma x + x^{\mathrm{H}}(K\Sigma - \Sigma K)x \\ &= \lambda x^{\mathrm{H}}\Sigma x + i\tau, \end{aligned} \quad (5.6)$$

where τ is real (here we have used the fact that the matrix $K\Sigma - \Sigma K$ is skew-Hermitian). Hence

$$\begin{aligned} \|H\Sigma - \Sigma K\|_2 &= \max_{\substack{\|u\|_2=1 \\ \|v\|_2=1}} |u^{\mathrm{H}}(H\Sigma - \Sigma K)v| \\ &\geq \max_{\|u\|_2=1} |u^{\mathrm{H}}(H\Sigma - \Sigma K)u| \\ &\geq |x^{\mathrm{H}}(H\Sigma - \Sigma K)x| \\ &\geq |\lambda| |x^{\mathrm{H}}\Sigma x| \quad [\text{by}(5.6)] \\ &= \|H - K\|_2 |x^{\mathrm{H}}\Sigma x| \geq \sigma_n \|H - K\|_2. \quad \blacksquare \end{aligned}$$

Lemma 5.3 allows us to establish the following theorem.

Theorem 5.4. *Let $A, \tilde{A} \in \mathbf{C}^{n \times n}$, and suppose that there are two nonsingular matrices P and Q such that $P^{-1}AP$ and $Q^{-1}\tilde{A}Q$ are Hermitian. Let the eigenvalues of A and $\tilde{A}$ (which are necessarily real) be $\lambda_1 \geq \cdots \geq \lambda_n$ and $\tilde{\lambda}_1 \geq \cdots \geq \tilde{\lambda}_n$. Then*

$$|\tilde{\lambda}_i - \lambda_i| \leq \kappa_2(P)\kappa_2(Q)\|\tilde{A} - A\|_2, \quad i = 1, \ldots, n. \quad (5.7)$$

Proof. Let
$$A_* = P^{-1}AP \quad \text{and} \quad \tilde{A}_* = Q^{-1}\tilde{A}Q.$$
Then
$$\begin{aligned}\|\tilde{A} - A\|_2 &= \|Q\tilde{A}_*Q^{-1} - PA_*P^{-1}\|_2 \\ &= \|Q(\tilde{A}_*Q^{-1}P - Q^{-1}PA_*)P^{-1}\|_2 \\ &\geq \|Q^{-1}\|_2^{-1}\|P\|_2^{-1}\|\tilde{A}_*(Q^{-1}P) - (Q^{-1}P)A_*\|_2.\end{aligned} \quad (5.8)$$

Let $U\Sigma V^{\mathrm{H}}$ be the singular value decomposition of $Q^{-1}P$. Then with σ_n denoting the smallest diagonal of Σ, we have from Lemma 5.3
$$\begin{aligned}\|\tilde{A}_*(Q^{-1}P) - (Q^{-1}P)A_*\|_2 &= \|(U^{\mathrm{H}}\tilde{A}_*U)\Sigma - \Sigma(V^{\mathrm{H}}A_*V)\|_2 \\ &\geq \sigma_n\|U^{\mathrm{H}}\tilde{A}_*U - V^{\mathrm{H}}A_*V\|_2 \\ &\geq \sigma_n|\tilde{\lambda}_i - \lambda_i|, \quad i = 1,\ldots,n\end{aligned}$$

Thus from (5.8) we get
$$|\tilde{\lambda}_i - \lambda_i| \leq \sigma_n^{-1}\|P\|_2\|Q^{-1}\|_2\|\tilde{A} - A\|_2, \quad i = 1,\ldots,n. \quad (5.9)$$
Now
$$\sigma_n^{-1} = \|(Q^{-1}P)^{-1}\|_2 \leq \|P^{-1}\|_2\|Q\|_2.$$
Combining this with (5.9) we get (5.7). ∎

There is an analogue of Theorem 5.4, due to Sun and Zhang, for matrices that can be transformed into normal matrices.

Theorem 5.5. *Let $A, \tilde{A} \in \mathbf{C}^{n\times n}$. Assume that there are nonsingular matrices P and Q such that $P^{-1}AP$ and $Q^{-1}\tilde{A}Q$ are normal. Then*
$$\mathrm{md}_2(A, \tilde{A}) \leq \kappa(P)\kappa(Q)\|\tilde{A} - A\|_{\mathrm{F}}.$$

Proof. If we can establish the analogue of Lemma 5.3 for normal matrices, then the proof of Theorem 5.4 goes through *mutatis mutandis*. Specifically we must show that if M and N are normal matrices and $\Sigma = \mathrm{diag}(\sigma_1,\ldots,\sigma_n)$ with $\sigma_1 \geq \cdots \geq \sigma_n \geq 0$, then
$$\|M\Sigma - \Sigma N\|_{\mathrm{F}} \geq \sigma_n\|M - N\|_{\mathrm{F}}.$$
To show this, set
$$\delta = \|M\Sigma - \Sigma N\|_{\mathrm{F}}^2 - \sigma_n^{\,2}\|M - N\|_{\mathrm{F}}^2.$$

5. Some Further Results

and
$$\Omega = \Sigma - \sigma_n I.$$
Obviously, the diagonal elements of Ω are nonnegative. Moreover,

$$\delta = \|M(\Omega + \sigma_n I) - (\Omega + \sigma_n I)N\|_F^2 - \sigma_n^2 \|M - N\|_F^2$$
$$= \|M\Omega - \Omega N + \sigma_n(M - N)\|_F^2 - \sigma_n^2 \|M - N\|_F^2$$
$$= \|M\Omega - \Omega N\|_F^2 + 2\sigma_n \Re\{\text{trace}[(M\Omega - \Omega N)^H(M - N)]\}$$
$$= \|M\Omega - \Omega N\|_F^2$$
$$\quad + \sigma_n \text{trace}\{\Omega[(M - N)^H(M - N) + (M - N)(M - N)^H]\}$$
$$\geq 0,$$

which is the required inequality. ■

Notes and References

With the exception of Theorem 5.5, which is due to Sun [227, 1984] and Zhang [277, 1986], the results of this section are taken from a paper by Kahan [133, 1975]. Kahan writes as if Theorem 5.1 were due to Wilkinson [269, 1965], but although Wilkinson gives a brief discussion of nonsymmetric perturbations, he does not bound the imaginary parts.

In addition to the results of this section, Kahan shows that the matching distance of a non-Hermitian perturbation is proportional to $\log n \|E\|_2$ (Exercise 5.2).

Exercises

1. Let A be Hermitian and $\tilde{A}$ be normal. In the notation of Theorem 5.1 show that
$$\sqrt{\sum_{k=1}^n \nu_k^2} \leq \|E_\Im\|_F$$
and
$$\sqrt{\sum_{k=1}^n (\mu_k - \lambda_k)^2} \leq \|E_\Re\|_F.$$

Conclude that
$$\sqrt{\sum_{k=1}^n |(\mu_k + i\nu_k) - \lambda_k|^2} \leq \|E\|_F.$$

[Note: This is the bound that the Hoffman–Wielandt theorem would provide, except that the pairing of the eigenvalues of A and $\tilde{A}$ is explicit.]

2. (Kahan [133]). Use the fact [132] that if $\lambda(Z)$ is real then $\|Z - Z^H\|_2 \leq (0.038 + \log_2 n)\|Z + Z^H\|_2$) to show that if A is Hermitian then

$$\mathrm{md}(A, \tilde{A}) \leq \|E_\Re\|_2 + (0.038 + \log_2 n)\|E_\Im\|_2.$$

Chapter V
Invariant Subspaces

We have already observed in Section II.4 that the problem of establishing perturbation bounds for eigenvectors is complicated by the fact that eigenvectors corresponding to multiple eigenvalues are not unique. This has the consequence that the eigenvectors corresponding to a tight cluster of eigenvalues will be ill conditioned. However, the *space* spanned by these eigenvectors is an invariant subspace, which need not be sensitive to perturbations in the matrix. It therefore makes sense to derive perturbation bounds for invariant subspaces, from which bounds for eigenvectors follow as a special case.

The first section of this chapter may be regarded as a continuation of the subsection on invariant subspaces in Chapter I. Here we introduce the notion of a simple invariant subspace — the analogue of a simple eigenpair — and establish its properties. In the next section we derive error and perturbation bounds for a simple invariant subspace of a general matrix. In the third section we present the Davis–Kahan theory for invariant subspaces of Hermitian matrices. The chapter concludes with a section on the singular value decomposition.

Throughout this chapter, A will denote a matrix of order n, except in the last section, where it will denote an $m \times n$ matrix.

1. The Theory of Simple Invariant Subspaces

1.1. Definition

Let $\mathcal{X}$ be an invariant subspace of A, and let the columns of X form a basis for $\mathcal{X}$. In Section I.3 we showed that there was a unique matrix L such that
$$AX = XL.$$
The matrix L is the representation of A on $\mathcal{X}$ with respect to the basis X, and the eigenvalues of L are eigenvalues of A.

Unfortunately, the matrix L need not characterize the invariant subspace $\mathcal{X}$. For example, if $A = I_n$, then any matrix $X \in \mathbf{C}^{n \times 2}$ with orthonormal columns spans an invariant subspace whose representation with respect to X is I_2. This shows that we cannot circumvent the problem of nonuniqueness of eigenvectors by passing to invariant subspaces: we need additional conditions to insure that the invariant subspaces are themselves unique.

The key is provided by the observation that if λ is a simple eigenvalue of A, then its eigenvector is unique up to a scalar multiple. The analogous requirement for an invariant subspace is that the eigenvalues of its representation L be distinct from the other eigenvalues of A. We will say that such an invariant subspace is simple. However, before making a formal definition, it will be convenient to establish some preliminary results.

We begin with a useful characterization of invariant subspaces.

Theorem 1.1. *Let the columns of X be linearly independent and let the columns of Y span $\mathcal{R}(X)^\perp$. Then $\mathcal{R}(X)$ is an invariant subspace of A if and only if*
$$Y^{\mathrm{H}} A X = 0. \tag{1.1}$$
In this case $\mathcal{R}(Y)$ is an invariant subspace of A^{H}.

Proof. Let $\mathcal{X} = \mathcal{R}(X)$. Then by definition $\mathcal{X}$ is an invariant subspace of A if and only if $A\mathcal{X} \subset \mathcal{X}$. But
$$A\mathcal{X} \subset \mathcal{X} \iff A\mathcal{X} \perp \mathcal{X}^\perp$$
$$\iff \mathcal{R}(AX) \perp \mathcal{R}(Y)$$
$$\iff Y^{\mathrm{H}} A X = 0,$$

1. Simple Invariant Subspaces 221

which establishes (1.1). Writing (1.1) in the form $X^H A^H Y = 0$, we see that $\mathcal{R}(Y)$ must be an invariant subspace of A^H. ∎

Just the invariant subspace $\mathcal{R}(X)$ is a generalization of the notion of a right eigenvector, so $\mathcal{R}(Y)$ is a generalization of a left eigenvector. Consequently we shall call $\mathcal{R}(Y)$ a LEFT INVARIANT SUBSPACE of A.

The condition (1.1) can be regarded as saying that A can be reduced to a block triangular form by a unitary similarity. To see this, let $\mathcal{X}_1$ be an invariant subspace of A, and the columns of X_1 form an orthonormal basis for $\mathcal{X}$. Let $(X_1 \; Y_2)$ be unitary. Then

$$(X_1 \; Y_2)^H A (X_1 \; Y_2) = \begin{pmatrix} X_1^H A X_1 & X_1^H A Y_2 \\ Y_2^H A X_1 & Y_2^H A Y_2 \end{pmatrix}.$$

By (1.1) the matrix $Y_2^H A X_1$ is zero. Consequently, if we set

$$L_1 = X_1^H A X_1, \qquad L_2 = Y_2^H A Y_2,$$

and

$$H = X_1^H A Y_2,$$

then

$$(X_1 \; Y_2)^H A (X_1 \; Y_2) = \begin{pmatrix} L_1 & H \\ 0 & L_2 \end{pmatrix}. \tag{1.2}$$

It is easy to see that

$$A X_1 = X_1 L_1,$$

so that L_1 is the representation of A on $\mathcal{X}$ with respect to X. Thus the eigenvalues of L_1 are the eigenvalues of A associated with $\mathcal{X}$. The complementary set of eigenvalues are those of L_2. All this suggests the following definition.

Definition 1.2. *Let $\mathcal{X}$ be an invariant subspace of A, and let (1.2) be its* REDUCED FORM *with respect to the unitary matrix $(X_1 \; Y_2)$. Then $\mathcal{X}$ is a* SIMPLE INVARIANT SUBSPACE *if*

$$\mathcal{L}(L_1) \cap \mathcal{L}(L_2) = \emptyset.$$

A key fact about simple invariant subspaces is that they have complements in $\mathbf{C}^n$ that are also an invariant subspaces (to see that this is not true in general, let

$$A = \begin{pmatrix} 0 & 1 \\ 0 & 0 \end{pmatrix}$$

and consider the invariant subspace spanned by $\mathbf{1}_1$). However, before we can prove this fact we must digress to establish the properties of a certain linear operator.

1.2. The Operator $\mathbf{T} = X \mapsto AX - XB$

In the sequel we shall have to solve SYLVESTER'S EQUATION, which is of the form

$$AX - XB = C, \tag{1.3}$$

where A and B are square matrices of orders n and m, so that X and C are $n \times m$ matrices. We will be concerned with conditions under which (1.3) has a unique solution. Equivalently, if we define the linear operator $\mathbf{T}: \mathbf{C}^{n \times m} \to \mathbf{C}^{n \times m}$ by

$$\mathbf{T} = X \mapsto AX - XB,$$

then the problem becomes one of determining when $\mathbf{T}$ is nonsingular.

Theorem 1.3. *The linear operator* $\mathbf{T} = X \mapsto AX - XB$ *is nonsingular if and only if*

$$\mathcal{L}(A) \cap \mathcal{L}(B) = \emptyset.$$

Proof. First suppose that $\lambda \in \mathcal{L}(A) \cap \mathcal{L}(B)$. Let $Ap = \lambda p$ and $q^\mathrm{H} B = \lambda q^\mathrm{H}$ ($p, q \neq 0$). Let $X = pq^\mathrm{H}$. Then

$$\mathbf{T}(X) = Apq^\mathrm{H} - pq^\mathrm{H} B = \lambda pq^\mathrm{H} - \lambda pq^\mathrm{H} = 0.$$

Thus $\mathbf{T}$ annihilates the nonzero matrix X and must be singular.

Conversely, assume that $\mathcal{L}(A) \cap \mathcal{L}(B) = \emptyset$. We must show that the system $AX - XB = C$ has a unique solution. Let the Schur decomposition of B be $T = V^\mathrm{H} BV$. Then with $Y = XV$ and $D = CV$, the equation $AX - XB = C$ is equivalent to

$$AY - YT = D. \tag{1.4}$$

1. SIMPLE INVARIANT SUBSPACES

We will show that this equation has a unique solution. Partition $Y = (y_1 \ldots y_n)$ and $D = (d_1 \ldots d_n)$ by columns. Since T is upper triangular, the first column in the relation (1.4) is $Ay_1 - \tau_{11} y_1 = d_1$ or

$$(A - \tau_{11} I) y_1 = d_1 \qquad (1.5)$$

Since $\tau_{11} \in \mathcal{L}(B)$, the matrix $A - \tau_{11} I$ is nonsingular. Hence y_1 is the unique solution of (1.5).

Now suppose that $y_1, \ldots, y_{k-1}$ are uniquely determined. The kth column of (1.4) is

$$Ay_k - \sum_{i=1}^{k} \tau_{ik} y_i = d_k$$

or

$$(A - \tau_{kk} I) y_k = d_k + \sum_{i=1}^{k-1} \tau_{ik} y_i. \qquad (1.6)$$

Since $\tau_{kk} \in \mathcal{L}(B)$, the matrix $A - \tau_{kk} I$ is nonsingular. Hence y_k is the unique solution of (1.6). ∎

A corollary of this result is a characterization of the eigenvalues of **T**.

Corollary 1.4.

$$\mathcal{L}(\mathbf{T}) = \mathcal{L}(A) - \mathcal{L}(B).$$

Proof. If $\nu \in \mathcal{L}(\mathbf{T})$, then there is an X such that $AX - XB = \nu X$, or $(A - \nu I)X - XB = 0$; i.e., the operator $X \mapsto (A - \nu I)X - XB$ is singular. It follows that $\mathcal{L}(A - \nu I)$ and $\mathcal{L}(B)$ have a common element, which is to say that $\nu = \lambda - \mu$ for some $\lambda \in \mathcal{L}(A)$ and $\mu \in \mathcal{L}(B)$. Hence $\mathcal{L}(\mathbf{T}) \subset \mathcal{L}(A) - \mathcal{L}(B)$. The inclusion in the other direction follows by reversing the above argument. ∎

1.3. The Spectral Resolution

We are now in a position to show that a simple invariant subspace has a complementary subspace. The following theorem exhibits the complement as the column space of a matrix constructed from a reduced form of the invariant subspace.

Theorem 1.5. *Let the simple invariant subspace $\mathcal{X}_1$ have the reduced form (1.2) with respect to the orthogonal matrix $(X_1 \ Y_2)$. Then there are matrices X_2 and Y_1 such that*

$$(X_1 \ X_2)^{-1} = (Y_1 \ Y_2)^{\mathrm{H}},$$

and

$$A = X_1 L_1 Y_1^{\mathrm{H}} + X_2 L_2 Y_2^{\mathrm{H}}, \tag{1.7}$$

where

$$L_i = Y_i^{\mathrm{H}} A X_i, \quad i = 1, 2.$$

Proof. We begin by reducing the matrix

$$\begin{pmatrix} L_1 & H \\ 0 & L_2 \end{pmatrix}$$

from (1.2) to block diagonal form by a similarity transformation. Specifically, we will show that there is a matrix Q such that

$$\begin{pmatrix} I & -Q \\ 0 & I \end{pmatrix} \begin{pmatrix} L_1 & H \\ 0 & L_2 \end{pmatrix} \begin{pmatrix} I & Q \\ 0 & I \end{pmatrix} = \begin{pmatrix} L_1 & 0 \\ 0 & L_2 \end{pmatrix}. \tag{1.8}$$

This is equivalent to showing that there is a matrix Q such that

$$L_1 Q - Q L_2 = -H.$$

Since $\mathcal{X}$ is simple, $\mathcal{L}(L_1) \cap \mathcal{L}(L_2) = \emptyset$. Hence by Theorem 1.3, Q exists (and is unique).

It follows from (1.2) and (1.8) that

$$\begin{pmatrix} L_1 & 0 \\ 0 & L_2 \end{pmatrix} = \begin{pmatrix} I & -Q \\ 0 & I \end{pmatrix} \begin{pmatrix} X_1^{\mathrm{H}} \\ Y_2^{\mathrm{H}} \end{pmatrix} A(X_1 \ Y_2) \begin{pmatrix} I & Q \\ 0 & I \end{pmatrix}$$
$$= \begin{pmatrix} Y_1^{\mathrm{H}} \\ Y_2^{\mathrm{H}} \end{pmatrix} A(X_1 \ X_2), \tag{1.9}$$

where

$$X_2 = Y_2 + X_1 Q$$

1. SIMPLE INVARIANT SUBSPACES

and
$$Y_1 = X_1 - Y_2 Q^{\mathrm{H}}.$$
The fact that $(X_1\ X_2)^{-1} = (Y_1\ Y_2)^{\mathrm{H}}$ follows from the fact that (1.9) is a similarity transformation. Hence we may write (1.9) in the form
$$A = (X_1\ X_2) \begin{pmatrix} L_1 & 0 \\ 0 & L_2 \end{pmatrix} \begin{pmatrix} Y_1^{\mathrm{H}} \\ Y_2^{\mathrm{H}} \end{pmatrix},$$
from which (1.7) follows directly. ∎

From (1.7) we see that $AX_1 = X_1 L_1$. More important, $AX_2 = X_2 L_2$, which implies that $\mathcal{X}_2 = \mathcal{R}(X_2)$ is an invariant subspace of A. Since $(X_1\ X_2)$ is nonsingular, together $\mathcal{X}_1$ and $\mathcal{X}_2$ span $\mathbf{C}^n$. Thus we have the following corollary.

Corollary 1.6. *Let $\mathcal{X}_1$ be a simple invariant subspace of A. Then A has a complementary invariant subspace $\mathcal{X}_2$. Moreover, the spaces $\mathcal{Y}_1 = \mathcal{X}_2^{\perp}$ and $\mathcal{Y}_2 = \mathcal{X}_1^{\perp}$ are the corresponding complementary pair of left invariant subspaces.*

We will call (1.7) the SPECTRAL RESOLUTION of A along $\mathcal{X}_1$ and $\mathcal{X}_2$. It is instructive to write the spectral resolution in a different way. Let
$$P_i = X_i Y_i^{\mathrm{H}}, \qquad i = 1, 2.$$
Then it is easily verified that

1. $P_i^2 = P_i$ $(i = 1, 2)$,
2. $P_1 P_2 = P_2 P_1 = 0$, (1.10)
3. $A = P_1 A P_1 + P_2 A P_2$.

As we saw in Section I.2, the first condition says that $P_i \mathcal{X}_i = \mathcal{X}_i$ ($i = 1, 2$). The second condition says that $P_1 \mathcal{X}_2 = 0$. Hence if we decompose any vector z into the sum
$$z = x_1 + x_2, \qquad x_1 \in \mathcal{X}_1, x_2 \in \mathcal{X}_2,$$
then $x_1 = P_1 z$ and $x_2 = (I - P_1)z = P_2 z$. For this reason we say that P_1 is a the projection onto $\mathcal{X}_1$ along $\mathcal{X}_2$. We will call it the SPECTRAL PROJECTION of the simple invariant subspace $\mathcal{X}_1$.

When $\dim(\mathcal{X}_1) = 1$, that is when $x_1 = X_1$ is an eigenvector, the spectral projection is $P_1 = xy^H$ and $\|P_1\|_2 = \|y_1\|_2$. We have already seen that this quantity is a condition number for the eigenvalue λ_1 [(IV.2.8)]. The quantity $\|P_1\|$ will play an analogous role for L_1. Hence it is of interest to know the singular values of P_1.

Theorem 1.7. *Let $\mathcal{X}$ be a simple invariant subspace of A and let P be its spectral projection. Let $\mathcal{Y}$ be the corresponding left invariant subspace, and let $\theta_1 \geq \theta_2 \geq \cdots \geq 0$ be the canonical angles between $\mathcal{X}$ and $\mathcal{Y}$. Then $\theta_1 < \frac{\pi}{2}$ and*

$$\mathcal{S}(P) = \{\sec\theta_1, \sec\theta_2, \ldots\}. \tag{1.11}$$

Proof. We will adopt the notation of Theorem 1.5, with $\mathcal{X}_1 = \mathcal{X}$, etc. Since $Y_1 = X_1 + Y_2 Q$, we have

$$Y_1^H Y_1 = I + Q^H Q.$$

It follows that if $\rho_1, \rho_2, \ldots$ are the singular values of Q, then

$$\mathcal{S}(Y_1) = \left\{\sqrt{1 + \rho_1^2}, \sqrt{1 + \rho_2^2}, \ldots\right\}. \tag{1.12}$$

Clearly the columns of $Y_1(I + Q^H Q)^{-\frac{1}{2}}$ form an orthonormal basis for $\mathcal{Y}_1$. Since the columns of Y_2 form an orthonormal basis for $\mathcal{X}_1^\perp$, it follows from Corollary I.5.4 that the sines of the canonical angles of $\mathcal{X}_1$ and $\mathcal{Y}_1$ are the singular values of

$$Y_2^H Y_1 (I + Q^H Q)^{-\frac{1}{2}} = Q(I + Q^H Q)^{-\frac{1}{2}}.$$

Hence

$$\sin\theta_i = \frac{\rho_i}{\sqrt{1 + \rho_i^2}} < 1. \tag{1.13}$$

It follows that the canonical angles must be less than $\frac{\pi}{2}$. Finally, (1.11) follows from (1.12) and (1.13) and the fact that X_1 in the expression $P = X_1 Y_1^H$ has orthonormal columns. ∎

Although we shall not use the fact in the sequel, it is worth noting that a spectral resolution can be defined for more than two complementary subspaces. An extreme example is given by the spectral decomposition (I.3.1), in which each invariant subspace is spanned by a single eigenvector. This example also shows that although the simplicity of an invariant subspace is sufficient for a spectral resolution, it is not necessary.

1. Simple Invariant Subspaces

Notes and References

The theory developed in this section is constructive, in the sense that given a basis for an invariant subspace, one can construct the associated spectral resolution. From a pedagogical point of view, the approach has the advantage that one can develop the theory for a simple *eigenpair* — something students grasp readily — and then generalize it by replacing lower case letters with capital letters [214].

The disadvantage of the approach is that it deals only with simple invariant subspaces, whereas the set of all invariant subspaces of a matrix has a far richer structure. For a detailed exposition see the book by Gohberg, Lancaster, and Rodman [87, 1986].

In spite of its simplicity, Theorem 1.1 is the key to perturbation theory for invariant subspaces, since it furnishes an equation that an invariant subspace must satisfy. To obtain perturbation bounds all one has to do is solve the equation.

Equation (1.3) is known variously as Sylvester's equation and Rosenblum's equation [188, 1956]. The proof of the existence of a solution (Theorem 1.3) is constructive and can serve as a basis for efficient algorithms for solving the equation [11, 89]. Integral representations of the solution have been given by Rosenblum (Exercise 1.4) and Bhatia, Davis, and McIntosh [31] (Exercise 1.5).

The possibility of spectral resolutions into more than two blocks is treated in the exercises below. The ultimate spectral resolution is the Jordan canonical form (cf. I.3.4), in which the blocks are as small and simple as possible. However, the transformations which produce the Jordan form may be too ill conditioned to make it usable. This has led some algorithmists to seek resolutions in which the blocks are nearly as small as possible, given a bound on the condition of the transformations (e.g., see [95, 189, 18, 128]).

Exercises

1. Given a (not necessarily orthonormal) basis for an invariant subspace of A, describe in detail how to compute its spectral resolution.

THE FOLLOWING EXERCISES CONCERN THE SOLUTION OF SYLVESTER'S EQUATION $AX - XB = C$ AND THE ASSOCIATED OPERATOR $\mathbf{T} = X \mapsto AX - XB$.

2. Assuming A and B are diagonalizable, show that $\mathbf{T}$ has a complete system of eigenvectors. Use this fact to give an alternative proof of Theorem 1.3.

3. (Matrix representation of $\mathbf{T}$). Let $X = (x_1 \ldots x_m)$ and define
$$\mathrm{vec}(X) = \begin{pmatrix} x_1 \\ \vdots \\ x_m \end{pmatrix}.$$
Show that
$$\mathrm{vec}[\mathbf{T}(X)] = (I_m \otimes A - B \otimes I_n)\mathrm{vec}(X),$$
where $\otimes$ is the Kronecker product defined in Exercise I.3.26.

4. (Rosenblum [188]). Let $\mathcal{G}$ be a simple closed curve containing $\mathcal{L}(B)$ and excluding $\mathcal{L}(A)$. Show that
$$X = -\frac{1}{2\pi i} \int_{\mathcal{G}} (A - \zeta I)^{-1} C (B - \zeta I)^{-1} \, d\zeta.$$

5. (Bhatia, Davis, and McIntosh [31]). Let A and B be normal with $\mathcal{L}(A) \cap \mathcal{L}(B) = \emptyset$. Let $A = A_\Re + i A_\Im$, where $A_\Re$ and $A_\Im$ are Hermitian, and similarly for B. For $t = (\tau_1 \ \tau_2)^{\mathrm{T}} \in \mathbf{R}^2$, let
$$U(t) = e^{i(\tau_1 A_\Re + \tau_2 A_\Im)} \quad \text{and} \quad V(t) = e^{i(\tau_1 B_\Re + \tau_2 B_\Im)}.$$
Show that if ϕ_δ is any function integrable on $\mathbf{R}^2$ satisfying
$$\int_{\mathbf{R}^2} e^{-it^{\mathrm{T}} x} \phi_\delta(x) \, dx = \frac{1}{\tau_1 + i\tau_2}, \qquad \|t\|_2 \geq \delta,$$
then
$$X = \int_{\mathbf{R}^2} U(-t) C V(t) \phi_\delta(t) \, dt.$$

———◇———

THESE EXERCISES DEVELOP THE PROPERTIES OF SPECTRAL RESOLUTIONS WITH MORE THAN TWO BLOCKS.

6. Let the distinct eigenvalues of A be $\lambda_1, \ldots \lambda_k$. Show that there are matrices $X = (X_1 \ldots X_k)$ and $Y = (Y_1 \ldots Y_k)$ such that $X^{-1} = Y^{\mathrm{H}}$ and
$$Y^{\mathrm{H}} A X = \mathrm{diag}(L_1, \ldots, L_k),$$
where $\mathcal{L}(L_i) = \{\lambda_i\}$. Conclude that
$$A = X_1 L_1 Y_1^{\mathrm{H}} + \cdots + X_k L_k Y_k^{\mathrm{H}}.$$

7. Let $P_i = X_i Y_i^H$ ($i = 1, \ldots, k$). Show that the P_i are (oblique) projections satisfying $P_i P_j = 0$ ($i \neq j$) and

$$A = P_1 A P_1 + \cdots + P_k A P_k.$$

8. Let $\phi(A)$ be defined as in Exercise II.2.20. Show that

$$\phi(A) = X_1 \phi(L_1) Y_1^H + \cdots + X_k \phi(L_k) Y_k^H.$$

[Note: This exercise shows that the evaluation of $\phi(A)$ may be reduced to the evaluation of $\phi(L_i)$ ($i = 1, \ldots, k$). Since the orders of the L_i may be much less than the order of A, this reduction may save a great deal of work.]

———◇———

2. Perturbation of Invariant Subspaces

In this section we will treat two closely related problems. The first is the problem of assessing the accuracy of an approximate invariant subspace in terms of a residual. Specifically, let the columns of X_1 form an orthonormal basis for an approximate invariant subspace of A. Let $L = X_1^H A X_1$, and let $R = AX_1 - X_1 L_1$ be the associated residual. If $R = 0$, then $\mathcal{R}(X_1)$ is an invariant subspace of A. This suggests that if R is sufficiently small there will be an invariant subspace $\tilde{\mathcal{X}}_1$ of A that approaches $\mathcal{R}(X_1)$ as R approaches zero. The problem is to bound the difference in terms of R.

The second problem is our usual perturbation problem. Let $\mathcal{X}_1$ be a simple invariant subspace of A and let $\tilde{A} = A + E$. Show that for sufficiently small E there is an invariant subspace $\tilde{\mathcal{X}}_1$ of $\tilde{A}$, that approaches $\mathcal{X}_1$ as E approaches zero, and bound their difference in terms of E.

The two problems are closely related. For example, if the orthonormal columns of X_1 span an invariant subspace of A and we set $M = X_1^H \tilde{A} X_1$, then $R = \tilde{A}X - XM = EX$. Thus for any unitarily invariant norm, $\|R\| \leq \|E\|$, and we may use any residual bound to determine how near $\mathcal{R}(X_1)$ is to an invariant subspace of $\tilde{A}$. In fact, this is the general approach we will take in this section — first establish a residual bound, then derive a perturbation bound from it.

2.1. The Approximation Problem

Let $(X_1\ Y_2)$ be a unitary matrix and let

$$(X_1\ Y_2)^{\mathrm{H}} A (X_1\ Y_2) = \begin{pmatrix} L_1 & H \\ G & L_2 \end{pmatrix}. \qquad (2.1)$$

By Theorem 1.1 the space $\mathcal{R}(X_1)$ is an invariant subspace if and only if $G = Y_2^{\mathrm{H}} A X_1$ is zero. The problem treated in this section is to determine how near $\mathcal{R}(X_1)$ is to an invariant subspace of A, when G is small.

The solution is conceptually simple, although fussy to realize. Let

$$\hat{X}_1 = (X_1 + Y_2 P)(I + P^{\mathrm{H}} P)^{-\frac{1}{2}} \qquad (2.2)$$

and

$$\hat{Y}_2 = (Y_2 - X_1 P^{\mathrm{H}})(I + P P^{\mathrm{H}})^{-\frac{1}{2}}, \qquad (2.3)$$

where P is a matrix to be determined so that $\mathcal{R}(\hat{X}_1)$ is an invariant subspace of A. It is easy to see that $(\hat{X}_1\ \hat{Y}_2)$ is unitary. Hence by Theorem 1.1, $\mathcal{R}(\hat{X}_1)$ is an invariant subspace of A if and only if $\hat{Y}_2^{\mathrm{H}} A \hat{X}_1 = 0$. Writing this condition out in terms of (2.2) and (2.3) yields the following nonlinear equation for P:

$$P L_1 - L_2 P = G - PHP. \qquad (2.4)$$

If we define

$$\mathbf{T} = P \mapsto P L_1 - L_2 P, \qquad (2.5)$$

then this equation can be written

$$\mathbf{T}(P) = G - PHP. \qquad (2.6)$$

The conditions under which this equation has a solution are given in the following theorem, in which $\|\cdot\|$ represents a consistent family of norms.

Theorem 2.1. *Let $(X_1\ Y_2)$ be unitary. Let L_1, L_2, G and H be defined by (2.1) and set*

$$\gamma = \|G\|, \qquad \eta = \|H\|.$$

2. PERTURBATION OF INVARIANT SUBSPACES

Assume that $\mathcal{L}(L_1) \cap \mathcal{L}(L_2) = \emptyset$, so that the operator $\mathbf{T}$ defined by (2.5) is nonsingular, and set

$$\delta = \text{sep}(L_1, L_2) \stackrel{\text{def}}{=} \inf_{\|P\|=1} \|\mathbf{T}(P)\| > 0.$$

Then if
$$\frac{\gamma \eta}{\delta^2} < \frac{1}{4}, \qquad (2.7)$$
there is a unique solution P of (2.6) satisfying

$$\|P\| \leq \frac{2\gamma}{\delta + \sqrt{\delta^2 - 4\gamma\eta}} < 2\frac{\gamma}{\delta}. \qquad (2.8)$$

If $\hat{X}_1$ and $\hat{Y}_2$ are defined by (2.2) and (2.3), then $\mathcal{R}(\hat{X}_1)$ and $\mathcal{R}(\hat{Y}_2)$ are simple right and left invariant subspaces of A. The representation of A with respect to $\hat{X}_1$ is

$$\hat{L}_1 = (I + P^{\text{H}} P)^{\frac{1}{2}} (L_1 + HP)(I + P^{\text{H}} P)^{-\frac{1}{2}}. \qquad (2.9)$$

The representation of A with respect to $\hat{Y}_2$ is

$$\hat{L}_2 = (I + PP^{\text{H}})^{-\frac{1}{2}} (L_2 - PH)(I + PP^{\text{H}})^{\frac{1}{2}}. \qquad (2.10)$$

Proof. The existence of a P satisfying (2.8) is a consequence of Theorem 2.11 at the end of this section. By construction $\mathcal{R}(\hat{X}_1)$ and $\mathcal{R}(\hat{Y}_2)$ are invariant subspaces. We will establish their simplicity later. It remains only to establish the representations (2.9) and (2.10).

The representation of A with respect to $\hat{X}_1$ is $\hat{X}_1^{\text{H}} A \hat{X}_1$. From (2.2)

$$\hat{X}_1^{\text{H}} A \hat{X}_1 = (I + P^{\text{H}} P)^{\frac{1}{2}} (L_1 + HP + P^{\text{H}} G + P^{\text{H}} L_2 P)(I + P^{\text{H}} P)^{\frac{1}{2}}. \qquad (2.11)$$

From (2.4),
$$L_2 P = PL_1 + PHP - G.$$

If this value is substituted into (2.11), the result, after some simplification, is (2.9). The representation (2.10) follows similarly. ∎

We now turn to an extended discussion of the theorem. The first point to consider is the interpretation of P and its norm. By Corollary I.5.4 the singular values of the matrix

$$Y_2^{\text{H}} \hat{X}_1 = P(I + P^{\text{H}} P)^{-\frac{1}{2}}$$

are the sines of the canonical angles $\theta_1, \theta_2, \ldots$ between $\mathcal{R}(X_1)$ and the invariant subspace $\mathcal{R}(\hat{X}_1)$. If $\pi_1, \pi_2, \ldots$ are the singular values of P, then
$$\frac{\pi_i}{\sqrt{1+\pi_i^2}} = \sin\theta_i.$$
Hence
$$\pi_i = \tan\theta_i.$$
Thus, loosely speaking, the theorem bounds the tangents of the canonical angles between $\mathcal{R}(X_1)$ and an invariant subspace of A. In particular, since $\sin\theta \leq \tan\theta$ ($0 \leq \theta \leq \frac{\pi}{2}$), if $\|\cdot\|$ is the spectral norm, then
$$\|P_1 - \hat{P}_1\|_2 \leq 2\frac{\gamma_2}{\delta_2},$$
where P_1 and $\hat{P}_1$ are the orthogonal projections onto $\mathcal{R}(X_1)$ and $\mathcal{R}(\hat{X}_1)$. In terms of the Frobenius norm
$$\|P_1 - \hat{P}_1\|_F \leq 2\sqrt{2}\frac{\gamma_F}{\delta_F}$$
(see Theorem I.5.5).

There are three numbers — γ, η, and δ — that determine the existence and size of P. The number γ is closely related to the residual
$$R = AX_1 - X_1 L_1.$$
In fact, since $L_1 = X_1^H A X_1$, it follows that
$$(X_1 \; Y_2)^H R = \begin{pmatrix} 0 \\ G \end{pmatrix}.$$

Hence if $\|\cdot\|$ is unitarily invariant, $\gamma = \|R\|$. Moreover, by Theorem IV.1.15, R is optimal in the sense that $\|R\|$ is minimal among all residuals of the form $AX_1 - X_1 L$.

The number $\eta = \|H\|$ is of only secondary importance in the bound (2.8). However, it features prominently in the condition $\gamma\eta/\delta^2 < 1/4$, which insures the existence of $\hat{X}_1$. Here it plays a role analogous to the quantity η introduced at the beginning of the second subsection of Section IV.1. If η is small, the eigenvalues of L_1 and L_2 are effectively

2. PERTURBATION OF INVARIANT SUBSPACES

uncoupled. On the other hand, if η is large, the eigenvalues of L_1 and L_2 look like a single cluster compared to $\|A\|$, and the residual must be very small for our theorem to hold.

Before we go on to discuss the meaning of the number δ, it will be worth our while to recast part of the theorem in the less cluttered form of a residual bound. The key is to note that if we set $S = X_1^H A - L_1 X_1^H$, then $\|S\| = \|H\|$ for any unitarily invariant norm. In the following corollary we change our notation slightly.

Corollary 2.2. *Let $\|\cdot\|$ be a unitarily invariant norm. Let $(X\ Y)$ be unitary and $\mathcal{X} = \mathcal{R}(X)$. Let*

$$L = X^H A X \quad \text{and} \quad M = Y^H A Y.$$

Finally, let

$$R = AX - XL \quad \text{and} \quad S = X^H A - LX^H.$$

If

$$\frac{\|R\|\|S\|}{\text{sep}(L,M)^2} < \frac{1}{4},$$

then there is a simple invariant subspace $\hat{\mathcal{X}}$ of A such that

$$\|\tan[\Theta(\mathcal{X},\hat{\mathcal{X}})]\| < 2\frac{\|R\|}{\text{sep}(L,M)}.$$

We now turn to the number $\delta = \text{sep}(L_1, L_2)$, which is the thing that makes the whole theory work. As the name "sep" indicates, it is related to the separation of $\mathcal{L}(L_1)$ and $\mathcal{L}(L_2)$.

Theorem 2.3. *For any square matrices L and M,*

$$\text{sep}(L, M) \leq \min |\mathcal{L}(L) - \mathcal{L}(M)|. \tag{2.12}$$

Proof. As above, set $\mathbf{T} = P \mapsto PL - MP$. If $\text{sep}(L, M) = 0$, the inequality (2.12) holds trivially. Otherwise, $\mathbf{T}$ is nonsingular, and

$$\text{sep}^{-1}(L, M) = \sup_{\|Q\|=1} \mathbf{T}^{-1}(Q) = \|\mathbf{T}^{-1}\|.$$

Now by Theorem II.2.6 the spectral radius of $\mathbf{T}^{-1}$ is bounded by $\|\mathbf{T}^{-1}\|$. By Corollary 1.4, $\mathcal{L}(\mathbf{T}) = \mathcal{L}(L) - \mathcal{L}(M)$. Hence

$$\operatorname{sep}^{-1}(L, M) \geq \rho(\mathbf{T}^{-1}) = \max |\mathcal{L}(L) - \mathcal{L}(M)|^{-1},$$

which is equivalent to (2.12). ∎

Theorem 2.3 and the bound (2.8) imply that if some of the eigenvalues of L_1 and L_2 are close then the invariant subspace $\mathcal{R}(\hat{X}_1)$ may be distant from $\mathcal{R}(X_1)$. However, the converse need not be true, since $\operatorname{sep}(L_1, L_2)$ can be small when the eigenvalues of L_1 and L_2 are well separated, as the following example shows.

Example 2.4. Let

$$L = \begin{pmatrix} 0 & 1 \\ 0 & 0 \end{pmatrix}$$

and

$$M = \begin{pmatrix} 0 & 1 \\ \epsilon & 0 \end{pmatrix}.$$

Then $\mathcal{L}(L) = \{0\}$ and $\mathcal{L}(M) = \{\pm\epsilon^{\frac{1}{2}}\}$, so that $\delta_\lambda = \min |\mathcal{L}(L) - \mathcal{L}(M)| = \epsilon^{\frac{1}{2}}$. On the other hand, $\operatorname{sep}(L, M) = \epsilon$. Thus $\operatorname{sep}(L, M)$ can be arbitrarily smaller than the distance between the eigenvalues of L and those of M, in the sense that $\lim_{\epsilon \to 0} \operatorname{sep}(L, M)/\delta_\lambda = 0$.

The distance δ_λ between the eigenvalues of L and those of M is necessarily a continuous function of the elements of L and M; however, it need not be analytic. For example, perturbing M in the above example so that it becomes equal to L changes the $(2,2)$-element of M by ϵ. However it changes δ_λ by $\epsilon^{\frac{1}{2}}$. The function $\operatorname{sep}(L, M)$ is better behaved.

Theorem 2.5.

$$\operatorname{sep}(L, M) - \|E\| - \|F\| \leq \operatorname{sep}(L+E, M+F) \leq \operatorname{sep}(L, M) + \|E\| + \|F\|.$$

Remark 2.6. If the norm $\|\cdot\|$ is unitarily invariant, then we may replace $\|E\|$ and $\|F\|$ by $\|E\|_2$ and $\|F\|_2$.

2. Perturbation of Invariant Subspaces

Proof. For the first inequality,

$$\begin{aligned} \operatorname{sep}(L+E, M+F) &= \inf_{\|P\|=1} \|(L+E)P - P(M+F)\| \\ &\geq \inf_{\|P\|=1} \{\|LP - PM\| - \|EP\| - \|PF\|\} \\ &\geq \inf_{\|P\|=1} \{\|LP - PM\|\} - \|E\| - \|F\| \\ &= \operatorname{sep}(L, M) - \|E\| - \|F\|. \end{aligned}$$

The second inequality is established similarly. ∎

Thus a perturbation in L or M cannot induce a larger perturbation in $\operatorname{sep}(L, M)$. This stability of the function sep will be important in establishing the perturbation bounds of the next subsection.

The representations (2.9) and (2.10) imply that

$$\mathcal{L}(A) = \mathcal{L}(L_1 + HP) \cup \mathcal{L}(L_2 - PH).$$

By (2.8)

$$\|PH\|, \|HP\| < 2\frac{\gamma\eta}{\delta}, \qquad (2.13)$$

and it follows from Theorem 2.5 that

$$\operatorname{sep}(\hat{L}_1, \hat{L}_2) \geq \delta - 4\frac{\gamma\eta}{\delta} > 0,$$

the last inequality following from (2.7). This implies that $\mathcal{L}(\hat{L}_1) \cap \mathcal{L}(\hat{L}_2) = \emptyset$, which shows that $\mathcal{R}(\hat{X}_1)$ and $\mathcal{R}(\hat{Y}_2)$ are *simple* invariant subspaces.

An important consequence of the approximation theorem is a new class of residual bounds for the eigenvalues of A. Specifically, we have

$$\|(L_1 + PH) - L_1\| = \|PH\| \leq 2\frac{\gamma\eta}{\delta}.$$

Since L_1 is known and $\mathcal{L}(L_1 + PH) \subset \mathcal{L}(A)$, we may use the perturbation techniques of the last chapter to bound the distance between a subset of the eigenvalues of A and the eigenvalues of L_1.

Comparing these bounds with the bounds from Theorem IV.1.13, we see that the approaches are different and give different results. In the case of residual bounds, we apply perturbation theory to a manufactured perturbation $A + E$ to relate the eigenvalues of L_1 (M in Theorem IV.1.13) to those of A. In the approximation theorem, we apply perturbation theory to L_1 to relate its eigenvalues to those of $L_1 + HP$, which are a subset of the eigenvalues of A. Moreover the perturbations are of different sizes: γ in the case of the residual bounds and bounded by $\gamma\eta/\delta$ in the approximation theorem. Which is better will depend on the application.

2.2. Perturbation Theorems

The key to obtaining perturbation theorems for invariant subspaces is to combine the approximation theorem with continuity of the measure sep. Specifically, we have the following theorem, whose proof is left as an exercise.

Theorem 2.7. *Let $(X_1\ Y_2)$ be unitary and suppose that $\mathcal{R}(X_1)$ is a simple invariant subspace of A, so that*

$$(X_1\ Y_2)^{\mathrm{H}} A (X_1\ Y_2) = \begin{pmatrix} L_1 & H \\ 0 & L_2 \end{pmatrix}.$$

Given a perturbation E, let

$$(X_1\ Y_2)^{\mathrm{H}} E (X_1\ Y_2) = \begin{pmatrix} E_{11} & E_{12} \\ E_{21} & E_{22} \end{pmatrix}.$$

Let $\|\cdot\|$ represent a consistent family of norms and set

$$\tilde{\gamma} = \|E_{21}\|,$$
$$\tilde{\eta} = \|H\| + \|E_{12}\|,$$
$$\tilde{\delta} = \mathrm{sep}(L_1, L_2) - \|E_{11}\| - \|E_{22}\|.$$

If $\tilde{\delta} > 0$ and

$$\frac{\tilde{\gamma}\tilde{\eta}}{\tilde{\delta}^2} < \frac{1}{4},$$

there is a unique matrix P satisfying

$$\|P\| \le \frac{2\tilde{\gamma}}{\tilde{\delta} + \sqrt{\tilde{\delta}^2 - 4\tilde{\gamma}\tilde{\eta}}} < 2\frac{\tilde{\gamma}}{\tilde{\delta}}$$

2. Perturbation of Invariant Subspaces

such that the columns of

$$\tilde{X}_1 = (X_1 + Y_2 P)(I + P^{\rm H} P)^{-\frac{1}{2}}$$

and

$$\tilde{Y}_2 = (Y_2 - X_1 P^{\rm H})(I + P P^{\rm H})^{-\frac{1}{2}}$$

form orthonormal bases for simple right and left invariant subspaces of $\tilde{A} = A + E$. The representation of $\tilde{A}$ with respect to $\tilde{X}_1$ is

$$\tilde{L}_1 = (I + P^{\rm H} P)^{\frac{1}{2}}[L_1 + E_{11} + (H + E_{12})P](I + P^{\rm H} P)^{-\frac{1}{2}},$$

and the representation of $\tilde{A}$ with respect to $\tilde{Y}_2$ is

$$\tilde{L}_2 = (I + P P^{\rm H})^{-\frac{1}{2}}[L_2 + E_{22} - P(H + E_{12})](I + P P^{\rm H})^{\frac{1}{2}}.$$

The comments following Theorem 2.1 apply here. In particular the singular values of P are the tangents of the canonical angles between $\mathcal{R}(X_1)$ and $\mathcal{R}(\tilde{X}_1)$.

The expressions for $\tilde{L}_1$ and $\tilde{L}_2$ are somewhat awkward to interpret. This is because of the way we have chosen to express $\tilde{X}_1$ and $\tilde{Y}_2$. If we choose different expressions, we will obtain different bases for the perturbed invariant subspaces and hence different representations for A on those subspaces.

A good choice, it turns out, is to express $\tilde{X}_1$ in terms of the spectral resolution of A. Specifically, let $(X_1\ X_2)^{-1} = (Y_1\ Y_2)^{\rm H}$ and

$$(Y_1\ Y_2)^{\rm H} A (X_1\ X_2) = \begin{pmatrix} L_1 & 0 \\ 0 & L_2 \end{pmatrix}, \qquad (2.14)$$

as in (1.9). If we seek $\tilde{X}_1$ and $\tilde{Y}_2$ in the forms

$$\tilde{X}_1 = X_1 + X_2 P$$

and

$$\tilde{Y}_2 = Y_2 - Y_1 P^{\rm H},$$

then we have the following theorem.

Theorem 2.8. *Let A have the spectral resolution (2.14) and set*

$$(Y_1\ Y_2)^{\mathrm{H}} E (X_1\ X_2) = \begin{pmatrix} F_{11} & F_{12} \\ F_{21} & F_{22} \end{pmatrix}.$$

Let $\|\cdot\|$ represent a consistent family of norms, and set

$$\tilde{\gamma} = \|F_{21}\|,$$
$$\tilde{\eta} = \|F_{12}\|,$$
$$\tilde{\delta} = \mathrm{sep}(L_1, L_2) - \|F_{11}\| - \|F_{22}\|.$$

If $\tilde{\delta} > 0$ and

$$\frac{\tilde{\gamma}\tilde{\eta}}{\tilde{\delta}^2} < \frac{1}{4},$$

there is a unique matrix P satisfying

$$\|P\| \leq \frac{2\tilde{\gamma}}{\tilde{\delta} + \sqrt{\tilde{\delta}^2 - 4\tilde{\gamma}\tilde{\eta}}} < 2\frac{\tilde{\gamma}}{\tilde{\delta}},$$

such that the columns of

$$\tilde{X}_1 = X_1 + X_2 P$$

and

$$\tilde{Y}_2 = Y_2 - Y_1 P^{\mathrm{H}}$$

form bases for simple right and left invariant subspace of $\tilde{A} = A + E$. The representation of $\tilde{A}$ with respect to $\tilde{X}_1$ is

$$\tilde{L}_1 = L_1 + F_{11} + F_{12} P k$$

and the representation of $\tilde{A}$ with respect to $\tilde{Y}_2$ is

$$\tilde{L}_2 = L_2 + F_{22} - P F_{12}.$$

Proof. Apply the approximation theorem to eliminate $F_{21} E$ in the matrix

$$(Y_1\ Y_2)^{\mathrm{H}} (A+E)(X_1\ X_2) = \begin{pmatrix} L_1 + F_{11} & F_{12} \\ F_{21} & L_2 + F_{22} \end{pmatrix}. \quad\blacksquare$$

2. Perturbation of Invariant Subspaces

Remark 2.9. Since X_1 and Y_2 are the same in Theorems 2.7 and 2.8, we have $E_{21} = F_{21}$. Hence if P_1 denotes the matrix P produced by Theorem 2.7 and P_2 denotes the matrix P produced by Theorem 2.8, then

$$P_1(L_1 + E_{11}) - (L_2 + E_{22})P_1 = E_{21} - P_1(H + E_{12})P_1,$$

and

$$P_2(L_1 + F_{11}) - (L_2 + F_{22})P_2 = E_{21} - P_2(F_{12})P_2.$$

Since these two equations differ in terms of order $O(\|E\|^2)$, we have the remarkable result:

$$P_1 = P_2 + O(\|E\|^2).$$

It is instructive to consider the difference between Theorem 2.7 and Theorem 2.8 from a different point of view. There is no unique basis for the subspace $\tilde{\mathcal{X}}_1$. However if we chose a matrix Z whose columns span a subspace that is acute to $\tilde{\mathcal{X}}_1$, the normalization

$$Z^{\mathrm{H}} \tilde{X}_1 = I,$$

along with $\mathcal{R}(\tilde{X}_1) = \tilde{\mathcal{X}}_1$, uniquely determines $\tilde{X}_1$. In Theorem 2.7 we require the normalization

$$X_1^{\mathrm{H}} \tilde{X}_1 = I + O(\|E\|^2), \qquad (2.15)$$

whereas in Theorem 2.8 we require

$$Y_1^{\mathrm{H}} \tilde{X}_1 = I + O(\|E\|^2). \qquad (2.16)$$

Which is the better theorem? It depends on the application. If the angles between the invariant subspaces themselves are the chief concern, it does not make a great deal of difference, since the matrices P produced by the two theorems are the same up to second order terms; the difference is in the way they are used to get a basis for $\mathcal{R}(X_1)$.

On the other hand, if the representations of A on the perturbed subspaces are of interest, then Theorem 2.8 is the more natural. In the expression $\tilde{L}_1 = L_1 + F_{11} + F_{12}P$, the matrix $F_{12}P$ is of order $\|E\|^2$. Since $F_{11} = Y_1^{\mathrm{H}} E X_1$, we have

$$\tilde{L}_1 = L_1 + Y_1^{\mathrm{H}} E X_1 + O(\|E\|^2).$$

Since $Y^H X_1 = I$, this equation is completely analogous to the relation
$$\tilde{\lambda} = \lambda + y^H E x + O(\|E\|^2),$$
which we derived in Theorem IV.2.3. Moreover, if we write $\tilde{L}_1 = Y_1^H(A+E)X_1 + O(\|E\|^2)$ and compare this expression with (IV.2.7), we see that the matrix $Y_1^H(A+E)X_1$ is a generalization of the Rayleigh quotient, which we will call the GENERALIZED RAYLEIGH QUOTIENT.

Furthermore, since $X_1^H X_1 = I$, for any unitarily invariant norm
$$\|\tilde{L}_1 - L_1\| \lesssim \|Y_1\|_2 \|E\|.$$

Thus $\|Y_1\|_2$ is a condition number for L_1. This number is also the norm of the spectral projection associated with the invariant subspace $\mathcal{R}(X_1)$, which justifies the statement that the condition number of an eigenvalue is the norm of its spectral projection. Moreover, by Theorem 1.7, $\|Y_1\|_2 = \sec\theta_1$, where θ_1 is the largest canonical angle between $\mathcal{R}(X_1)$ and $\mathcal{R}(Y_1)$. Hence,

> the condition number of L_1 (with respect to the normalization (2.16)) is the secant of the largest canonical angle between $\mathcal{R}(X_1)$ and $\mathcal{R}(Y_1)$.

2.3. Eigenvectors

When the concern is with eigenvectors, the preceding results simplify considerably, since the operator **T**—in an obvious specialization of the above notation— becomes the matrix $\lambda_1 I - L_2$. Hence for both theorems,
$$p \cong (\lambda_1 I - L_2)^{-1} Y_2^H E x_1.$$
For Theorem 2.7,
$$\tilde{x}_1 \cong x_1 + Y_2(\lambda_1 I - L_2)^{-1} Y_2^H E x_1,$$
while for Theorem 2.8
$$\tilde{x}_1 \cong x_1 + X_2(\lambda_1 I - L_2)^{-1} Y_2^H E x_1.$$
The matrix
$$(\lambda I - A)^{\#} = X_2(\lambda_1 I - L_2)^{-1} Y_2^H$$

2. Perturbation of Invariant Subspaces

is called the GROUP INVERSE or DRAZIN GENERALIZED INVERSE of $\lambda_1 I - L_2$ (see Exercises III.1.23–III.1.24). Since for Theorem 2.8

$$\|\tilde{x}_1 - x_1\| \lesssim \|(\lambda I - A)^{\#}\| \|E\|,$$

the number $\|(\lambda I - A)^{\#}\|$ is sometimes said to be a condition number for the eigenvector x_1; but as we have seen above this depends on the application. If the angles between the eigenvectors are the concern, then $\delta^{-1} = \|(\lambda_1 I - L_2)^{-1}\|$ is the condition number of the problem. But if we are interested in the difference between the eigenvectors when $y_1^H \tilde{x}_1 \cong 1$, then $\|(\lambda I - A)^{\#}\|$ is the condition number. Here is an example of the latter application.

Example 2.10. *A square, nonnegative matrix A is said to be* STOCHASTIC *if $A\mathbf{1} = \mathbf{1}$; i.e., its rows sum to one. Clearly, $\mathbf{1}$ is a right eigenvector of A corresponding to the eigenvalue one. Moreover, if A is irreducible (see Example 2.7 for the definition), then one is a simple eigenvalue and hence has a unique, positive left eigenvector p that satisfies $p^T \mathbf{1} = 1$. With this normalization, the components of p can be regarded as probabilities.*

Now if we perturb A in such a way that $\tilde{A}$ remains an irreducible stochastic matrix, then we will want to keep the normalization $\tilde{y}^H \mathbf{1} = 1$, so that we can continue to regard the components of $\tilde{y}$ as probabilities. In this case the Drazin generalized inverse provides the condition number for the problem.

The two theorems give us two bounds for eigenvalues — the first a bound for the eigenvalue itself, the second for the Rayleigh quotient. For the eigenvalue we have

$$|\tilde{\lambda} - \lambda| \leq \left(1 + 2\frac{\tilde{\eta}}{\tilde{\delta}}\right) \|E\|.$$

For the Rayleigh quotient we have

$$|\tilde{\lambda} - (\lambda + y_1^H E x_1)| \leq 2\frac{\tilde{\gamma}^2}{\tilde{\delta}}.$$

2.4. Solution of a Nonlinear Equation

In this subsection we will prove a general theorem that can be used to establish the existence of P in Theorem 2.1. We state and prove it for a Banach space, which the reader may take to be $\mathbf{C}^{m\times n}$.

Theorem 2.11. Let T be a bounded linear operator on a Banach space $\mathcal{B}$. Assume that T has a bounded inverse, and set

$$\delta = \|T^{-1}\|^{-1}.$$

Let $\varphi : \mathcal{B} \to \mathcal{B}$ be a function that satisfies

$$\|\varphi(x)\| \leq \eta \|x\|^2$$

and

$$\|\varphi(x) - \varphi(y)\| \leq 2\eta \max\{\|x\|, \|y\|\} \|x - y\|$$

for some $\eta \geq 0$. For any $g \in \mathcal{B}$, let

$$\gamma = \|g\|.$$

If

$$\rho = 4\frac{\gamma\eta}{\delta^2} < 1,$$

then the sequence defined by $x_0 = 0$ and

$$x_{k+1} = T^{-1}[g + \varphi(x_k)], \qquad k = 0, 1, \ldots \qquad (2.17)$$

converges to the unique solution of

$$Tx = g + \varphi(x)$$

that satisfies

$$\|x\| \leq \frac{2\gamma}{\delta + \sqrt{\delta^2 - 4\gamma\eta}} < 2\frac{\gamma}{\delta}.$$

Moreover,

$$\|x_k - x\| \leq \frac{\rho^k}{1-\rho} \frac{\gamma}{\delta}.$$

2. Perturbation of Invariant Subspaces

Proof. We first construct an upper bound on $\|x_k\|$. From (2.17),

$$\|x_{k+1}\| \leq \|T^{-1}\|(\|g\| + \|\varphi(x_k)\|) \leq \frac{\gamma}{\delta} + \frac{\eta}{\delta}\|x_k\|^2.$$

Consequently if we set $\xi_0 = 0$ and

$$\xi_{k+1} = \frac{\gamma}{\delta} + \frac{\eta}{\delta}\xi_k^2, \qquad k = 0, 1, \ldots,$$

then $\|x_k\| \leq \xi_k$.

Now the sequence $\xi_0, \xi_1, \ldots$ is clearly increasing. Moreover, if $\rho < 1$, the function

$$\phi(\xi) = \frac{\gamma}{\delta} + \frac{\eta}{\delta}\xi^2,$$

has a smallest fixed point

$$\xi_* = \frac{2\gamma}{\delta + \sqrt{\delta^2 - 4\gamma\eta}}.$$

If $\xi_k \leq \xi_*$, then $\xi_{k+1} = \phi(\xi_k) \leq \phi(\xi_*) = \xi_*$. Hence all the ξ_k are bounded by ξ_*, and the sequence $\{\xi_k\}$ must converge to ξ_*. Thus

$$\|x_k\| \leq \|\xi_*\| < 2\frac{\gamma}{\delta}.$$

We next show that the sequence $\{x_k\}$ converges. We have

$$\|x_{k+1} - x_k\| \leq \|T^{-1}\|\|\varphi(x_k) - \varphi(x_{k-1})\|$$
$$\leq 2\delta^{-1}\eta \max\{\|x_k\|, \|x_{k-1}\|\}\|x_k - x_{k-1}\|$$
$$\leq \rho\|x_k - x_{k-1}\|.$$

Hence

$$\|x_{k+1} - x_k\| \leq \rho^k\|x_1 - x_0\| \leq \rho^k\frac{\gamma}{\delta}.$$

It follows that $\{x_k\}$ is a Cauchy sequence and must have a limit x. Moreover,

$$\|x - x_k\| \leq \sum_{i=k}^{\infty}\|x_{i+1} - x_i\| \leq \sum_{i=k}^{\infty}\rho^i\frac{\gamma}{\delta} = \frac{\rho^k}{1-\rho}\frac{\gamma}{\delta}. \qquad \blacksquare$$

Notes and References

Although various results for the perturbation of eigenvectors are found scattered in the literature [274, 69, 269], the modern approach via invariant subspaces crystallized in the sixties and early seventies.

For Hermitian matrices, a little note by Swanson [232, 1961] contains in embryonic form much of what was to follow. The first clear statement that the problem really concerned invariant subspaces is due to Davis [50, 51, 1963, 1965]. The importance of Sylvester's equation emerged in the famous paper of Davis and Kahan [53, 1970], whose content more than justifies its impenetrability.

For nonnormal matrices the problem is complicated by two facts. First, there is no orthonormal system of eigenvectors. Second, the differences among the eigenvalues of a nonnormal matrix are not Lipschitz continuous. In his thesis, Varah [250, 1967] (see also [251]) ameliorated the first difficulty by working with spectral resolutions, whose transformations are in general better conditioned than the matrix of eigenvectors; however, his bounds have the distance between the spectra raised to a power in the denominator. Ruhe [190, 1970] proposed replacing this difference with the smallest singular value of a power of $A - \lambda I$ for suitable λ. The use of an orthogonal reduction to block triangular form to circumvent the first difficulty and the introduction of the function sep to circumvent the second is due to Stewart [200, 1971], who proved his theorems for closed operators in a Hilbert space. The exposition in this chapter is based on a later survey paper [202, 1973]. Lower bounds on the function sep have been given by Sun [226, 1984]

Yamamoto [275, 1980] exploits a different nonlinear equation and eigenpair to get component-wise bounds (Exercise 2.6).

The fact that the generalized Rayleigh quotient $Y_1 \tilde{A} X_1$ provides a first-order approximation to the representation $\tilde{L}_1$ appears to have first been noted by Stewart [202, 1973], although it is readily derivable from standard perturbation expansions, such as are found in Kato [135, Ch.II]. The observation that a change in normalization [e.g., from (2.15) to (2.16)] leaves the multiplier P essentially unaffected may be found in [215,]. For eigenvectors, nonlinear normalizations are common; for example, one frequently requires that $\tilde{x}^H \tilde{x} = 1$. This complicates the asymptotic theory for complex matrices, since the normalization may not be analytic. Meyer and Stewart [155, 1988] have treated this problem in detail.

Owing to their structure, the perturbation theory for stochastic matrices

2. Perturbation of Invariant Subspaces 245

can be developed independently of the theory here [194, 106, 47, 155] (see Exercise 2.7).

Exercises

THE FOLLOWING EXERCISES DEVELOP SOME OF THE ELEMENTARY PROPERTIES OF THE FUNCTION sep, WHICH WE SUPPOSE TO BE DEFINED WITH RESPECT TO A CONSISTENT FAMILY OF NORMS $\|\cdot\|$.

1. Show that if X and Y are nonsingular, then

$$\frac{\operatorname{sep}(A,B)}{\kappa(X)\kappa(Y)} \leq \operatorname{sep}(XAX^{-1}, YBY^{-1}) \leq \kappa(X)\kappa(Y)\operatorname{sep}(A,B).$$

2. Show that if X and Y are unitary and $\|\cdot\|$ is unitarily invariant, then

$$\operatorname{sep}(X^{\mathrm{H}}AX, Y^{\mathrm{H}}AY) = \operatorname{sep}(A, B).$$

3. Show that if $A = \operatorname{diag}(A_1, \ldots, A_k)$ and $B = \operatorname{diag}(B_1, \ldots, B_l)$, then

$$\operatorname{sep}_{\mathrm{F}}(A, B) = \min\{\operatorname{sep}_{\mathrm{F}}(A_i, B_j) : i = 1, \ldots, k; j = 1, \ldots, l\}.$$

4. Show that if A and B are diagonalizable, then

$$\operatorname{sep}_{\mathrm{F}}(A, B) \geq \frac{|\mathcal{L}(A) - \mathcal{L}(B)|}{\kappa_2(X)\kappa_2(Y)},$$

where X and Y are matrices of the eigenvectors of A and B.

5. Let $\|\cdot\|_r$ and $\|\cdot\|_s$ be consistent norms satisfying

$$\sigma\|P\|_r \leq \|P\|_s \leq \tau\|P\|_s$$

for all $P \in \mathbf{C}^{n \times n}$. Show that

$$\operatorname{sep}_r(A, B) \geq \frac{\sigma}{\tau}\operatorname{sep}_s(A, B).$$

Use this fact to bound sep_2 in terms of $\operatorname{sep}_{\mathrm{F}}$.

—— ◇ ——

6. (Yamamoto [275]). Let $(\hat{x}, \hat{\lambda})$ be a simple eigenpair of A, with $\|\hat{x}\|_2 = 1$. Let $(x, \lambda) = (\hat{x} - h, \hat{\lambda} - \eta)$ be an approximate eigenpair. Set $\epsilon = \max\{\|h\|, |\eta|\}$. Show that if ϵ is sufficiently small then the matrix

$$\begin{pmatrix} A - \lambda I & x \\ x^{\mathrm{H}} & 0 \end{pmatrix}$$

is nonsingular and

$$\begin{pmatrix} A - \lambda I & x \\ x^{\mathrm{H}} & 0 \end{pmatrix} \begin{pmatrix} h \\ \eta \end{pmatrix} = - \begin{pmatrix} Ax - \lambda x \\ \frac{1}{2}(1 - \|x\|_2^2) \end{pmatrix} + O(\epsilon^2).$$

Analyze this equation to obtain an approximation theorem for eigenvectors.

7. Let A and $\tilde{A}$ be stochastic matrices, each having one as a simple eigenvalue. Let y^{T} and $\tilde{y}^{\mathrm{T}}$ be the corresponding left eigenvectors, normalized so that $y^{\mathrm{T}} \mathbf{1} = \tilde{y}^{\mathrm{T}} \mathbf{1} = 1$. Show that

$$\tilde{y}^{\mathrm{T}} - y^{\mathrm{T}} = y^{\mathrm{T}} E (I - \tilde{A})^{\#}.$$

3. Hermitian Matrices

We now turn to the the perturbation of invariant subspaces of Hermitian matrices. In the next two subsections, we will apply the theory of the last section to Hermitian matrices — first the approximation theorem and then the perturbation theorem. In the third subsection we will develop part of the elegant Davis-Kahan theory of invariant subspaces. Finally, we will develop two residual bounds that in some cases are improvements on the bound of Theorem IV.4.14.

Throughout this section A will be a Hermitian matrix of order n, as will the error matrix E.

3.1. The Approximation Theorem

When A is Hermitian, several things simplify in the approximation theorem. In the first place, $H = G^{\mathrm{H}}$. It follows that any unitary similarity that reduces G to zero also reduces H to zero and hence that $\mathcal{R}(\hat{Y}_2)$ is the invariant subspace complementary to $\mathcal{R}(\hat{X}_1)$.

3. Hermitian Matrices

If $\|\cdot\|$ is unitarily invariant, then $\gamma = \|G\| = \|G^H\| = \|H\| = \eta$, and the condition $\eta\gamma/\delta^2 < 1/4$ becomes

$$\frac{\gamma}{\delta} < \frac{1}{2}.$$

The most striking simplification occurs when we take $\|\cdot\|$ to be the Frobenius norm.

Theorem 3.1. *If L and M are Hermitian, then*

$$\operatorname{sep}_F(L, M) = \min |\mathcal{L}(L) - \mathcal{L}(M)|.$$

Proof. As usual, let $\mathbf{T} = P \mapsto PL - MP$. For any matrix $P = (p_1 \ p_2 \ \ldots \ p_l)$ let

$$\operatorname{vec}(P) = \begin{pmatrix} p_1 \\ p_2 \\ \vdots \\ p_l \end{pmatrix}.$$

Then

$$\operatorname{vec}(PL) = \vec{L}\operatorname{vec}(P),$$

where

$$\vec{L} = \begin{pmatrix} \lambda_{11}I_m & \lambda_{12}I_m & \cdots & \lambda_{1l}I_m \\ \lambda_{21}I_m & \lambda_{22}I_m & \cdots & \lambda_{2l}I_m \\ \vdots & \vdots & & \vdots \\ \lambda_{l1}I_m & \lambda_{l2}I_m & \cdots & \lambda_{ll}I_m \end{pmatrix}.$$

Similarly,

$$\operatorname{vec}(MP) = \vec{M}\operatorname{vec}(P),$$

where

$$\vec{M} = \operatorname{diag}(M, M, \ldots, M).$$

Hence

$$\operatorname{vec}[\mathbf{T}(P)] = (\vec{L} - \vec{M})\operatorname{vec}(P).$$

Since $\vec{L}$ and $\vec{M}$ are Hermitian, the linear operator T is Hermitian. Since $\|P\|_F = \|\operatorname{vec}(P)\|_2$,

$$\operatorname{sep}_F(L, M) = \inf_{\|P\|_F=1} \mathbf{T}(P) = \min \mathcal{L}(\mathbf{T}) = \min|\mathcal{L}(L) - \mathcal{L}(M)|,$$

the last equality following from Corollary 1.4. ∎

Thus for Hermitian A, the number δ_F in the approximation theorem truly measures the distance between the eigenvalues of L_1 and those of L_2.

Since L_1 and $\hat{L}_1 = (I+P^HP)^{\frac{1}{2}}(L_1+HP)(I+P^HP)^{-\frac{1}{2}}$ are Hermitian, we may use Theorem IV.5.5 to bound the eigenvalues of $\hat{L}_1$.

Theorem 3.2. *In the notation of the approximation theorem, let $\|\cdot\|$ be the Frobenius norm, so that*

$$\delta = \min |\mathcal{L}(L_1) - \mathcal{L}(L_2)|.$$

Let the eigenvalues of L_1 be $\lambda_1 \geq \cdots \geq \lambda_k$ and those of $\hat{L}_1$ be $\hat{\lambda}_1 \geq \cdots \geq \hat{\lambda}_k$. Then

$$\sqrt{\sum_{i=1}^k (\lambda_i - \hat{\lambda}_i)^2} \leq 2\left(1 + 2\frac{\gamma^2}{\delta^2}\right)\frac{\gamma^2}{\delta}. \tag{3.1}$$

Proof. By Theorem IV.5.5 we have

$$\sqrt{\sum_{i=1}^k (\lambda_i - \hat{\lambda}_i)^2} \leq \kappa_2[(I+P^HP)^{\frac{1}{2}}]\|H\|_F\|P\|_F. \tag{3.2}$$

Now $\|H\|_F\|P\|_F \leq 2\gamma^2/\delta$. Moreover, $\|(I+P^HP)^{-\frac{1}{2}}\|_2 \leq 1$. Finally,

$$\|(I+P^HP)^{\frac{1}{2}}\|_2 \leq (1+\|P\|_F^2)^{\frac{1}{2}} \leq 1 + \frac{\|P\|_F^2}{2} \leq 1 + 2\frac{\gamma^2}{\delta^2}.$$

Combining these inequalities yields (3.2). ∎

Note that as $\gamma \to 0$, the constants 2 in the bound (3.1) can be replaced by functions that approach 1 [cf., (2.8)]. Consequently, the right-hand side of (3.1) is bounded by a quantity that is asymptotic to γ^2/δ.

3.2. Generalized Rayleigh Quotients

So far as invariant subspaces are concerned, the comments of the last subsection apply to the perturbation theorems. However, the representations of A on the perturbed subspaces provide new perturbation

3. HERMITIAN MATRICES

theorems for eigenvalues. Since A is Hermitian, it is most natural to work with Theorem 2.8, taking $X_i = Y_i$ ($i = 1, 2$). We will also take $\|\cdot\|$ to be the Frobenius norm, so that and $\tilde{\gamma}_i = \tilde{\eta}_i$ and δ_{F} is the distance between the spectra of L_1 and L_2.

The proof of the following theorem is similar to that of Theorem 3.2 and is left as an exercise.

Theorem 3.3. *Let the Hermitian matrix A have the spectral resolution*

$$(X_1\ X_2)^{\mathrm{H}} A (X_1\ X_2) = \begin{pmatrix} L_1 & 0 \\ 0 & L_2 \end{pmatrix},$$

where L_1 is $k \times k$, and set

$$(Y_1\ Y_2)^{\mathrm{H}} E (X_1\ X_2) = \begin{pmatrix} F_{11} & F_{21}^{\mathrm{H}} \\ F_{21} & F_{22} \end{pmatrix}.$$

Let

$$\tilde{\gamma} = \|F_{21}\|_{\mathrm{F}}$$

and

$$\tilde{\delta} = \mathrm{sep}_{\mathrm{F}}(L_1, L_2) - \|F_{11}\|_2 - \|F_{22}\|_2.$$

Let $\lambda_1 \geq \cdots \geq \lambda_k$ be the eigenvalues of L_1 and $\check{\lambda}_1 \geq \cdots \geq \check{\lambda}_k$ be the eigenvalues of the generalized Rayleigh quotient $\check{L}_1 = X_1^{\mathrm{H}}(A + E) X_1$. Then if $\tilde{\delta} > 0$ and $\tilde{\gamma}/\tilde{\delta} < 1/2$, there are eigenvalues $\tilde{\lambda}_{j_1}, \ldots, \tilde{\lambda}_{j_k}$ of $\tilde{A}$ satisfying

$$\sqrt{\sum_{i=1}^{k} (\tilde{\lambda}_{j_i} - \check{\lambda}_i)^2} \leq 2\left(1 + 2\frac{\tilde{\gamma}^2}{\tilde{\delta}^2}\right) \frac{\tilde{\gamma}^2}{\tilde{\delta}}.$$

This theorem shows that the eigenvalues of the Rayleigh quotient are, up to terms in $\|E\|^2$, eigenvalues of $\tilde{A}$.

3.3. Direct Bounds

A great deal of the complexity of the preceding theory is due to the necessity of establishing the existence of the perturbed invariant subspace. For Hermitian matrices the existence is often obvious. For example, in the usual notation, suppose that $\min |\mathcal{L}(L_1) - \mathcal{L}(L_2)| = \delta$ and $\|E\|_2 < \frac{\delta}{2}$. Then by Corollary IV.4.6, $\min |\mathcal{L}(\tilde{L}_1) - \mathcal{L}(\tilde{L}_2)| > 0$, and there are unique complementary invariant subspaces associated with $\tilde{L}_1$ and $\tilde{L}_2$. Thus we may assume the existence of the perturbed invariant subspace and proceed directly to bounds on the canonical angles between the original and its perturbation. This general approach is due to Davis and Kahan.

The first "$\sin\Theta$" theorem is so called because it bounds the sum of squares of the sines of the canonical angles between an invariant subspace of A and an approximation.

Theorem 3.4. *Let A have the spectral resolution*

$$\begin{pmatrix} X_1^H \\ X_2^H \end{pmatrix} A(X_1\ X_2) = \mathrm{diag}(L_1, L_2),$$

where $(X_1\ X_2)$ is unitary with $X_1 \in \mathbf{C}^{n\times k}$. Let $Z \in \mathbf{C}^{n\times k}$ have orthonormal columns, and for any Hermitian M of order k, let

$$R = AZ - ZM.$$

If

$$\delta = \min|\mathcal{L}(L_2) - \mathcal{L}(M)| > 0, \qquad (3.3)$$

then

$$\|\sin\Theta[\mathcal{R}(X_1), \mathcal{R}(Z)]\|_F \le \frac{\|R\|_F}{\delta}.$$

Proof. From the definition of R and the fact that $X_2^H A = L_2 X_2^H$, we have

$$X_2^H R = L_2 X_2^H Z - X_2^H Z M. \qquad (3.4)$$

By Theorem 3.1,

$$\|X_2^H Z\|_F \le \frac{\|R\|_F}{\delta}.$$

By Theorem I.5.5, $\|X_2^H Z\|_F = \|\sin\Theta[\mathcal{R}(X_1), \mathcal{R}(Z)]\|_F$. ∎

Thus the theorem bounds the error in the approximate subspace in terms of the residual R. Since Z and M are arbitrary, we may use it to assess the accuracy of the vector from any approximate eigenpair (z, μ), provided we can find a lower bound the distance from μ to $n-1$ eigenvalues of A.

3. Hermitian Matrices

As k becomes large, the conclusion of the theorem becomes less and less satisfactory because $\|\sin\Theta[\mathcal{R}(X_1),\mathcal{R}(Z)]\|_F$ can be large even when the individual canonical angles are small. What we need in this case is a bound on $\|\sin\Theta[\mathcal{R}(X_1),\mathcal{R}(Z)]\|_2$, which we can obtain if we are willing to place further restrictions on the spectra of L_2 and M. We begin with a lemma.

Lemma 3.5. *Let $\|\cdot\|$ be a consistent norm. Let A and B be square with $\|A\| \leq \alpha$ and $\|B^{-1}\|^{-1} \geq \alpha + \delta$, where $\delta > 0$. If*

$$AX - XB = C,$$

then

$$\|X\| \leq \frac{\|C\|}{\delta}.$$

Proof. By consistency $\|AX\| \leq \alpha\|X\|$ and $\|XB\| \geq (\alpha+\delta)\|X\|$. Hence

$$\|C\| \geq \|BX\| - \|AX\| \geq (\alpha+\delta)\|X\| - \alpha\|X\| \geq \delta\|X\|. \blacksquare$$

We are now in position to establish a second $\sin\Theta$ theorem.

Theorem 3.6. *In the notation of Theorem 3.4, suppose that*

$$\mathcal{L}(M) \subset [\alpha,\beta] \tag{3.5}$$

and that for some $\delta > 0$,

$$\mathcal{L}(L_2) \subset \mathbf{R} \setminus [\alpha-\delta,\beta+\delta]. \tag{3.6}$$

Then for any unitarily invariant norm

$$\|\sin\Theta[\mathcal{R}(X_1),\mathcal{R}(Z)]\| \leq \frac{\|R\|}{\delta}.$$

Remark 3.7. The matrices L_1 and M may be switched in (3.5) and (3.6).

Proof. By translating the spectra of A and M, we may assume without loss of generality that $\alpha = -\beta$. The result now follows on applying Lemma 3.5 to (3.4). $\blacksquare$

In some applications, the columns of the matrix Z may not be orthonormal. The following theorem shows that with an appropriate correction factor, the above bounds continue to hold.

Theorem 3.8. *In Theorems 3.4 and 3.6, let* $\inf_2(Z)$ *(i.e., the smallest singular value of Z) be positive. Then*

$$\|\sin\Theta[\mathcal{R}(X_1),\mathcal{R}(Z)]\| \leq \frac{\|R\|}{\delta\inf_2(Z)}.$$

Proof. Let $Z = QR$ be the QR factorization of Z. Then the proofs of the $\sin\Theta$ theorems show that

$$\|X_2^H Z\| = \|X_2^H QR\| \leq \frac{\|R\|}{\delta}.$$

But

$$\|X_2^H QR\| \geq \inf_2(Z)\|X_2^H Q\| = \inf_2(Z)\|\sin\Theta[\mathcal{R}(X_1),\mathcal{R}(Z)]\|. \blacksquare$$

We conclude this subsection with a bound on the tangents of the canonical angles. Here we must restrict M to be equal to $Z^H A Z$ and impose further restrictions on the disposition of the spectrum. The idea of the proof is to prove the theorem for the norms $\|\cdot\|_{\Phi_j}$ associated with Fan's symmetric gauge functions

$$\Phi_j(x) = \max_{1 \leq i_1 < \cdots < i_j \leq n} \{|\xi_{i_1}| + \cdots + |\xi_{i_j}|\}.$$

We begin with a lemma.

Lemma 3.9. *Let R have singular values $\sigma_1 \geq \sigma_2 \geq \cdots \geq \sigma_k$. If R_j is any leading principle submatrix of R, then*

$$\mathrm{trace}(R_j) \leq \sum_{i=1}^{j} \sigma_i.$$

Proof. By Theorem I.4.4 the sum of the singular values of R_j is less than or equal to $\sum_{i=1}^{j} \sigma_i$. By Lemma II.3.4 the trace of R_j is less than or equal to the sum of its singular values. $\blacksquare$

We may now prove the $\tan\Theta$ theorem.

3. Hermitian Matrices

Theorem 3.10. *In the notation of the* $\sin \Theta$ *theorems, let*

$$M = Z^{\mathrm{H}} A Z,$$

and assume that

$$\mathcal{L}(M) \subset [\alpha, \beta]$$

while for some $\delta > 0$

$$\mathcal{L}(L_2) \subset (-\infty, \alpha - \delta]$$

(or $\mathcal{L}(L_2) \subset [\beta + \delta, \infty)$*). Then*

$$\| \tan \Theta[\mathcal{R}(X_1), \mathcal{R}(Z)] \| \leq \frac{\|R\|}{\delta}.$$

Proof. Some preliminary transformation will make the proof easier. First, we may assume without loss of generality that L_1 and L_2 are of the same order. For if the order of L_1 is less than that of L_2, we may augment the reduced form of A to $\mathrm{diag}(L_1, \nu I, L_2)$, where $\nu \in [\alpha, \beta]$. This will make no difference in the final bounds.

Next by passing to canonical bases, we may assume that

$$Z = \begin{pmatrix} I \\ 0 \end{pmatrix}$$

and

$$X = (X_1 \ X_2) = \begin{pmatrix} \Gamma & -\Sigma \\ \Sigma & \Gamma \end{pmatrix},$$

where $\Gamma = \mathrm{diag}(\cos \theta_i)$ and $\Sigma = \mathrm{diag}(\sin \theta_i)$ consist of the cosines (in ascending order) and the sines (in descending order) of the canonical angles between $\mathcal{R}(Z)$ and $\mathcal{R}(X_1)$. In this coordinate system partition A in the form

$$A = \begin{pmatrix} A_{11} & A_{12} \\ A_{21} & A_{22} \end{pmatrix}.$$

Since $M = A_{11}$, we have

$$R = \begin{pmatrix} A_{11} & A_{12} \\ A_{21} & A_{22} \end{pmatrix} \begin{pmatrix} I \\ 0 \end{pmatrix} - \begin{pmatrix} I \\ 0 \end{pmatrix} A_{11} = \begin{pmatrix} 0 \\ A_{21} \end{pmatrix}. \qquad (3.7)$$

Note that by the simplification of the preceding paragraph all the above submatrices are square and of the same order.

Since
$$(-\Sigma \quad \Gamma)A = L_2(-\Sigma \quad \Gamma),$$
on multiplying (3.7) by $(-\Sigma \quad \Gamma)$ we have
$$\Gamma A_{21} = \Sigma A_{11} - L_2 \Sigma.$$
The ith diagonal element of this relation is
$$\cos\theta_i \alpha_{ii}^{(21)} = \sin\theta_i(\alpha_{ii}^{(11)} - \lambda_{ii}^{(22)}) \geq \delta \sin\theta_i,$$
the last inequality following from the fact that $\alpha_{ii}^{(11)} \in [\alpha, \beta]$ and $\lambda_{ii}^{(22)} \in (-\infty, \alpha - \delta]$. Since $\delta > 0$, the cosine of θ_i cannot be zero. Hence $\alpha_{ii}^{(21)} \geq \delta \tan\theta_i$. It follows from Lemma 3.9 that
$$\|R\|_{\Psi_j} \geq \sum_{i=1}^{j} \alpha_{ii}^{(21)} \geq \delta \sum_{i=1}^{j} \tan\theta_i = \|\tan\Theta\|_{\Psi_j},$$
where $\Theta = \text{diag}(\theta_i)$. The theorem now follows from Fan's theorem (Theorem II.3.17). ∎

Although the hypothesis on the situation of the spectra of M and L_2 may appear unnecessarily restrictive, it is necessary (Exercise 3.3). However, it answers to the frequently occurring case where Z is an approximation to an invariant subspace corresponding to the largest (or smallest) eigenvalues of A.

3.4. Residual Bounds for Eigenvalues

In Example IV.4.17 we saw that the residual bound for eigenvalues provided by Theorem IV.4.14 could be off by orders of magnitude. The problem is that the bound does not take into account the situation of the spectrum. In this subsection we will apply perturbation theory for invariant subspaces to derive new bounds that do just that.

Throughout this subsection X will be an $n \times k$ matrix with orthonormal columns. We will set
$$M = X^H A X$$

3. Hermitian Matrices

and
$$R = AX - XM.$$
We wish to show that there are k eigenvalues of A near the eigenvalues $\mu_1 \geq \cdots \geq \mu_k$ of M, and further show that the difference is proportional to $\|R\|^2$.

The basic idea is simple. We use one of the direct bounds from the last section to give us a matrix $\hat{X} = (X + YP)(I + P^H P)^{-\frac{1}{2}}$ whose columns span an invariant subspace. We then know from the approximation theorem that $\hat{M} = \hat{X}^H A \hat{X}$ is similar to $M + G^H P$. But G and P are both of order $\|R\|$, so that an application of perturbation theory for eigenvalues will give a bound of order $\|R\|^2$.

Since the direct bounds do not give us explicit relations between the subspaces, we must begin with a lemma that allows us to deduce the existence of P. Its proof, which uses the canonical bases of Theorem I.5.2, is left as an exercise.

Lemma 3.11. *Let $(X\ Y)$ be unitary. If $\mathcal{R}(X)$ and $\hat{\mathcal{X}}$ are acute, there is a matrix P such that $\mathcal{R}(X + YP) = \hat{\mathcal{X}}$. The singular values of $P(I + P^H P)^{-\frac{1}{2}}$ are the sines of the canonical angles between $\mathcal{R}(X)$ and $\hat{\mathcal{X}}$.*

The first residual bound is based on the second sin Θ theorem.

Theorem 3.12. *Suppose that*

> *there is a number $\delta > 0$ such that exactly*
> *$n - k$ of the eigenvalues of A lie outside the*   (3.8)
> *interval $[\mu_k - \delta, \mu_1 + \delta]$*

and
$$\rho \equiv \frac{\|R\|_2}{\delta} < 1.$$
Then there is an index j such that $\lambda_j, \ldots, \lambda_{j+k-1} \in (\mu_k - \delta, \mu_1 + \delta)$ and
$$|\mu_i - \lambda_{j+i-1}| \leq \frac{1}{1 - \rho^2} \frac{\|R\|_2^2}{\delta}, \qquad i = 1, \ldots, k. \tag{3.9}$$

Proof. Let $(X\ Y)$ be unitary. Then
$$\begin{pmatrix} X^H \\ Y^H \end{pmatrix} A (X\ Y) = \begin{pmatrix} M & G^H \\ G & N \end{pmatrix},$$

where $\|G\|_2 = \|R\|_2$. By Theorem 3.6 and Lemma 3.11, there is a matrix P satisfying

$$\|P(I+P^{\mathrm{H}}P)^{-\frac{1}{2}}\|_2 \leq \rho. \tag{3.10}$$

such that the columns of

$$\hat{X} = (X+YP)(I+P^{\mathrm{H}}P)^{-\frac{1}{2}}$$

span an invariant subspace of A. From (3.10) it follows that

$$\frac{\|P\|_2}{\sqrt{1+\|P\|_2^2}} \leq \rho,$$

and since $\rho < 1$,

$$\|P\|_2 \leq \frac{\rho}{\sqrt{1-\rho^2}}. \tag{3.11}$$

Let $\hat{Y} = (Y-XP^{\mathrm{H}})(I+PP^{\mathrm{H}})^{-\frac{1}{2}}$. Then $(\hat{X}\ \hat{Y})$ is unitary. Since the columns of $\hat{X}$ span an invariant subspace of A, we have $\hat{Y}^{\mathrm{H}}A\hat{X} = 0$. Hence

$$\begin{pmatrix} \hat{X}^{\mathrm{H}} \\ \hat{Y}^{\mathrm{H}} \end{pmatrix} A(\hat{X}\ \hat{Y}) = \begin{pmatrix} \hat{M} & 0 \\ 0 & \hat{N} \end{pmatrix}.$$

As in the proof of Theorem 2.1, it can be shown that

$$\hat{M} = (I+P^{\mathrm{H}}P)^{\frac{1}{2}}(M+G^{\mathrm{H}}P)(I+P^{\mathrm{H}}P)^{-\frac{1}{2}}.$$

The eigenvalues of $\hat{M}$ are eigenvalues of A. Since $\rho < 1$ it follows from the residual bound of Chapter IV (Theorem IV.4.14) that they lie in the interval $(\mu_k - \delta, \mu_1 + \delta)$, and hence are $\lambda_j, \ldots, \lambda_{j+k-1}$ for some index j. By the similarity bound of Theorem IV.5.3,

$$|\mu_i - \lambda_{j+i-1}| \leq \|(I+P^{\mathrm{H}}P)^{\frac{1}{2}}\|_2 \|(I+P^{\mathrm{H}}P)^{-\frac{1}{2}}\|_2 \|G\|_2 \|P\|_2, \quad i=1,\ldots,k.$$

The theorem now follows on noting that $\|(I+P^{\mathrm{H}}P)^{-\frac{1}{2}}\|_2 \leq 1$ and inserting the bound (3.11) for $\|P\|_2$. ∎

In the bound (3.9) the factor $(1-\rho^2)^{-1}$ is insignificant when ρ is even a little less than one. The factor $\|R\|_2^2/\delta$ is quadratic in $\|R\|_2$; however, as δ decreases the bound deteriorioates.

3. Hermitian Matrices

For $\rho < \frac{1}{2}$ the bound is less than the bound $\|R\|_2$ provided by Theorem IV.4.14. Moreover, the bound is asymptotically sharp, as the matrix

$$A = \begin{pmatrix} 0 & \epsilon \\ \epsilon & 1 \end{pmatrix}$$

from Example IV.4.17 shows.

The requirement (3.8) unfortunately does not allow the eigenvalues of M to be scattered through the spectrum of A. However if we pass to the Frobenius norm, then we can obtain a Hoffman–Wielandt type residual bound. Specifically, if there is a set $\mathcal{L}_2$ consisting of $n - k$ eigenvalues of A (counting multiplicities such that)

$$\delta = \min |\mathcal{L}(M) - \mathcal{L}_2| > 0, \qquad (3.12)$$

then Theorem 3.4 shows that there is a matrix P satisfying

$$\|P(I + P^{\mathrm{H}}P)^{\frac{1}{2}}\|_2 \leq \|P(I + P^{\mathrm{H}}P)^{\frac{1}{2}}\|_{\mathrm{F}} \leq \frac{\|R\|_{\mathrm{F}}}{\delta}$$

such that the columns of

$$\hat{X} = (X + YP)(I + P^{\mathrm{H}}P)^{-\frac{1}{2}}$$

span an invariant subspace of A. By the similarity bound of Theorem IV.5.5, the eigenvalues $\lambda_{j_1}, \ldots, \lambda_{j_k}$ of $\hat{M}$ may be ordered so that

$$\sqrt{\sum_{i=1}^{k}(\mu_i - \lambda_{j_i})^2} \leq \|(I + P^{\mathrm{H}}P)^{\frac{1}{2}}\|_2 \|(I + P^{\mathrm{H}}P)^{-\frac{1}{2}}\|_2 \|S\|_{\mathrm{F}} \|P\|_2.$$

Hence we have the following theorem.

Theorem 3.13. *With the above definitions, assume that A and M satisfy (3.12). If*

$$\rho_{\mathrm{F}} \equiv \frac{\|R\|_{\mathrm{F}}}{\delta} < 1,$$

then there are eigenvalues $\lambda_{j_1}, \ldots, \lambda_{j_k}$ of A such that

$$\sqrt{\sum_{i=1}^{k}(\mu_i - \lambda_{j_i})^2} \leq \frac{1}{1 - \rho_{\mathrm{F}}^2} \frac{\|R\|_{\mathrm{F}}^2}{\delta}.$$

V. INVARIANT SUBSPACES

Notes and References

The observation that for Hermitian matrices the function sepin the Frobenius norm reduces to the minimum difference between the eigenvalues was made by Stewart [200, 1971]. The knowledgeable reader will have noted that we have surreptitiously introduced the Kronecker or tensor product in the proof of Theorem 3.1.

The sin Θ and tan Θ theoremsare due to Davis and Kahan [53, 1970]. Earlier Davis [50, 51, 1963, 1965] established bounds on $\sin 2\Theta$ and $\tan 2\Theta$, which are also presented in this ground-breaking paper, along with much, much more, including Theorem 3.4. It should be noted that Davis and Kahan work with bounded operators in a Hilbert space, and some of their results extend to unbounded operators.

The residual bounds for eigenvalues are due to Stewart [218, 1989].

Exercises

THE FOLLOWING TWO EXERCISES SPECIALIZE THE RESULTS OF THIS SECTION TO EIGENVALUES AND EIGENVECTORS. NOTE THAT THE EIGENVALUE BOUNDS ARE A LITTLE SHARPER THAN THE ONES ONE WOULD GET FROM THE THEOREMS IN THE TEXT.

1. Let (z, μ) be an approximate eigenpair of A with $\|z\|_2 = 1$. Let $r = Ax - \mu x$. Suppose that there is a set $\mathcal{L}$ of $n - 1$ eigenvalues of A such that

$$\delta = \min |\mathcal{L} - \{\lambda\}| > 0.$$

Show that there is an eigenpair (x, λ) of A satisfying

$$\sin \angle(x, z) \leq \frac{\|r\|_2}{\delta}$$

and

$$|\mu - \lambda| \leq \min\left\{\|r\|_2, \frac{\|r\|_2^2}{\delta}\right\}.$$

2. Let (x, λ) be a simple eigenpair of A with $\|x\|_2 = 1$. Let $\tilde{A} = A + E$ and set $\epsilon = \|E\|_2$. Let δ be the distance from λ to $\mathcal{L}(A) \setminus \{\lambda\}$. Show that if $\epsilon < \delta$ and

$$\frac{\epsilon}{\delta - \epsilon} < \frac{1}{2},$$

then there is an eigenpair $(\tilde{x}, \tilde{\lambda})$ of $\tilde{A}$ satisfying

$$\tan \angle(x, \tilde{x}) < 2\frac{\epsilon}{\delta - \epsilon}$$

and

$$|\tilde{\lambda} - x^{\mathrm{H}}\tilde{A}x| < 2\frac{\epsilon^2}{\delta - \epsilon}.$$

———◇———

3. (Davis and Kahan [53]). By considering the matrices

$$A = \begin{pmatrix} 0 & 0 & 0 \\ 0 & 0 & \frac{1}{\sqrt{2}} \\ 0 & \frac{1}{\sqrt{2}} & 0 \end{pmatrix} \quad \text{and} \quad \tilde{A} = \begin{pmatrix} 0 & 0 & \frac{1}{\sqrt{2}} \\ 0 & 0 & \frac{1}{\sqrt{2}} \\ \frac{1}{\sqrt{2}} & \frac{1}{\sqrt{2}} & 0 \end{pmatrix},$$

show that the hypotheses on the situation of the eigenvalues in the $\tan\Theta$ theorem are necessary.

4. The Singular Value Decomposition

The perturbation theory for singular values and vectorscomplicated by two troublesome facts. The first is that we must deal with both right and left singular vectors. The second is that the singular values of a matrix are not differentiable functions of the matrix. For example, if $A = \alpha$ is a 1×1 complex matrix, then its singular value is $|\alpha|$, which is not an analytic function of α. In particular, if we seek a perturbation expansion for $\tilde{\alpha} = \alpha + \epsilon$, we cannot simply write

$$\tilde{\alpha} = \sqrt{(\alpha + \epsilon)^{\mathrm{H}}(\alpha + \epsilon)} \cong \alpha + \frac{1}{2}(\bar{\epsilon}\alpha + \bar{\alpha}\epsilon),$$

since the right-hand side of this expression may not be nonnegative. For the larger singular values this example presents no problem; but it shows that we must take care in dealing with singular values near zero.

In the next subsection we will consider a generalization of the $\sin\Theta$ theorem for subspaces spanned by singular vectors, which are sometimes called SINGULAR SUBSPACES Here we circumvent the problem by working with the Jordan–Wielandt matrix to get simultaneous bounds for spaces spanned by right and left singular vectors. In the following subsection, we derive a perturbation expansion based on the cross-product matrix.

Throughout this section A will be an $m \times n$ matrix with $m \geq n$.

4.1. Two $\sin \Theta$ Theorems

In this subsection we will establish $\sin \Theta$ theorems, due to Wedin, for spaces spanned by the singular vectors of A. To fix the notation, let

$$(U_1\ U_2\ U_3)^{\mathrm{H}} A (V_1\ V_2) = \begin{pmatrix} \Sigma_1 & 0 \\ 0 & \Sigma_2 \\ 0 & 0 \end{pmatrix}$$

be a partitioned singular value decomposition of A (here we do not place any constraints on the order in which the singular values appear), and let

$$(\tilde{U}_1\ \tilde{U}_2\ \tilde{U}_3)^{\mathrm{H}} \tilde{A} (\tilde{V}_1\ \tilde{V}_2) = \begin{pmatrix} \tilde{\Sigma}_1 & 0 \\ 0 & \tilde{\Sigma}_2 \\ 0 & 0 \end{pmatrix}$$

be a conformal partition of $\tilde{A} = A + E$. Let Φ be the matrix of canonical angles between $\mathcal{R}(U_1)$ and $\mathcal{R}(\tilde{U}_1)$, and let Θ be the matrix of canonical angles between $\mathcal{R}(V_1)$ and $\mathcal{R}(\tilde{V}_1)$ Finally, let

$$R = A\tilde{V}_1 - \tilde{U}_1 \tilde{\Sigma}_1 \quad \text{and} \quad S = A^{\mathrm{H}} \tilde{U}_1 - \tilde{V}_1 \tilde{\Sigma}_1. \tag{4.1}$$

The following theorem bounds the angles Φ and Θ in terms of the residuals R and S.

Theorem 4.1 (Wedin). *Suppose that there is a number $\delta > 0$ such that*

$$\min |\sigma(\tilde{\Sigma}_1) - \sigma(\Sigma_2)| \geq \delta \quad \text{and} \quad \min \sigma(\tilde{\Sigma}_1) \geq \delta. \tag{4.2}$$

Then

$$\sqrt{\|\sin \Phi\|_{\mathrm{F}}^2 + \|\sin \Theta\|_{\mathrm{F}}^2} \leq \frac{\sqrt{\|R\|_{\mathrm{F}}^2 + \|S\|_{\mathrm{F}}^2}}{\delta}.$$

Remark 4.2. The matrices U_i, V_i, $\tilde{U}_i$, and $\tilde{V}_i$ may be replaced by any matrices with orthonormal columns spanning the appropriate subspaces.

4. THE SINGULAR VALUE DECOMPOSITION

Proof. Consider the Jordan–Wielandt matrix

$$C = \begin{pmatrix} 0 & A \\ A^{\mathrm{H}} & 0 \end{pmatrix},$$

whose eigenvalues are $\pm\sigma_1, \ldots, \pm\sigma_n$ with $m - n$ additional zero eigenvalues. Let $\tilde{C}$ be the Jordan–Wielandt matrix for $\tilde{A}$.
It is easy to show that if

$$X = \frac{1}{\sqrt{2}} \begin{pmatrix} U_1 & U_1 \\ V_1 & -V_1 \end{pmatrix} \equiv (X_1 \ X_2)$$

then $\mathcal{R}(X)$ is an invariant subspace of C. The representation of C on this subspace is $\operatorname{diag}(\Sigma_1, -\Sigma_2)$. Similarly, if

$$\tilde{X} = \frac{1}{\sqrt{2}} \begin{pmatrix} \tilde{U}_1 & \tilde{U}_1 \\ \tilde{V}_1 & -\tilde{V}_1 \end{pmatrix} \equiv (\tilde{X}_1 \ \tilde{X}_2),$$

then $\mathcal{R}(\tilde{X})$ is an invariant subspace of $\tilde{C}$. The representation of $\tilde{C}$ on this subspace is $\operatorname{diag}(\tilde{\Sigma}_1, -\tilde{\Sigma}_2)$. Hence by Theorem 3.4, if we set

$$T = C\tilde{X} - \tilde{X}(\tilde{X}^{\mathrm{H}}\tilde{C}\tilde{X}),$$

then

$$\|\sin\Theta[\mathcal{R}(X), \mathcal{R}(\tilde{X})]\|_{\mathrm{F}} \leq \frac{\|T\|_{\mathrm{F}}}{\delta}. \qquad (4.3)$$

To arrive at the conclusion of our theorem, we must compute the left- and right-hand sides of (4.3). For the left-hand side, note that

$$P_X = XX^{\mathrm{H}} = \operatorname{diag}(UU^{\mathrm{H}}, VV^{\mathrm{H}}) = \operatorname{diag}(P_U, P_V),$$

and similarly for $P_{\tilde{X}}$. Hence by Theorem I.5.5

$$\|\sin\Theta[\mathcal{R}(X), \mathcal{R}(\tilde{X})]\|_{\mathrm{F}}^2 = \|P_X^\perp P_{\tilde{X}}\|_{\mathrm{F}}^2$$
$$= \|\operatorname{diag}(P_U^\perp P_{\tilde{U}}), \operatorname{diag}(P_V^\perp P_{\tilde{V}})\|_{\mathrm{F}}^2 \qquad (4.4)$$
$$= \|\sin\Phi\|_{\mathrm{F}}^2 + \|\sin\Theta\|_{\mathrm{F}}^2.$$

For the right-hand side, a straightforward computation shows that

$$T = \frac{1}{\sqrt{2}} \begin{pmatrix} R & -S \\ S & R \end{pmatrix}.$$

Hence
$$\|T\|_{\mathrm{F}}^2 = \|R\|_{\mathrm{F}}^2 + \|S\|_{\mathrm{F}}^2. \tag{4.5}$$
The theorem follows on combining (4.3), (4.4), and (4.5). ∎

The appearance of the condition $\sigma(\Sigma_1) \geq \delta$ seems strange at first, but it is necessary — as the following example shows.

Example 4.3. Let
$$A = \begin{pmatrix} \epsilon & 0 \\ 0 & 1 \\ 0 & 0 \end{pmatrix} \quad \text{and} \quad \tilde{A} = \begin{pmatrix} \epsilon & 0 \\ 0 & 1 \\ \epsilon & 0 \end{pmatrix}.$$

The $\tilde{u}_1$ makes an angle of 45 degrees with u_1, even though the singular value ϵ of A is well separated from the other singular value.

This example also points to a fundamental defect in Theorem 4.1. Although the vector v_1 is insensitive to perturbations in A, its bound is governed by the ill-conditioning of u_1. However, the problem can be circumvented by using the theorem to bound the perturbation in $\mathcal{R}(V_2)$. Since $\mathcal{R}(V_1)$ and $\mathcal{R}(V_2)$ are complementary subspaces, the same bound will serve for $\mathcal{R}(V_1)$.

By imposing further restrictions on the singular values, we may establish a bound on the 2-norm (actually on any unitarily invariant norm). The proof of the following theorem is a variant of the proof of Theorem 4.1 and is left as an exercise.

Theorem 4.4 (Wedin). *Suppose that there are numbers* $\alpha, \delta > 0$ *such that*
$$\min \sigma(\tilde{\Sigma}_1) \geq \alpha + \delta \quad \text{and} \quad \max \sigma(\Sigma_2) \leq \alpha. \tag{4.6}$$
Then
$$\max\{\|\sin \Phi\|_2, \|\sin \Theta\|_2\} \leq \frac{\max\{\|R\|_2, \|S\|_2\}}{\delta}.$$

The condition (4.6) restricts the bounds to subspaces associated with a group of the largest singular values. However, by the trick described in connection with Example 4.3, we can use the theorem indirectly to get bounds on the perturbation in $\mathcal{R}(V_2)$.

4. THE SINGULAR VALUE DECOMPOSITION

4.2. A Perturbation Expansion

In the last subsection we saw that the perturbation theory for singular vectors associated with small singular values presented some difficulties. Actually, small singular values themselves exhibit curious behavior — they tend to get larger (after all, they have nowhere to go but up). Since this fact has important consequences for applications to least squares problems and linear regression, we will develop a perturbation expansion that shows what is going on. The key is to smooth out the behavior of the small singular value by working with its square, or equivalently with the cross-product matrix $A^H A$.

We begin with a lemma that follows directly from Theorem 2.7 applied to Hermitian matrices.

Lemma 4.5. *In the Hermitian matrix*

$$\begin{pmatrix} \alpha & h^H \\ h & C \end{pmatrix}$$

let α and C, be constant and let h depend on a parameter ϵ in such a way that

$$\|h\|_2 = O(\epsilon)$$

as $\epsilon \to 0$. Let the quantities $\tilde{\alpha}$, $\tilde{C}$ and $\tilde{h}$ satisfy

$$|\tilde{\alpha} - \alpha|, \|\tilde{C} - C\|_2 = O(\epsilon)$$

and

$$\|\tilde{h} - h\|_2 = O(\epsilon^2).$$

If $\alpha I - C$ is nonsingular, then for all sufficiently small ϵ the matrix

$$\begin{pmatrix} \tilde{\alpha} & \tilde{h}^H \\ \tilde{h} & \tilde{C} \end{pmatrix} \quad (4.7)$$

has an eigenvector

$$\begin{pmatrix} 1 \\ (\alpha I - C)^{-1} h \end{pmatrix} + O(\epsilon^2) \quad (4.8)$$

corresponding to the eigenvector

$$\tilde{\alpha} + h^H (\alpha I - C)^{-1} h + O(\epsilon^3). \quad (4.9)$$

To apply the lemma, let A have the singular value decomposition

$$U^H A V = \begin{pmatrix} \Sigma \\ 0 \end{pmatrix}$$

(here, as above, we do not assume that the singular values appear on the diagonal of Σ in descending order). Partition

$$U = \begin{pmatrix} \overset{1}{u_1} & \overset{p-1}{U_2} & \overset{n-p}{U_3} \end{pmatrix}$$

and

$$V = \begin{pmatrix} \overset{1}{v_1} & \overset{p-1}{V_2} \end{pmatrix}.$$

Partition

$$U^H A V = \begin{pmatrix} \sigma_1 & 0 \\ 0 & \Sigma_2 \\ 0 & 0 \end{pmatrix}$$

conformally. Finally, given a perturbation $\tilde{A} = A + E$ of A, let

$$U^H E V = \begin{pmatrix} \gamma_{11} & g_{12}^H \\ g_{21} & G_{22} \\ g_{31} & G_{32} \end{pmatrix}$$

so that

$$U^H \tilde{A} V = \begin{pmatrix} \sigma_1 + \gamma_{11} & g_{12}^H \\ g_{21} & \Sigma_2 + G_{22} \\ g_{31} & G_{32} \end{pmatrix}. \qquad (4.10)$$

The following theorem contains the chief result of this subsection.

Theorem 4.6. *Let*
$$h = \sigma_1 g_{12} + \Sigma_2 g_{21}. \qquad (4.11)$$

If $\sigma_1 I - \Sigma_2$ is nonsingular (i.e., if σ_1 is a simple singular value of A), then as $E \to 0$ the matrix (4.10) has a right singular vector of the form

$$\begin{pmatrix} 1 \\ (\sigma_1^2 I - \Sigma_2^2)^{-1} h \end{pmatrix} + O(\|E\|_2^2) \qquad (4.12)$$

4. THE SINGULAR VALUE DECOMPOSITION

corresponding to a singular value $\tilde{\sigma}_1$ satisfying

$$\tilde{\sigma}_1^2 = (\sigma_1 + \gamma_{11})^2 + \|g_{21}\|_2^2 + \|g_{31}\|_2^2 + h^{\mathrm{H}}(\sigma_1^2 I - \Sigma_2^2)^{-1}h + O(\|E\|_2^3). \quad (4.13)$$

Proof. In Lemma 4.5 let $\alpha = \sigma_1^2$ and $A = \Sigma_2^2$, and let h be defined as in (4.11). Identify the elements of the matrix (4.7) with the elements of the partitioned cross product matrix $(U^{\mathrm{H}}AV)^{\mathrm{H}}(U^{\mathrm{H}}AV)$, so that

$$\tilde{\alpha} = (\sigma_1 + \gamma_{11})^2 + \|g_{21}\|_2^2 + \|g_{31}\|_2^2,$$
$$\tilde{h} = h + \gamma_{11}g_{12} + G_{22}^{\mathrm{H}}g_{21} + G_{32}^{\mathrm{H}}g_{31},$$
$$\tilde{A} = (\Sigma_2 + G_{22})^2 + g_{12}g_{12}^{\mathrm{H}} + G_{32}^{\mathrm{H}}G_{32}.$$

Then the conditions of the lemma are satisfied and (4.12) and (4.13) follow by making appropriate substitutions in (4.9) and (4.9). ∎

We have expanded the perturbed singular vector (4.12) in terms of the transformed matrix $U^{\mathrm{H}}\tilde{A}V$; in terms of the original matrix we get the expression

$$\tilde{v}_1 = v_1 + V_2(\sigma_1^2 I - \Sigma_2^2)^{-1}h + O(\|E\|_2^2).$$

The results can also be stated in terms of projections. Let P_1, P_2, and P_3 be the orthogonal projections onto the column spaces of u_1, U_2, and U_3, and let Q_1, Q_2, and Q_3 be the orthogonal projections onto the column spaces of v_1, V_2, and V_3. Then

$$\|U_i^{\mathrm{H}}EV_j\|_2 = \|P_iEQ_j\|,$$

so that an expression like $(\sigma_1 + \gamma_{11})^2 + \|g_{21}\|_2^2 + \|g_{31}\|_2^2$ can be written

$$\|P_1\tilde{A}Q_1\|_2^2 + \|P_2\tilde{A}Q_1\|_2^2 + \|P_3\tilde{A}Q_1\|_2^2 = \|\tilde{A}Q_1\|_2^2.$$

In particular, if σ_1 is large compared with E, then the second order terms in (4.13) are negligible compared to the first order terms, and we have
$$\tilde{\sigma}_1 = \sigma_1 + \gamma_{11} + O(\|E\|_2^2)$$
$$= \sigma_1 + u_1^{\mathrm{H}}Ev_1 + O(\|E\|_2^2)$$
$$= \|P_1\tilde{A}Q_1\|_2 + O(\|E\|_2^2).$$

Our expansion quantifies the observation, made at the beginning of this subsection, that small singular values tend to increase. For if $\sigma_1 = 0$, then $h = \Sigma_2 g_{21}$, and

$$h^{\mathrm{H}}(\sigma_1^2 I - \Sigma_2^2)^{-1}h = -\|g_{21}\|_2^2.$$

It follows that

$$\tilde{\sigma}_1^2 = \gamma_{11}^2 + \|g_{31}\|_2^2 + O(\|E\|_2^3)$$
$$= (u_1 E v_1)^2 + \|U_3^{\mathrm{H}} E v_1\|_2^2 + O(\|E\|_2^3)$$
$$= \|(P_1 + P_3)E Q_1\|_2^2 + O(\|E\|_2^3).$$

Notes and References

The Φ-Θ theorem for the 2-norm (Theorem 4.4) was proven by Wedin [259, 1972], who established the results for arbitrary unitarily invariant norms. The Φ-*Theta* theorem for the Frobenius norm (Theorem 4.1) is technically new, but the proof is a modification of a comment by Wedin on another way of proving the Φ-Θ theorem for the 2-norm.

Although we have stressed direct bounds in this section, the approach taken in Section 2 for invariant subspaces can be adapted to the singular value decomposition. Briefly, let $U = (U_1\ U_2)$ and $V = (V_1\ V_2)$ be unitary, and let

$$U^{\mathrm{H}} A V = \begin{pmatrix} S_1 & H^{\mathrm{H}} \\ G & S_2 \end{pmatrix}.$$

We seek

$$\hat{U} = (U_1\ U_2) \begin{pmatrix} I & -P^{\mathrm{H}} \\ P & I \end{pmatrix} \begin{pmatrix} (I + P^{\mathrm{H}} P)^{-\frac{1}{2}} & 0 \\ 0 & (I + P P^{\mathrm{H}})^{-\frac{1}{2}} \end{pmatrix}$$

and

$$\hat{V} = (V_1\ V_2) \begin{pmatrix} I & Q^{\mathrm{H}} \\ -Q & I \end{pmatrix} \begin{pmatrix} (I + Q^{\mathrm{H}} Q)^{-\frac{1}{2}} & 0 \\ 0 & (I + Q Q^{\mathrm{H}})^{-\frac{1}{2}} \end{pmatrix}$$

such that that $\hat{U}^{\mathrm{H}} A \hat{V}$ is block diagonal. This requirement leads to the equation

$$\mathbf{T}(Q, P) = (G, H) - \phi(Q, P),$$

4. THE SINGULAR VALUE DECOMPOSITION

where
$$\mathbf{T} = (Q, P) \mapsto (QS_1 - S_2 P, PS_1^H - S_2^H Q)$$
and
$$\phi = (Q, P) \mapsto (QGP, PHQ).$$

If we set $\|(Q,P)\|_F^2 = \|Q\|_F^2 + \|P\|_F^2$ and let $\|\mathbf{T}\|$ be the subordinate operator norm, then $\|\mathbf{T}^{-1}\|^{-1} = \min |\mathcal{S}(S_1) - \mathcal{S}(S_2)|$. Theorem 2.11 now applies to give conditions for the existence of P and Q and bounds on their norms. This development is due to Stewart [202, 1973].

The material on perturbation expansions in the second subsection is taken with small changes from a paper by Stewart [212, 1984]. In least squares problems with errors in the least squares matrix (errors in the variables, as they are known to the statistical community), the increase of small singular values manifests itself in a downward bias of the least squares solution (e.g., see [19, 213]). This has lead to the development of techniques to remove the bias [80, 92]. It should be noted that the solutions produced by these techniques differ from a least squares solution only in second order terms and higher [210].

Closely related to these perturbation expansions are characterization by Sun [230, 1988] of the behavior of a simple singular value when the elements of its matrix are analytic functions of several complex variables.

Exercises

1. Verify Remark 4.2.

2. Let $\sigma = \inf_2(A)$ be the smallest singular value of A with right singular vector v and similarly for $\tilde{\sigma}$ and $\tilde{v}$. Let δ be the distance between σ and the next largest singular value of A. Show that if $\|E\|_2 < \delta$ then

$$\sin \angle(v, \tilde{v}) \leq \frac{\|E\|_2}{\delta - \|E\|_2}.$$

[Hint: Work with the complementary spaces and regard A as a perturbation of $\tilde{A}$.]

THE FOLLOWING EXERCISES PRESENT WEDIN'S PROOF [259] OF THEOREM 4.4, WHICH IS VALID FOR ANY UNITARILY INVARIANT NORM $\|\cdot\|$. HERE, IN THE NOTATION OF THE FIRST SUBSECTION, WE SET $A_i = U_i \Sigma_i V_i^H$ ($i = 1, 2$) AND $\tilde{A}_i = \tilde{U}_i \tilde{\Sigma}_i \tilde{V}_i^H$ ($i = 1, 2$).

3. Show that
$$P_{A_1}^\perp P_{\tilde{A}_1} = (P_{A_1}^\perp E P_{\tilde{A}_1^H} + A_2 P_{A_1^H}^\perp P_{\tilde{A}_1^H})\tilde{A}_1^\dagger$$
and
$$P_{\tilde{A}_1^H} P_{A_1^H}^\perp = \tilde{A}_1^H(P_{\tilde{A}_1} E P_{A_1^H}^\perp + P_{\tilde{A}_1} P_{A_1}^\perp A_1).$$

4. Let R and S be defined by (4.1). Show that
$$\mu \equiv \max\{\|P_{A_1}^\perp E P_{\tilde{A}_1^H}\|, P_{\tilde{A}_1} E P_{A_1^H}^\perp\} \leq \max\{\|R\|, \|S\|\}.$$

5. Let μ be defined as in the last exercise. Show that under the hypotheses of Theorem 4.4
$$\|\sin\Phi\| \leq \frac{\mu + \alpha\|\sin\Theta\|}{\alpha + \delta}$$
and
$$\|\sin\Theta\| \leq \frac{\mu + \alpha\|\sin\Phi\|}{\alpha + \delta}.$$
Hence
$$\max\{\|\sin\Phi\|, \|\sin\Theta\|\} \leq \frac{\max\{\|R\|, \|S\|\}}{\delta}.$$

——◇——

6. Fill in the details of the sketch given in the notes and references for an approximation theorem for the singular value decomposition. Be sure to exhibit matrices whose singular values are those of A. Derive a perturbation theorem from the approximation theorem.

7. (A norm version of Theorem 4.6 [208]). Let A have singular values $\sigma_1 \geq \cdots \geq \sigma_n$ and $\tilde{A}$ have singular values $\tilde{\sigma}_1 \geq \cdots \geq \tilde{\sigma}_n$. Show that
$$\tilde{\sigma}_i^2 = (\sigma_i + \gamma_i)^2 + \eta_i^2, \qquad i = 1, \ldots, p,$$
where
$$|\gamma_i| \leq \|P_A E\|_2$$
and
$$\inf_2(P_A^\perp E) \leq \eta_i \leq \|P_A^\perp E\|_2$$

8. (Scaled null vectors [210]). Let $A \in \mathbf{C}^{m \times n}$ have rank n, and let $b = Ax$. Let $\tilde{A} = A + E$ and $\tilde{b} = B + e$. For T nonsingular, let σ_T be the smallest singular value of $(\tilde{A}T\ b)$ and
$$\begin{pmatrix} \hat{x}_T \\ -1 \end{pmatrix}$$

4. THE SINGULAR VALUE DECOMPOSITION

be the corresponding singular vector. Show that
$$T^{-1}\hat{x}_T = \tilde{A}^\dagger \tilde{b} + O(\|(E\ e)\|^2).$$

Chapter VI
Generalized Eigenvalue Problems

A MATRIX PENCIL is a family of matrices $A - \lambda B$, parameterized by a complex number λ. When A is square and $B = I$, the zeros of the function $\det(A - \lambda B)$ are the eigenvalues of A. Consequently, the problem of finding the nontrivial solutions of the equation

$$Ax = \lambda Bx$$

is called the GENERALIZED EIGENVALUE PROBLEM.

Although the generalized eigenvalue problem looks like a simple generalization of the usual eigenvalue problem, it exhibits some important differences. In the first place, it is possible for $\det(A - \lambda B)$ to be identically zero, independent of λ. For such SINGULAR PENCILS every scalar can be regarded as an eigenvalue.

Second, it is possible for B to be singular, in which case the problem has infinite eigenvalues. To see this, write the generalized eigenvalue problem in the reciprocal form

$$Bx = \lambda^{-1} Ax.$$

If B is singular with a null vector x, then $Bx = 0Ax$, so that x is an eigenvector of the reciprocal problem corresponding to eigenvalue

$\lambda^{-1} = 0$; i.e., $\lambda = \infty$. It might be thought that infinite eigenvalues are special, unhappy cases to be ignored in our perturbation theory, but that is a misconception. If we write the eigenvalue problem in the cross-product form

$$\beta Ax = \alpha Bx, \qquad (1)$$

then we see that infinite eigenvalues correspond to nonzero pairs (α, β) for which $\beta = 0$, a case that is not essentially different from the case $\alpha = 0$ (i.e., $\lambda = 0$). In this chapter we will deal with the problem of infinite eigenvalues by treating generalized eigenvalue problems in their cross-product forms.

Finally, there are difficult and unresolved problems connected with the scaling of generalized eigenvalue problems. In the ordinary eigenvalue problem, the fact that $B = I$ provides a natural scale: namely the size of A. For the generalized eigenvalue problem, we may scale both A and B, and the perturbation bounds we derive will be essentially different for different scalings. This is an open research problem, which will keep returning to haunt us.

In spite of the differences between the generalized and the ordinary eigenvalue problems, they have striking similarities, similarities we will stress as much as the differences. In fact, this chapter is a copy *en miniature* of the part of the book that concerns eigenvalue problems. The first section is devoted to the background—an algebraic introduction to the subject. We then turn to perturbation bounds for the eigenvalues of regular matrix pencils—the natural generalization of the ordinary eigenvalue problem—and then for their eigenspaces— the natural generalization of their eigenvectors. Finally, we consider both the eigenvalues and eigenspaces of definite matrix pencils, which generalize the Hermitian eigenvalue problem.

Although rectangular matrix pencils—matrix pencils $A - \lambda B$ with A and B rectangular—occur and have important applications, we shall consider only square matrix pencils, for which the perturbation theory is less immature.

Unless otherwise stated, A and B will be square matrices of order n throughout this chapter.

1. Background

1.1. Matrix Pairs

We have seen in the introduction to this chapter that the presence of infinite eigenvalues results from the asymmetrical treatment of A and B in the definition of a matrix pencil and its generalized eigenvalue problem. The solution to the problem is to recast it in the form (1), in which A and B play equivalent roles.

However, there is a technical problem here. If the pair (α, β) satisfies (1) then so does $\tau(\alpha, \beta)$ for any scalar τ. Consequently, if we are to regard (α, β) as a generalized eigenvalue, we must so regard its nonzero scalar multiples. This suggests that it is the subspace spanned by the vector $(\alpha, \beta)^{\mathrm{T}}$ that should be regarded as the generalized eigenvalue. To distinguish between the subspace and the pair, we make the following definition.

Definition 1.1. *Let* $(\alpha, \beta) \neq (0, 0)$. *Then*

$$\langle \alpha, \beta \rangle \stackrel{\text{def}}{=} \{\tau(\alpha, \beta)^{\mathrm{T}} : \tau \in \mathbf{C}\}.$$

In order to preserve the connection of the generalized eigenvalue problem with the ordinary eigenvalue problem, we will occasionally abuse the notation introduced in the above definition and write $\langle \lambda \rangle$ for $\langle \lambda, 1 \rangle$. For infinite λ, we define $\langle \infty \rangle = \langle 1, 0 \rangle$.

We are now in a position to define the generalized eigenvalue problem. Since the definition of matrix pencil treats A and B differently, we will drop the term and refer simply to pairs of matrices.

Definition 1.2. *A* MATRIX PAIR (A, B) *is* SINGULAR *if for all* (α, β)

$$\det(\beta A - \alpha B) = 0.$$

Otherwise the pair (A, B) *is said to be* REGULAR. *If* (A, B) *is regular and*

$$\beta A x = \alpha B x \qquad (1.1)$$

for $(\alpha, \beta) \neq (0, 0)$ *and* $x \neq 0$, *then* $\langle \alpha, \beta \rangle$ *is an* EIGENVALUE *of* (A, B) *with* (RIGHT) EIGENVECTOR x. *The corresponding solution* $y \neq 0$ *of the equation*

$$\beta y^{\mathrm{H}} A = \alpha y^{\mathrm{H}} B$$

is called a LEFT EIGENVECTOR.

Some examples may make these definitions clearer.

Example 1.3. Suppose that the null spaces of A and B intersect, and let $x \neq 0$ belong to the intersection. Then for any (α, β), we have $(\beta A - \alpha B)x = 0$, so that the pair (A, B) is singular.

Example 1.4. Let B be nonsingular. Then with $(\alpha, \beta) = (1, 0)$, we have
$$\det(\beta A - \alpha B) = -\det(B) \neq 0.$$
Consequently the pair (A, B) is regular.

In fact the eigenvalue problem for the pair in this example is equivalent to an ordinary eigenvalue problem. To see this, note that if $\langle \alpha, \beta \rangle$ is an eigenvalue of (A, B), then $\beta \neq 0$. It follows that (1.1) can be rewritten in the form $B^{-1}Ax = \lambda x$, where $\lambda = \alpha/\beta$. Conversely if $\lambda \in \mathcal{L}(B^{-1}A)$, then $\langle \lambda \rangle$ is an eigenvalue of (A, B). This observation — that the generalized eigenvalue problem with nonsingular B can be converted to an ordinary eigenvalue problem — is the basis of many numerical methods, which, however, can fail in the presence of rounding error when B is ill conditioned.

Example 1.5. The pair
$$A = \begin{pmatrix} 1 & 0 \\ 0 & 0 \end{pmatrix}, \quad B = \begin{pmatrix} 0 & 0 \\ 0 & 1 \end{pmatrix}$$
is obviously regular. Its eigenvalues are $\langle 1 \rangle$ and $\langle \infty \rangle$, and the corresponding eigenvectors are $\mathbf{1}_1$ and $\mathbf{1}_2$. We shall see that in spite of the infinite eigenvalue, the pair behaves well under perturbations, provided we make the proper definition of "well behaved."

When B is nonsingular, the eigenvalues of the pair (A, B) satisfy the characteristic equation
$$\det(A - \lambda B) = 0.$$

When B is singular, the characteristic equation will have degree less than n. For example, the pair in Example 1.5 has the characteristic polynomial $\det(A - \lambda B) \equiv \lambda$. The missing eigenvalue is the infinite

1. BACKGROUND

one. By transforming the problem we can make the infinite eigenvalues finite and restore the lost degrees in the characteristic equation. The proof of the following theorem is purely computational and is left as a exercise.

Theorem 1.6. *Let W be a 2×2 nonsingular matrix. Given the pair (A, B), set*

$$(C\ D) = (A\ B) \begin{pmatrix} \omega_{11} I & \omega_{12} I \\ \omega_{21} I & \omega_{22} I \end{pmatrix} \equiv (A\ B)(W \otimes I).$$

Given the pair $(\alpha, \beta) \neq (0, 0)$, define (γ, δ) by

$$\begin{pmatrix} \delta \\ -\gamma \end{pmatrix} = W^{-1} \begin{pmatrix} \beta \\ -\alpha \end{pmatrix}.$$

Then $\langle \alpha, \beta \rangle$ is an eigenvalue of (A, B) if and only if $\langle \gamma, \delta \rangle$ is an eigenvalue of (C, D).

If (A, B) is a regular pair, there are constants σ and τ such that $\tau A - \sigma B$ is nonsingular. If we set

$$W = \begin{pmatrix} \sigma & \tau \\ \tau & -\sigma \end{pmatrix},$$

then W is nonsingular. If C and D are defined as in Theorem 1.6, then the eigenvalues of (A, B) are in one-to-one correspondence with those of (C, D). But D is nonsingular, and hence by Example 1.4 the eigenvalues of the pair (C, D) are the eigenvalues of $D^{-1}C$. Thus we have established that

a regular matrix pencil of order n has n eigenvalues.

As with the ordinary eigenvalue problem, we will denote the set of eigenvalues of the pair (A, B) by $\mathcal{L}[(A, B)]$.

1.2. Triangular and Weierstrass Forms

As we saw in Chapter I, an important theme of matrix algebra is the reduction of matrices to simpler forms by means of appropriate transformations. The key word here is "appropriate." For the computation of projections, the appropriate transformation is premultiplication by a unitary matrix. For the eigenvalue problem it is similarity transformations. For the generalized eigenvalue problem it is equivalence transformations.

Definition 1.7. *If X and Y are nonsingular, then the pair (A, B) and $(Y^H AX, Y^H BX)$ are* EQUIVALENT.

Equivalence, like similarity, preserves eigenvalues while transforming eigenvectors in a simple manner. The proof of the following theorem is left as an exercise.

Theorem 1.8. *Let $\langle \alpha, \beta \rangle$ be an eigenvalue of the pair (A, B) with eigenvector x. Then $\langle \alpha, \beta \rangle$ is an eigenvalue of the equivalent pair $(Y^H AX, Y^H BX)$ with eigenvector $X^{-1}x$.*

The first application of this observation is a reduction to the equivalent of the Schur form.

Theorem 1.9. *Let (A, B) be a regular pair. Then there are unitary matrices U and V such that the components of the equivalent pair $(S, T) = (V^H AU, V^H BU)$ are triangular. The quantities $\langle \sigma_{ii}, \tau_{ii} \rangle$ ($i = 1, \ldots, n$) are the eigenvalues of (A, B) and may be made to appear in any order on the diagonals of S and T.*

Proof. Let $\langle \sigma, \tau \rangle$ be the first eigenvalue in some prespecified order of the eigenvalues of (A, B), and let $\tau Ax = \sigma Bx$ ($x \neq 0$). Since (A, B) is regular, not both Ax and Bx can be zero — say $Ax \neq 0$. Let $U = (u\; U_*)$ be a unitary matrix with u proportional to x, and let $V = (v\; V_*)$ be a unitary matrix with v proportional to Ax. Then

$$V^H AU = \begin{pmatrix} v^H Au & v^H AU_* \\ V_*^H Au & V_*^H AU_* \end{pmatrix} = \begin{pmatrix} \sigma_{11} & s_*^H \\ 0 & A_* \end{pmatrix}$$

1. BACKGROUND 277

is block triangular. Since $\tau Au = \sigma Bu$, we must have $V_*^H Bu = 0$. Hence

$$V^H BU = \begin{pmatrix} \tau_{11} & t_*^H \\ 0 & B_* \end{pmatrix}$$

is also block triangular.

This completes one step of the reduction. The reduction continues inductively *à la* Schur (cf. Theorem I.3.3). ∎

We note in passing that singular pairs can also be reduced to triangular form by unitary equivalences. The proof is a minor variant of the above proof.

The computational consequences of this theorem are the same as for the Schur theorem: it provides a target for iterative generalized eigenvalue algorithms to aim for. However, it does not have the broad theoretical implications of Schur's theorem. This is because the transformations involved do not preserve Hermitian matrices. Consequently, we cannot read off the theory of Hermitian pairs from the theorem; instead we must develop it directly, as we will do in the next subsection.

However, the triangular reduction has one important consequence for simple eigenvalues.

Corollary 1.10. *Let $\langle \alpha, \beta \rangle$ be a simple eigenvalue of the regular pair (A, B) with right eigenvector x and left eigenvector y. Then*

$$\langle \alpha, \beta \rangle = \langle y^H Ax, y^H Bx \rangle.$$

Proof. It is sufficient to consider (A, B) in the triangular form

$$\left[\begin{pmatrix} \alpha & a^H \\ 0 & A_* \end{pmatrix}, \begin{pmatrix} \beta & b^H \\ 0 & B_* \end{pmatrix} \right].$$

In this case x is a multiple of 1_1. Moreover, the first component of y is nonzero; otherwise, y would be a left eigenvector of $\langle A_*, B_* \rangle$, contradicting the simplicity of $\langle \alpha, \beta \rangle$. Hence $y^H x \neq 0$ and

$$\langle y^H Ax, y^H Bx \rangle = \langle \alpha y^H x, \beta y^H x \rangle = \langle \alpha, \beta \rangle. \quad \blacksquare$$

We now turn to the further reduction of the Schur form to block diagonal form. Let (A, B) be a pair in triangular form and partition

$$A = \begin{pmatrix} A_{11} & A_{12} \\ 0 & A_{22} \end{pmatrix}, \quad B = \begin{pmatrix} B_{11} & B_{12} \\ 0 & B_{22} \end{pmatrix}.$$

We wish to find matrices P and Q such that

$$\begin{pmatrix} I & Q \\ 0 & I \end{pmatrix} \begin{pmatrix} A_{11} & A_{12} \\ 0 & A_{22} \end{pmatrix} \begin{pmatrix} I & P \\ 0 & I \end{pmatrix} = \begin{pmatrix} A_{11} & 0 \\ 0 & A_{22} \end{pmatrix}$$

and

$$\begin{pmatrix} I & Q \\ 0 & I \end{pmatrix} \begin{pmatrix} B_{11} & B_{12} \\ 0 & B_{22} \end{pmatrix} \begin{pmatrix} I & P \\ 0 & I \end{pmatrix} = \begin{pmatrix} B_{11} & 0 \\ 0 & B_{22} \end{pmatrix}.$$

With a little manipulation, this requirement yields the pair of equations

$$A_{11}P + QA_{22} = -A_{12},$$
$$B_{11}P + QB_{22} = -B_{12}.$$

which may be called the GENERALIZED SYLVESTER EQUATIONS If we set

$$\mathbf{T} = (P, Q) \mapsto (A_{11}P + QA_{22}, B_{11}P + QB_{22}), \qquad (1.2)$$

then our problem becomes one of determining when the linear operator $\mathbf{T}$ is nonsingular. It turns out that a separation condition, analogous to the condition of Theorem V.1.3 for the ordinary eigenvalue problem, is necessary and sufficient for $\mathbf{T}$ to be nonsingular.

Theorem 1.11. *Let (A_{11}, B_{11}) and (A_{22}, B_{22}) be regular pairs, and let $\mathbf{T}$ be defined by (1.2). Then $\mathbf{T}$ is nonsingular if and only if*

$$\mathcal{L}[(A_{11}, B_{11})] \cap \mathcal{L}[(A_{22}, B_{22})] = \emptyset. \qquad (1.3)$$

Proof. Suppose that $\mathcal{L}[(A_{11}, B_{11})] \cap \mathcal{L}[(A_{22}, B_{22})] = \emptyset$. We will show that for any (R, S), the equation $\mathbf{T}(P, Q) = (R, S)$ has a solution, which implies that $\mathbf{T}$ is nonsingular.

We may assume without loss of generality that A_{11}, A_{22}, B_{11}, and B_{22} are upper triangular. For by Theorem 1.8, there are unitary matrices U_i, V_i ($i = 1, 2$) such that the pairs $(V_i^H A_{ii} U_i, V_i^H B_{ii} U_i)$ are upper triangular. Then the equation $\mathbf{T}(P, Q) = (R, S)$ is equivalent to

$$(V_1^H A_{11} U_1)(U_1^H P U_2) + (V_1^H Q V_2)(V_2^H A_{22} U_2) = V_1^H R U_2,$$
$$(V_1^H B_{11} U_1)(U_1^H P U_2) + (V_1^H Q V_2)(V_2^H B_{22} U_2) = V_1^H S U_2.$$

1. BACKGROUND

Hence with the substitutions $A_{11} \leftarrow V_1^H A_{11} U_1$, $P \leftarrow U_1^H P U_2$, etc., the problem reduces to one in which the pairs (A_{ii}, B_{ii}) ($i = 1, 2$) are triangular.

We shall now show how to solve the equations

$$\begin{aligned} A_{11} P + Q A_{22} &= R, \\ B_{11} P + Q B_{22} &= S \end{aligned} \quad (1.4)$$

column by column beginning with the first columns of P and Q. Suppose that the columns $p_1, p_2, \ldots, p_{k-1}$ and $q_1, q_2, \ldots, q_{k-1}$ have already been computed (n.b., k may be equal to one). From (1.4) and the upper triangularity of A_{ii} and B_{ii}, it follows that the kth columns of P and Q must satisfy

$$\begin{aligned} A_{11} p_k + \alpha_{kk}^{(22)} q_k &= r_k - \sum_{i=1}^{k-1} \alpha_{ik}^{(22)} q_i, \\ B_{11} p_k + \beta_{kk}^{(22)} q_k &= s_k - \sum_{i=1}^{k-1} \beta_{ik}^{(22)} q_i. \end{aligned} \quad (1.5)$$

Multiply the first equation by $\beta_{kk}^{(22)}$, the second by $\alpha_{kk}^{(22)}$, and subtract to get

$$(\beta_{kk}^{(22)} A_{11} - \alpha_{kk}^{(22)} B_{11}) p_k = \beta_{kk}^{(22)} (r_k - \sum_{i=1}^{k-1} \alpha_{ik}^{(22)} q_i) - \alpha_{kk}^{(22)} (s_k - \sum_{i=1}^{k-1} \beta_{ik}^{(22)} q_i). \quad (1.6)$$

Since $\mathcal{L}[(A_{11}, B_{11})] \cap \mathcal{L}[(A_{22}, B_{22})] = \emptyset$, the matrix $\beta_{kk}^{(22)} A_{11} - \alpha_{kk}^{(22)} B_{11}$ is nonsingular. Hence equation (1.6) may be solved for p_k. Since (A_{22}, B_{22}) is a regular pair, not both $\alpha_{kk}^{(22)}$ and $\beta_{kk}^{(22)}$ are zero. Hence one of the equations (1.5) may be solved for q_k, and it is easily verified that this solution is consistent with the other equation. This completes the computation of p_k and q_k from $p_1, p_2, \ldots, p_{k-1}$ and $q_1, q_2, \ldots, q_{k-1}$.

For the converse, suppose that $\langle \lambda \rangle \in \mathcal{L}[(A_{11}, B_{11})] \cap \mathcal{L}[(A_{22}, B_{22})]$, and assume without loss of generality that $\lambda \neq \infty$ (otherwise reverse the roles of A and B in what follows). Then there are nonzero vectors x and y such that

$$A_{11} x = \lambda B_{11} x, \quad y^H A_{22} = \lambda y^H B_{22}.$$

Let

$$P = x y^H B_{22}, \quad Q = -B_{11} x y^H.$$

Then
$$\mathbf{T}(P,Q) = [(\lambda - \lambda)B_{11}xy^H B_{22}, (1-1)B_{11}xy^H B_{22}] = (0,0),$$
which shows that **T** is singular. ∎

One consequence of this theorem is that we can reduce any regular pair to any block diagonal form, in which the diagonal block-pairs do not have common eigenvalues. In particular, we have the following corollary.

Corollary 1.12. *Let the regular pair (A,B) have distinct eigenvalues. Then there are nonsingular matrices X and Y such that the pair $(Y^H AX, Y^H BX)$ is diagonal. The columns of X are the right eigenvectors of (A,B), and the columns of Y are its left eigenvectors.*

A more interesting consequence is WEIERSTRASS'S CANONICAL FORM.

Theorem 1.13 (Weierstrass). *Any regular pair is equivalent to a form*
$$[\mathrm{diag}(J, I), \mathrm{diag}(I, N)], \qquad (1.7)$$
where J and N are in Jordan canonical form and N is nilpotent (i.e., has only zero eigenvalues).

Proof. Assume that the eigenvalues of the pair have been ordered so that its triangular form can be partitioned
$$\left[\begin{pmatrix} A_{11} & A_{12} \\ 0 & A_{22} \end{pmatrix}, \begin{pmatrix} B_{11} & B_{12} \\ 0 & B_{22} \end{pmatrix}\right],$$
where the diagonals of B_{11} are nonzero and the diagonals of B_{22} are zero. Since $\mathcal{L}[(A_{11}, B_{11})] \cap \mathcal{L}[(A_{22}, B_{22})] = \emptyset$, we may further reduce the pair to the form
$$[\mathrm{diag}(A_{11}, A_{22}), \mathrm{diag}(B_{11}, B_{22})].$$
Since the pair is regular and the diagonals of B_{22} are zero, the diagonals of A_{22} are nonzero; i.e., A_{22} is nonsingular. Hence we may further reduce the pair to the form
$$\left[\mathrm{diag}(A_{11}B_{11}^{-1}, I), \mathrm{diag}(I, B_{22}A_{22}^{-1})\right].$$
The reduction is completed by reducing $A_{11}B_{11}^{-1}$ and $B_{22}A_{22}^{-1}$ to their Jordan canonical forms. ∎

1. BACKGROUND

1.3. Definite Pairs

The natural generalization of the Hermitian eigenvalue problem is to pairs of Hermitian matrices. Unfortunately, the property of being a Hermitian pair is not in itself enough to guarantee that the pair has nice properties, as the following example shows.

Example 1.14. *Let*

$$A = \begin{pmatrix} 1 & 0 \\ 0 & -1 \end{pmatrix}, \qquad B = \begin{pmatrix} 0 & 1 \\ 1 & 0 \end{pmatrix}.$$

Then the pair (A, B) *is Hermitian, but the eigenvalues of the pair are clearly* $\langle \pm i \rangle$.

Even more pathological cases can occur. For example, any matrix can be written in the form $B^{-1}A$, where A and B are Hermitian. Clearly, we must impose additional conditions if we are to have a workable theory. One possibility, which accounts for many applications, is to require that B be positive definite. This condition is justified by the following theorem

Theorem 1.15. *In the Hermitian pair* (A, B) *let* B *be positive definite. Then there is a nonsingular matrix* X *satisfying* $X^H B X = I$ *such that* $X^H A X = \Lambda$, *where* Λ *is real and diagonal.*

Proof. Since B is positive definite, it has a positive definite square root $B^{\frac{1}{2}}$. Then the pair (A, B) is equivalent to the pair $(B^{-\frac{1}{2}} A B^{-\frac{1}{2}}, I)$. Let $B^{-\frac{1}{2}} A B^{-\frac{1}{2}} = U \Lambda U^H$ be the spectral decomposition of $B^{-\frac{1}{2}} A B^{-\frac{1}{2}}$. Then $X = B^{-\frac{1}{2}} U$ is easily seen to be the required matrix. ∎

The construction used in the proof of this theorem can be used to establish a min-max characterization of the eigenvalues of (A, B) in the spirit of Fischer's theorem (Corollary IV.4.7). In fact, the following corollary, whose proof is left as an exercise, is what Fischer originally established.

Corollary 1.16 (Fischer). *Let the eigenvalues of* (A, B) *be ordered so that* $\lambda_1 \geq \lambda_2 \geq \cdots \geq \lambda_n$. *Then*

$$\lambda_i = \max_{\dim(\mathcal{X}) = i} \min_{\substack{x \in \mathcal{X} \\ x \neq 0}} \frac{x^H A x}{x^H B x},$$

and
$$\lambda_i = \min_{\dim(\mathcal{X})=n-i+1} \max_{\substack{x \in \mathcal{X} \\ x \neq 0}} \frac{x^{\mathrm{H}} A x}{x^{\mathrm{H}} B x}.$$

Although the condition that B be positive definite covers many cases occurring in practice, we can make do with an even weaker condition, which includes cases in which neither A nor B is positive definite.

Definition 1.17. *The Hermitian pair (A, B) is a* DEFINITE PAIR *if*

$$\gamma(A, B) \stackrel{\text{def}}{=} \min_{\substack{x \in \mathbf{C}^n \\ \|x\|_2=1}} |x^{\mathrm{H}}(A + iB)x| \equiv \min_{\substack{x \in \mathbf{C}^n \\ \|x\|_2=1}} \sqrt{(x^{\mathrm{H}} A x)^2 + (x^{\mathrm{H}} B x)^2} > 0. \tag{1.8}$$

The basic fact about definite pairs is that they can be transformed into a pair in which B is positive definite. Specifically, we have the following result.

Theorem 1.18. *Let (A, B) be a definite pair, and for $\phi \in \mathbf{R}$ let*

$$\begin{aligned} A_\phi &= A \cos \phi - B \sin \phi, \\ B_\phi &= A \sin \phi + B \cos \phi. \end{aligned} \tag{1.9}$$

Then there is a $\phi \in [0, 2\pi)$ such that B_ϕ is positive definite and

$$\gamma(A, B) = \lambda_{\min}(B_\phi),$$

where $\lambda_{\min}(B_\phi)$ is the smallest eigenvalue of B_ϕ.

Proof. Let $\mathcal{F}$ be the field of values of $A+iB$ (see Definition 3.10). Then $\gamma(A, B) = \min_{h \in \mathcal{F}} \|h\|_2$. Let the minimum be attained at the point $h = x_0^{\mathrm{H}}(A+iB)x_0$. Since $\mathcal{F}$ is a bounded, convex set (see Theorem 3.11), it is contained in the half plane $\mathcal{H}$, whose boundary passes perpendicularly through h.

Let $\mathcal{F}_\phi$, H_ϕ, and h_ϕ be the quantities corresponding to the pair (A_ϕ, B_ϕ). Since $A_\phi + iB_\phi = e^{i\phi}(A + iB)$, these quantities are just the original quantities rotated through the angle ϕ. Choose ϕ so that $\mathcal{H}_\phi$ lies in the upper half plane; i.e., so that h_ϕ lies along the imaginary

… 1. BACKGROUND

axis. Then $x_0^{\text{H}} A_\phi x_0 = \Re h_\phi = 0$. Moreover, since no point of $\mathcal{F}_\phi$ lies below $\mathcal{H}$, we must have

$$0 < \gamma(A,B) = x_0^{\text{H}} B_\phi x_0 = \min_{\|x\|=1} x^{\text{H}} B_\phi x = \lambda_{\min}(B),$$

which proves that B_ϕ is positive definite. ∎

If we now combine Theorems 1.15 and 1.18, we have the following corollary.

Corollary 1.19. *Let (A,B) be a definite pair. Then (A,B) is regular. Moreover, there is a nonsingular matrix X such that $X^{\text{H}} A X$ and $X^{\text{H}} B X$ are diagonal.*

1.4. Metrics and Their Limitations

A novel feature of the perturbation theory for the generalized eigenvalue problem is that two matrices — A and B — vary instead of one. Moreover, when we consider the perturbation of eigenvalues, we must introduce a distance between pairs $\langle \alpha, \beta \rangle$ and $\langle \tilde{\alpha}, \tilde{\beta} \rangle$. This subsection is devoted to defining the metrics that will be used in the remainder of the chapter.

We will first consider metrics for eigenvalues. Since we have chosen to regard eigenvalues of matrix pairs as two dimensional subspaces, it is natural to use the metrics of Section II.4. Of these metrics, one, the chordal metric, has an especially natural geometric interpretation.

Definition 1.20. *The* CHORDAL DISTANCE *between $\langle \alpha, \beta \rangle$ and $\langle \gamma, \delta \rangle$ is the number*

$$\chi(\langle \alpha, \beta \rangle, \langle \gamma, \delta \rangle) \stackrel{\text{def}}{=} \rho_{\text{g},2}(\langle \alpha, \beta \rangle, \langle \gamma, \delta \rangle),$$

where $\rho_{\text{g},2}$ is the gap metric in the 2-norm (see Definition II.4.3).

By definition the chordal distance is a metric. It is also easily computed. In terms of α, β, γ, and δ it has the form

$$\chi(\langle \alpha, \beta \rangle, \langle \gamma, \delta \rangle) = \frac{|\alpha \delta - \beta \gamma|}{\sqrt{|\alpha|^2 + |\beta|^2}\sqrt{|\gamma|^2 + |\delta|^2}}.$$

If we set $\lambda = \alpha/\beta$ and $\mu = \gamma/\delta$, then we have

$$\chi(\langle\lambda\rangle, \langle\mu\rangle) = \frac{|\lambda - \mu|}{\sqrt{1 + |\lambda|^2}\sqrt{1 + |\mu|^2}}.$$

From the latter form it is seen that

$$\chi(\langle\lambda\rangle, \langle\infty\rangle) = \frac{1}{\sqrt{1 + |\lambda|^2}} \leq 1.$$

Thus, the chordal metric regularizes the point at infinity by making it no more than unit distance from any other point.

The name "chordal metric" comes from the following considerations. In $\mathbf{R}^3$ let the x-y plane represent the complex numbers. For any complex number λ draw a line between λ and the point $(0, 0, 1)$ and let $s(\lambda)$ be the intersection, other than $(0, 0, 1)$, of the line with the unit sphere centered at the origin (the Riemann sphere). Then it can be shown that

$$\chi(\lambda, \mu) = \frac{1}{2}\|s(\lambda) - s(\mu)\|_2. \tag{1.10}$$

In other words, the chordal distance between λ and μ is one half the length of the chord joining the projections of λ and μ onto the Riemann sphere.

For numbers less than one in magnitude, the chordal metric behaves essentially like the ordinary Euclidean metric. In particular

$$|\lambda|, |\mu| \leq 1 \implies \chi(\langle\lambda\rangle, \langle\infty\rangle) \leq |\lambda - \mu| \leq 2\chi(\langle\lambda\rangle, \langle\infty\rangle).$$

Moreover, as $\lambda, \mu \to 0$, we have $\chi(\langle\lambda\rangle, \langle\mu\rangle) \cong |\mu - \lambda|$.

On the other hand, for large numbers the chordal metric behaves counter-intuitively. For example, as $\lambda \to \infty$ we have

$$\chi(\langle\lambda\rangle, \langle 2\lambda\rangle) \cong \frac{1}{|\lambda|}.$$

Thus large numbers can have very small chordal differences, even when they have large relative errors. ,

Let us now consider metrics for matrix pairs. Let (A, B) be a matrix pair and let $(\tilde{A}, \tilde{B}) = (A + E, B + F)$. A natural way to define the

1. BACKGROUND

distance between the pairs is to apply a norm to the difference $(\tilde{A}\ \tilde{B}) - (A\ B)$, i.e., to the matrix $(E\ F)$. For example we might say that the distance between (A, B) and $(\tilde{A}, \tilde{B})$ is $\|(E\ F)\|_2$ or $\sqrt{\|E\|_2^2 + \|F\|_2^2}$ or, depending on the application, some other combination.

In many respects this is the most natural approach; however, it has an important drawback. Since the generalized eigenvalue problem is homogeneous in A and B, we do not feel there is any substantial difference between the the pair (A, B) and any nonzero multiple $(\tau A, \tau B)$. But with the above approach, these two pairs have positive distance, unless $\tau = 1$. Consequently, we will also use other, less discriminating metrics.

Let us start with definite pairs. We will take the same approach as we did with eigenvalues and define our metric over equivalence classes of pairs, which in some sense represent the same problem.

Definition 1.21. *Let (A, B) be a definite pair. Then $\langle A, B \rangle_D$ (D for definite) is the set of pairs (C, D) such that there exists a real multiplier μ for which one of the following conditions holds:*

1. *$C = \mu A$ and $D = \mu B$ ($\mu \neq 0$),*

2. *$A = \mu B$ and $C = \mu D$,*

3. *$B = \mu A$ and $D = \mu C$.*

The first case in the above definition corresponds to the case where the pair (C, D) is proportional to the pair (A, B), and hence the pairs have the same eigenvalues. It the second case both pairs have the single eigenvalue $\langle \mu \rangle$, and in the third they have the eigenvalue $\langle \mu^{-1} \rangle$. It is easily verified that the operator $\langle \cdot, \cdot \rangle_D$ divides the set of definite pairs into equivalence classes.

We now define

$$\rho(\langle A, B \rangle_D, \langle C, D \rangle_D) \stackrel{\text{def}}{=} \max_{x \neq 0} \chi(\langle x^H A x, x^H B x \rangle, \langle x^H C x, x^H D x \rangle). \quad (1.11)$$

Theorem 1.22. *The function ρ defined by (1.11) is a metric on the space of definite pairs $\langle A, B \rangle_D$.*

Proof. The only property of a metric that is difficult to verify is that

$$\rho(\langle A,B\rangle_\mathrm{D}, \langle C,D\rangle_\mathrm{D}) = 0 \implies \langle A,B\rangle_\mathrm{D} = \langle C,D\rangle_\mathrm{D}.$$

We will establish this implication, leaving the other properties as an exercise.

Suppose that $\rho(\langle A,B\rangle_\mathrm{D}, \langle C,D\rangle_\mathrm{D}) = 0$. Then for all x

$$x^\mathrm{H} A x x^\mathrm{H} D x = x^\mathrm{H} B x x^\mathrm{H} C x. \tag{1.12}$$

Let us first dispose of cases two and three in Definition 1.21. Suppose that $A = \mu B$. Since (A,B) is definite, it follows that $x^\mathrm{H} B x \neq 0$ for all x. Hence from (1.12), we have $x^\mathrm{H} C x = \mu x^\mathrm{H} D x$ for all x. Equivalently $C = \mu D$, which shows that $\langle A,B\rangle = \langle C,D\rangle$. The third case is treated similarly.

Turning to the first case, it is easy to see that the first relation in Definition 1.21 and the equation (1.12) remain invariant under substitutions of the form

$$A \leftarrow A\cos\phi - B\sin\phi,$$
$$B \leftarrow A\sin\phi + B\cos\phi,$$
$$C \leftarrow C\cos\phi - C\sin\phi,$$
$$D \leftarrow C\sin\phi + D\cos\phi.$$

Hence by Theorem 1.18, we may assume B is positive definite. The same relations are also invariant under congruence transformations. Hence by Theorem 1.15 we may assume that $B = I$ and that $A = \operatorname{diag}(\alpha_1 I, \ldots, \alpha_m I)$, where the α_i are distinct numbers. Since A is not a multiple of B, we must have $m > 1$. The general proof is sufficiently well illustrated by the case $m = 2$.

Let

$$C = \begin{pmatrix} C_{11} & C_{12} \\ C_{21} & C_{22} \end{pmatrix}, \qquad D = \begin{pmatrix} D_{11} & D_{12} \\ D_{21} & D_{22} \end{pmatrix}$$

be conformal partitions of C and D. Consider the vector

$$x = (x_1^\mathrm{H} \; \tau x_2^\mathrm{H})^\mathrm{H},$$

1. BACKGROUND

where τ is real. From (1.12)

$$(\alpha_1 x_1^H x_1 + \alpha_2 x_2^H x_2 \tau^2) \times$$
$$[x_1^H D_{11} x_1 + (x_1^H D_{12} x_2 + x_2^H D_{21} x_1)\tau + x_2^H D_{22} x_2 \tau^2] =$$
$$(x_1^H x_1 + x_2^H x_2 \tau^2)[x_1^H C_{11} x_1 + (x_1^H C_{12} x_2 + x_2^H C_{21} x_1)\tau + x_2^H C_{22} x_2 \tau^2].$$

Equating powers of τ, we get

$$\alpha_j x_j^H D_{jj} x_j = x_j^H C_{jj} x_j, \qquad j = 1, 2, \qquad (1.13)$$

$$x_1^H x_1 x_2^H (\alpha_1 D_{22} - C_{22}) x_2 = x_2^H x_2 x_1^H (C_{11} - \alpha_2 D_{11}) x_1, \qquad (1.14)$$

and

$$\alpha_j (x_1^H D_{12} x_2 + x_2^H D_{21} x_1) = x_1^H C_{12} x_2 + x_2^H D_{21} x_1, \qquad j = 1, 2. \quad (1.15)$$

From (1.13)
$$C_{11} = \alpha_1 D_{11}, \quad C_{22} = \alpha_2 D_{22}.$$

Hence from (1.14) we obtain

$$\frac{x_2^H D_{22} x_2}{x_2^H x_2} = \frac{x_1^H D_{11} x_1}{x_1^H x_1} = \mu$$

for some real μ. Since this relation holds generally in x_1 and x_2, we must have

$$D_{11} = D_{22} = \mu I.$$

Since $\alpha_1 \neq \alpha_2$, we have from (1.15),

$$x_1^H D_{12} x_2 + x_2^H D_{21} x_1 = 0. \qquad (1.16)$$

Taking first $x_2 = D_{12}^H x_1$ and then $x_1 = D_{21}^H x_2$ in (1.16), we obtain

$$D_{12} = 0, \quad D_{21} = 0.$$

Hence
$$C = \mu A,$$

and
$$D = \mu I = \mu B. \quad \blacksquare$$

The notation $\rho(\langle A, B\rangle_\mathrm{D}, \langle C, D\rangle_\mathrm{D})$ is a little clumsy, and in the sequel we will write
$$\rho_\mathrm{D}[(A, B), (C, D)].$$
However, the reader should keep in mind that the function ρ_D so defined is a pseudo-metric, not a metric.

Turning now to general, regular matrix pairs, we will say that the pair (A, B) is LEFT EQUIVALENT to (C, D) if there is a nonsingular matrix Y such that $Y^\mathrm{H}(A, B) = (C, D)$. This is clearly an equivalence relation and we shall denote the equivalence classes by $\langle A, B\rangle_\mathrm{L}$.

We wish to define a metric over the equivalence classes $\langle A, B\rangle_\mathrm{L}$. The key observation is that (A, B) is left equivalent to (C, D) if and only if the row spaces of the matrices $(A\ B)$ and $(C\ D)$ are the same. Consequently we can define a metric by using any one of the metrics in Section II.4. For definiteness we will choose the gap metric in the 2-norm, or equivalently we will make the following definition.

Definition 1.23. *Let (A, B) and (C, D) be regular matrix pairs. Then*
$$\rho(\langle A, B\rangle_\mathrm{L}, \langle C, D\rangle_\mathrm{L}) = \sin\theta_1,$$
where θ_1 is the largest canonical angle between the row spaces of $(A\ B)$ and $(C\ D)$.

Again, we will usually write $\rho_\mathrm{L}[(A, B), (C, D)]$ for $\rho(\langle A, B\rangle_\mathrm{L}, \langle C, D\rangle_\mathrm{L})$. Note that there is a natural function ρ_R obtained by considering the pair $(A^\mathrm{H}, B^\mathrm{H})$.

Let us now step back from the technical details and take a broader view. The justification for the metrics we have introduced is convenience and elegance. The metrics are convenient because they regularize singular cases. For example, all eigenvalues, finite and infinite, are treated uniformly. The elegance can only be judged by the results, but the reader is invited to look ahead to the statement of Theorem 3.2.

On the other hand, convenience and elegance exact a toll. In Chapter III we discussed the loss of information entailed in using norms, and the same caveats apply here. But there is more.

In the next section we will show that if x is an eigenvector of the definite pair (A, B) corresponding to the simple eigenvalue $\langle x^\mathrm{H}Ax, x^\mathrm{H}Bx\rangle$, then $\langle x^\mathrm{H}\tilde{A}x, x^\mathrm{H}\tilde{B}x\rangle$ is a first order approximation of the eigenvalue of

1. BACKGROUND

the perturbed pair $(\tilde{A}, \tilde{B})$. This implies that up to first order terms, the errors in, say, A do not cross over to affect the second component β of the eigenvalue. But the metric ρ_D confounds errors A and B, while the chordal metric confounds the resulting perturbations in α and β.

Moreover, the resulting bounds are not scale invariant. Replacing the pair (A, B) with $(\tau A, B)$ will give essentially different bounds — bounds that change nonlinearly with τ. For example, our theorems will not reduce to the usual perturbation theorems when $B = I$; to recover them we must let $\tau \to 0$.

However, the situation is not entirely bleak. For nicely scaled problems the perturbation bounds we derive may be quite satisfactory. Moreover, some of our theorems will give bounds on perturbations of the components of $\langle \alpha, \beta \rangle$, bounds that the analyst may use in any way he sees fit.

When it comes to general matrix pairs, the situation is even less satisfactory. The metric defined above has all the drawbacks discussed above. Moreover, it is asymmetric; we obtain essentially different theorems for the pairs (A, B) and (A^H, B^H). However, when A and B are Hermitian this last objection does not apply.

Notes and References

Matrix pairs arise naturally in the study of systems of ordinary differential equations of the form

$$A\frac{dx}{dt} = Bx,$$

where the simultaneous diagonalization of A and B by an equivalence represents a transformation which uncouples the system. Generalizations to higher order systems lead to λ-MATRICES of the form $A_0 + A_1\lambda + \cdots + A_k\lambda^k$, for treatments of which see [81, 86].

Weierstrass [263, 1867] established his canonical form (Theorem 1.13) by working with a pair of bilinear forms, as was customary at the time. Jordan [125, 1874] gave another proof, which included singular pencils. Later Kronecker [139, 1890] extended these results to rectangular pencils (for details see [81]). For modern computational treatments see [60, 272].

The generalized Schur form of Theorem 1.9 is due to Stewart [201, 1972], as is the condition for the generalized Sylvester equations to be nonsingular (Theorem 1.11).

Definite pairs in which one or the other of the components is positive definite constitute the majority of applications. It is not generally appreciated that Fischer [74, 1905] proved his min-max characterization (Corollary 1.16) for such pairs — not simply for eigenvalues of Hermitian matrices.

Theorem 1.18 characterizing definite pairs is one of a number of interrelated theorems, whose history and interconnections have been admirably surveyed by Uhlig [243]. The particular theorem given here is due to Crawford [48, 1976]. It should be noted that in the definition (1.8) of $\gamma(A,B)$ the minimum is taken over all vectors $x \in \mathbf{C}^n$. It might be hoped that for symmetric pairs one could let x range over $\mathbf{R}^n$. Although one can when $n > 2$, one cannot when $n = 2$, as the pair of Example 1.14 shows.

The chordal metric was first used in the perturbation theory for matrix pairs by Stewart [204, 1975]. The metric ρ_D was introduced by Sun [222, 1982], as were the metrics ρ_L and ρ_R [220, 1979] (see also [67]).

Exercises

1. Show that if A and B are nonsingular and $\langle \lambda \rangle$ is an eigenvalue of (A,B) then $\langle \lambda^{-1} \rangle$ is an eigenvalue of (A^{-1}, B^{-1}).

2. Let $A = I$ and

$$B = \begin{pmatrix} 1+\epsilon & 1 \\ 1 & 1-\epsilon \end{pmatrix},$$

where ϵ is small. Then the pair (A,B) has eigenvalues $\lambda_1 \cong \langle 1, 1+\frac{1}{2}\epsilon^2 \rangle$ and $\lambda_2 \cong \langle 1, -\frac{1}{2}\epsilon^2 \rangle$. Show that λ_1 is insensitive to perturbations in B, but is an ill-conditioned eigenvalue of $B^{-1}A$.

3. (Moler and Stewart [161]). Let (A,B) be a real, regular matrix pair. Show that (A,B) is orthogonally equivalent to (S,T), where T is upper triangular and S is block upper triangular with 1×1 or 2×2 blocks on its diagonal.

4. Show that the Weierstrass canonical form (1.7) is essentially unique.

5. Let A and B be positive definite. Show that if $x^H A x \leq x^H B x$ for all $x \neq 0$ then $x^H A^{-1} x \geq x^H B^{-1} x$.

6. Verify that the chordal metric may be defined by (1.10).

7. Show that

$$|\lambda|, |\mu| \leq 1 \implies \chi(\langle \lambda \rangle, \langle \infty \rangle) \leq |\lambda - \mu| \leq 2\chi(\langle \lambda \rangle, \langle \infty \rangle).$$

2. Regular Matrix Pairs

8. Let (A, B) be as in Example 1.14. Show that

$$\min_{\substack{x \in \mathbf{R}^n \\ \|x\|_2 = 1}} \sqrt{(x^T A x)^2 + (x^T B x)^2} = 1,$$

even though (A, B) is not definite.

2. Regular Matrix Pairs

In this section we will treat the perturbation of the eigenvalues of regular matrix pairs. We begin with first order perturbation theory, which exhibits the typical behavior of a simple generalized eigenvalue. We then turn to a generalization of the Gerschgorin theorem, which is the most useful tool for bounding perturbations of generalized eigenvalues. Next comes a generalization of Theorem IV.3.3, which bounds the spectral variation in terms of the condition of the eigenvalues. Finally, we develop the perturbation theory of eigenspaces.

Throughout this section, (A, B) will be a regular matrix pair of order n and

$$(\tilde{A}, \tilde{B}) = (A + E, B + F)$$

will be a perturbation of (A, B).

2.1. Continuity, First Order Theory

The first thing that must be established is the continuity of the eigenvalues of matrix pairs. We will use Theorem 1.6 to reduce the continuity of the generalized eigenvalues to that of the ordinary eigenvalue problem. Here it is not critical how we measure the size of the perturbation in (A, B), and to fix on a single measure we will set

$$\epsilon = \sqrt{\|E\|_2^2 + \|F\|_2^2}.$$

Theorem 2.1. *Let (A, B) be a regular pair, and let its eigenvalues be $\langle \lambda_1 \rangle, \ldots, \langle \lambda_n \rangle$. Then there is an ordering $\langle \tilde{\lambda}_1 \rangle, \ldots, \langle \tilde{\lambda}_n \rangle$ of the eigenvalues of $(\tilde{A}, \tilde{B})$ such that*

$$\lim_{\epsilon \to 0} \chi(\langle \tilde{\lambda}_i \rangle, \langle \lambda_i \rangle) = 0, \qquad i = 1, \ldots, n.$$

Proof. By Theorem 1.6 there is a 2×2 matrix W such that the matrix D in the pair $(C\ D) = (A\ B)(W \otimes I)$ is nonsingular. Let $\mu_1, \ldots, \mu_n$ be the eigenvalues of $D^{-1}C$. Let $(\tilde{C}\ \tilde{D}) = (\tilde{A}\ \tilde{B})(W \otimes I)$. For ϵ sufficiently small, $\tilde{D}$ is nonsingular. By the continuity of the ordinary eigenvalue problem, we know there is an ordering of the eigenvalues $\tilde{\mu}_1, \ldots, \tilde{\mu}_n$ of $\tilde{D}^{-1}\tilde{C}$ such that $\lim_{\epsilon \to 0} \tilde{\mu}_i = \mu_i$.

Let $(\beta_i; -\alpha_i) = (\lambda - 1)W^{-T}$. Then by Theorem 1.6, $\langle \lambda_i \rangle = \langle \alpha_i, \beta_i \rangle$ is an eigenvalue of (A, B). If we set $(\tilde{\beta}_i - \tilde{\alpha}_i) = (\tilde{\lambda} - 1)W^{-T}$ and $\langle \tilde{\lambda}_i \rangle = \langle \tilde{\alpha}_i, \tilde{\beta}_i \rangle$, then $\langle \tilde{\lambda}_i \rangle$ is an eigenvalue of $(\tilde{A}, \tilde{B})$. Since $(\tilde{\alpha}_i, \tilde{\beta}_i)$ converges to (α_i, β_i), it follows that $\langle \tilde{\lambda}_i \rangle$ converges to $\langle \lambda_i \rangle$ in the chordal metric. ∎

Throughout this book, we have presented first order perturbation expansions whenever they exist. Although these expansions are usually corollaries of more general results, in many cases the general results themselves were conjectured by looking at first order expansions. The reason is that the expansions often tell ninety percent of the story and yet are free of the clutter that accompanies rigorous upper bounds. Since research into the perturbation of matrix pairs is still in a state of flux, it is appropriate to begin with first order perturbation theory.

Let $\langle \alpha, \beta \rangle$ be a simple eigenvalue of $\langle A, B \rangle$. In order to derive a first order perturbation expansion, we must first show that one exists. In one sense this is trivial. By Theorem 1.6, we may assume that B is nonsingular. Hence for ϵ sufficiently small $\tilde{B}$ is nonsingular, and to the eigenvalue $\lambda = \alpha/\beta$ of $B^{-1}A$ there corresponds an eigenvalue $\tilde{\lambda} = \lambda + O(\epsilon)$ of $\tilde{B}^{-1}\tilde{A}$, which is differentiable in the elements of A and B. It follows that $\langle \tilde{\lambda} \rangle$ is the required expansion.

However, when we look at the individual components of $\langle \alpha, \beta \rangle$ we find that their perturbations are not unique. For if $\langle \tilde{\alpha}, \tilde{\beta} \rangle$ is an $O(\epsilon)$ perturbation of $\langle \alpha, \beta \rangle$ in the chordal metric and $\phi(\epsilon) = O(\epsilon)$, then $\langle \tilde{\alpha} + \alpha \phi(\epsilon), \tilde{\beta} + \beta \phi(\epsilon) \rangle$ differs from $\langle \tilde{\alpha}, \tilde{\beta} \rangle$ by $O(\epsilon^2)$. This follows directly from the formula

$$\chi(\langle \tilde{\alpha}, \tilde{\beta}, \rangle, \langle \tilde{\alpha} + \alpha\phi(\epsilon), \tilde{\beta} + \beta\phi(\epsilon) \rangle)$$
$$= \frac{|\phi(\epsilon)(\tilde{\alpha}\beta - \tilde{\beta}\alpha)|}{\sqrt{|\tilde{\alpha}|^2 + |\tilde{\beta}|^2}\sqrt{|\tilde{\alpha} + \alpha\phi(\epsilon)|^2 + |\tilde{\beta} + \beta\phi(\epsilon)|^2}},$$

in which the denominator is $O(\epsilon^2)$. Fortunately, Corollary 1.10 provides a canonical choice for $\langle \tilde{\alpha}, \tilde{\beta} \rangle$.

2. REGULAR MATRIX PAIRS

Theorem 2.2. *Let $\langle \alpha, \beta \rangle$ be a simple eigenvalue of the regular pair (A, B) with right and left eigenvectors x and y. Let $\langle \tilde{\alpha}, \tilde{\beta} \rangle$ be the corresponding eigenvalue of the $O(\epsilon)$ perturbation $(\tilde{A}, \tilde{B})$. Then*

$$\langle \tilde{\alpha}, \tilde{\beta} \rangle = \langle y^H \tilde{A} x, y^H \tilde{B} x \rangle + O(\epsilon^2). \tag{2.1}$$

Proof. Applying the perturbation theory for the ordinary eigenvalue problem first to $B^{-1}A$ and then to AB^{-1} (after a transformation, if necessary, to make B nonsingular), we find that we may take for the eigenvectors corresponding to $\langle \tilde{\alpha}, \tilde{\beta} \rangle$ the vectors $\tilde{x} = x + u$ and $\tilde{y} + v$, where $u, v = O(\epsilon)$. By Corollary 1.10,

$$\langle \tilde{\alpha}, \tilde{\beta} \rangle = \langle \tilde{y}^H \tilde{A} \tilde{x}, \tilde{y}^H \tilde{B} \tilde{x} \rangle =$$
$$\langle y^H \tilde{A} x + y^H A u + v^H A x + O(\epsilon^2), y^H \tilde{B} x + y^H A u + v^H A x + O(\epsilon^2) \rangle.$$

Since (A, B) is regular, at least one of α or β must be nonzero, say $\beta \neq 0$. Then

$$y^H A u + v^H A x = \alpha \frac{y^H B u + v^H B x}{\beta}$$

and

$$y^H B u + v^H B x = \beta \frac{y^H B u + v^H B x}{\beta}.$$

Thus $(y^H A u + v^H A x, y^H B u + v^H B x)$ is an order ϵ perturbation of $(y^H \tilde{A} x, y^H \tilde{A} x)$ that lies along (α, β). By the observation made just before the theorem, deleting these terms introduces an $O(\epsilon^2)$ error. ∎

From this theorem we may derive approximate error bounds for the perturbation of a simple eigenvalue. Specifically, if we set $\alpha = y^H A x$ and $\beta = y^H B x$, then

$$\chi(\langle \alpha, \beta \rangle, \langle \tilde{\alpha}, \tilde{\beta} \rangle) \cong \frac{|\alpha y^H F x - \beta y^H E x|}{\sqrt{|\alpha|^2 + |\beta|^2}\sqrt{|\alpha + y^H E x|^2 + |\beta + y^H F x|^2}}$$
$$\cong \frac{|\alpha y^H F x - \beta y^H E x|}{|\alpha|^2 + |\beta|^2}.$$

To turn this approximation into a bound, note that

$$|\alpha y^H F x - \beta y^H E x| = \left| y^H (E \ F) \begin{pmatrix} \beta x \\ -\alpha x \end{pmatrix} \right|$$
$$\leq \sqrt{|\alpha|^2 + |\beta^2|} \|x\|_2 \|y\|_2 \|(E \ F)\|_2.$$

Hence if we set
$$\nu = \frac{\|x\|_2 \|y\|_2}{\sqrt{|\alpha|^2 + |\beta|^2}}, \qquad (2.2)$$
then
$$\chi(\langle \alpha, \beta \rangle, \langle \tilde{\alpha}, \tilde{\beta} \rangle) \lesssim \nu \|(E\ F)\|_2. \qquad (2.3)$$

The number ν defined by (2.2) is completely analogous to the number ν defined by (IV.2.8) for the ordinary eigenvalue problem; that is, it serves the role of a condition number for its eigenvalue. Unfortunately we cannot obtain the usual bound for the ordinary eigenvalue problem $Ax = \lambda x$ by replacing B by I and F by 0. However, if we replace A by τA, λ by $\tau \lambda$, and E by τE, then the bound (2.3) becomes
$$\chi(\langle \tau \lambda \rangle, \langle \tau \tilde{\lambda} \rangle) \lesssim \nu_\tau \|\tau E\|_2,$$
where
$$\nu_\tau = \frac{\|x\|_2 \|y\|_2}{\sqrt{|\tau y^H A x|^2 + |y^H x|^2}}.$$

But as $\tau \to 0+$, we have $\chi(\langle \tau \lambda \rangle, \langle \tau \tilde{\lambda} \rangle) \cong \tau |\tilde{\lambda} - \lambda|$. Moreover, the condition number ν_τ approaches $\|x\|_2 \|y\|_2 / |y^H x|$. Hence we have
$$|\tilde{\lambda} - \lambda| \lesssim \frac{\|x\|_2 \|y\|_2}{|y^H x|} \|E\|_2,$$
which is the usual bound.

The approximate bound (2.3) is about as good as we will see for the generalized eigenvalue problem. However, the trouble we had to take to retrieve the Rayleigh quotient bound is a reminder that it suffers from the limitations noted at the end of the last section. When in doubt, one should return to explicit forms, like the approximation (2.1) provided by Theorem 2.2.

2.2. Gerschgorin Theory

In this subsection we will generalize Gerschgorin's theorem and apply it to the perturbation of multiple eigenvalues. As in Section IV.2, we approach the theorem through a generalization of the Bauer–Fike theorem.

2. REGULAR MATRIX PAIRS

Theorem 2.3. Let (A, B) and $(\tilde{A}, \tilde{B})$ be regular pairs, and let $\|\cdot\|$ be a consistent matrix norm. If $\langle \tilde{\alpha}, \tilde{\beta} \rangle \in \mathcal{L}[(\tilde{A}, \tilde{B})]$ is not an eigenvalue of (A, B), then

$$\|(\tilde{\beta}A - \tilde{\alpha}B)^{-1}(\tilde{\beta}E - \tilde{\alpha}F)\| \geq 1. \qquad (2.4)$$

Proof. Since $\langle \tilde{\alpha}, \tilde{\beta} \rangle \notin \mathcal{L}[(A, B)]$, the matrix $\tilde{\beta}A - \tilde{\alpha}B$ is nonsingular. Let $\tilde{x}$ be an eigenvector of $(\tilde{A}, \tilde{B})$ corresponding to $\langle \tilde{\alpha}, \tilde{\beta} \rangle$. Since

$$0 = (\tilde{\beta}\tilde{A} - \tilde{\alpha}\tilde{B})\tilde{x} = (\tilde{\beta}A - \tilde{\alpha}B)\tilde{x} + (\tilde{\beta}E - \tilde{\alpha}F)\tilde{x},$$

we have

$$(\tilde{\beta}A - \tilde{\alpha}B)^{-1}(\tilde{\beta}E - \tilde{\alpha}F)\tilde{x} = \tilde{x},$$

from which the theorem follows on taking norms. ∎

We may now state and prove the generalization of the Gerschgorin theorem.

Theorem 2.4. Let (A, B) be a regular pair. Let

$$\mathcal{D}_i = \{\, \langle \alpha, \beta \rangle : |\beta\alpha_{ii} - \alpha\beta_{ii}| \leq \sum_{j \neq i} |\beta\alpha_{ij} - \alpha\beta_{ij}| \,\}, \qquad i = 1, \ldots, n.$$

Then

$$\mathcal{L}[(A, B)] \subset \bigcup_{i=1}^{n} \mathcal{D}_i.$$

Moreover, if the union k of the regions $\mathcal{D}_i$ is disjoint from the others and is not equal to the space $\mathbf{C}_1^2$ of all $\langle \alpha, \beta \rangle$, then the union contains exactly k eigenvalues of (A, B).

Proof. Let $D_A = \text{diag}(\alpha_{11}, \ldots, \alpha_{nn})$ and $D_B = \text{diag}(\beta_{11}, \ldots, \beta_{nn})$. In Theorem 2.3 make the substitutions

$$A \leftarrow D_A, \qquad B \leftarrow D_B,$$
$$\tilde{A} \leftarrow A, \qquad \tilde{B} \leftarrow B,$$

and $\|\cdot\| \leftarrow \|\cdot\|_\infty$. Then it is easily verified that the inequality (2.4) is equivalent to saying that each eigenvalue of (A, B) is in some $\mathcal{D}_i$.

The statement about isolated disks follows from the continuity of the eigenvalues as in the ordinary Gerschgorin theorem — namely, if we introduce the pairs $(A_\tau, B_\tau) = [D_A + \tau(A - D_A), D_B + \tau(A - D_B)]$, then the corresponding regions $\mathcal{D}_i^{(\tau)}$ increase with τ. The only tricky point is to insure that the pair (A_τ, B_τ) is regular for $0 \leq \tau \leq 1$.

We argue as follows. Assume without loss of generality that the disjoint disks are $\mathcal{D}_1^{(\tau)}, \ldots, \mathcal{D}_k^{(\tau)}$. Then $\bigcup_{i=1}^k \mathcal{D}_i^{(\tau)}$ and $\bigcup_{i=k+1}^n \mathcal{D}_i^{(\tau)}$ are disjoint closed sets. Since $\mathbf{C}_1^2$ is connected, there must be a point $\langle \alpha, \beta \rangle \notin \bigcup_{i=1}^k \mathcal{D}_i^{(\tau)} \cup \bigcup_{i=k+1}^n \mathcal{D}_i^{(\tau)}$. Then $\langle \alpha, \beta \rangle$ is not an eigenvalue of (A_τ, B_τ), which is therefore regular. ∎

The first comment to be made about this theorem is that it becomes uninteresting when some $(\alpha_{ii}, \beta_{ii}) = (0,0)$, since in this case $\mathcal{D}_i$ includes all pairs $\langle \alpha, \beta \rangle$. In the sequel we will tacitly exclude this case (see Exercise 2.1).

The regions $\mathcal{D}_i$ are difficult to compute, since (α, β) appears on both sides of the bound. However, by expanding the regions, we may remove this dependence.

Corollary 2.5. *Let*

$$a_i = (\alpha_{i1}, \ldots, \alpha_{i,i-1}, 0, \alpha_{i,i+1}, \ldots, \alpha_{in})^\mathrm{T}$$

and

$$b_i = (\beta_{i1}, \ldots, \beta_{i,i-1}, 0, \beta_{i,i+1}, \ldots, \beta_{in})^\mathrm{T}$$

be the rows of $A - D_A$ and $B - D_B$. Let

$$\rho_i = \sqrt{\frac{\|a_i\|_1^2 + \|b_i\|_1^2}{|\alpha_{ii}|^2 + |\beta_{ii}|^2}},$$

and let

$$\mathcal{G}_i = \{ \langle \alpha, \beta \rangle : \chi(\langle \alpha, \beta \rangle, \langle \alpha_{ii}, \beta_{ii} \rangle) \leq \rho_i \}.$$

Then

$$\mathcal{D}_i \subset \mathcal{G}_i, \quad i = 1, \ldots, n.$$

Proof. We have

$$\sum_{j \neq i} |\beta \alpha_{ij} - \alpha \beta_{ij}| = \|\beta a_i - \alpha b_i\|_1$$

$$\leq (\beta \; \alpha) \begin{pmatrix} \|a_i\|_1 \\ \|b_i\|_1 \end{pmatrix}$$

$$\leq \sqrt{|\alpha|^2 + |\beta|^2} \sqrt{\|a_i\|_1^2 + \|b_i\|_1^2},$$

2. REGULAR MATRIX PAIRS

the last inequality following from the Cauchy inequality. Thus the inequality

$$|\beta\alpha_{ii} - \alpha\beta_{ii}| \leq \sum_{j \neq i} |\beta\alpha_{ij} - \alpha\beta_{ij}|$$

implies the inequality

$$|\beta\alpha_{ii} - \alpha\beta_{ii}| \leq \sqrt{|\alpha|^2 + |\beta|^2} \sqrt{\|u_i\|_1^2 + \|b_i\|_1^2}.$$

The corollary now follows on dividing by $\sqrt{|\alpha_{ii}|^2 + |\beta_{ii}|^2}\sqrt{|\alpha|^2 + |\beta|^2}$. ∎

This corollary is actually our principal Gerschgorin theorem. It should be noted that by the same kind of limit argument we used in the last subsection, we can recover the usual Gerschgorin theorem, for the ordinary eigenvalue problem.

The technique of diagonal similarities, which we used so successfully in Section IV.2, can be applied to the generalization of Gerschgorin's theorem. To vary the application, we will show how to apply the technique to a multiple eigenvalue of a diagonalizable pair. Since we have already showed in Section IV.2 how to take into account terms of order higher than the first, we will not bound their contribution here.

Let (A, B) be a diagonalizable pair; that is, suppose there exist nonsingular matrices $X = (x_1 \ldots x_n)$ and $Y = (y_1 \ldots y_n)$ such that

$$(Y^{\mathrm{H}}AX, Y^{\mathrm{H}}BX) = [\mathrm{diag}(\alpha_1, \ldots, \alpha_n), \mathrm{diag}(\beta_1, \ldots, \beta_n)].$$

Assume that $\langle \alpha_1, \beta_1 \rangle = \cdots = \langle \alpha_k, \beta_k \rangle$, and that these eigenvalues are distinct from the others, so that

$$\delta = \min_{k < i \leq n} \chi(\langle \alpha_1, \beta_1 \rangle, \langle \alpha_i, \beta_i \rangle) \neq 0.$$

Set

$$\nu_i = \frac{\|x_i\|_2 \|y_i\|_2}{\sqrt{|\alpha_i|^2 + |\beta_i|^2}},$$

and let

$$\nu = \max_{1 \leq i \leq k} \nu_i$$

and

$$\nu' = \max_{k < i \leq n} \nu_i.$$

VI. GENERALIZED EIGENVALUE PROBLEMS

Now consider the pair $(\tilde{A}, \tilde{B}) = (A + E, B + F)$. Write (for $n = 6$)

$$Y^{\mathrm{H}}(A + E)X =$$

$$\begin{pmatrix}
\alpha_{11} + \gamma_{11} & \gamma_{12} & \gamma_{13} & \gamma_{14} & \gamma_{15} & \gamma_{16} \\
\gamma_{21} & \alpha_{22} + \gamma_{22} & \gamma_{23} & \gamma_{24} & \gamma_{25} & \gamma_{26} \\
\gamma_{31} & \gamma_{32} & \alpha_{33} + \gamma_{33} & \gamma_{34} & \gamma_{35} & \gamma_{36} \\
\gamma_{41} & \gamma_{42} & \gamma_{43} & \alpha_{44} + \gamma_{44} & \gamma_{45} & \gamma_{46} \\
\gamma_{51} & \gamma_{52} & \gamma_{53} & \gamma_{54} & \alpha_{55} + \gamma_{55} & \gamma_{56} \\
\gamma_{61} & \gamma_{62} & \gamma_{63} & \gamma_{64} & \gamma_{65} & \alpha_{66} + \gamma_{66}
\end{pmatrix}$$

and

$$Y^{\mathrm{H}}(B + F)X =$$

$$\begin{pmatrix}
\beta_{11} + \eta_{11} & \eta_{12} & \eta_{13} & \eta_{14} & \eta_{15} & \eta_{16} \\
\eta_{21} & \beta_{22} + \eta_{22} & \eta_{23} & \eta_{24} & \eta_{25} & \eta_{26} \\
\eta_{31} & \eta_{32} & \beta_{33} + \eta_{33} & \eta_{34} & \eta_{35} & \eta_{36} \\
\eta_{41} & \eta_{42} & \eta_{43} & \beta_{44} + \eta_{44} & \eta_{45} & \eta_{46} \\
\eta_{51} & \eta_{52} & \eta_{53} & \eta_{54} & \beta_{55} + \eta_{55} & \eta_{56} \\
\eta_{61} & \eta_{62} & \eta_{63} & \eta_{64} & \eta_{65} & \beta_{66} + \eta_{66}
\end{pmatrix}.$$

Let $\epsilon = \max\{\|E\|_2, \|F\|_2\}$, so that

$$\epsilon \|x_j\|_2 \|y_i\|_2 \geq |\gamma_{ij}|, |\eta_{ij}|.$$

For definiteness let us suppose $k = 3$ and $n = 6$. Let

$$T = \operatorname{diag}(\tau, \tau, \tau, 1, 1, 1).$$

2. Regular Matrix Pairs

Then

$$TY^H(A+E)XT^{-1} =$$

$$\begin{pmatrix} \alpha_{11}+\gamma_{11} & \gamma_{12} & \gamma_{13} & \tau\gamma_{14} & \tau\gamma_{15} & \tau\gamma_{16} \\ \gamma_{21} & \alpha_{22}+\gamma_{22} & \gamma_{23} & \tau\gamma_{24} & \tau\gamma_{25} & \tau\gamma_{26} \\ \gamma_{31} & \gamma_{32} & \alpha_{33}+\gamma_{33} & \tau\gamma_{34} & \tau\gamma_{35} & \tau\gamma_{36} \\ \tau^{-1}\gamma_{41} & \tau^{-1}\gamma_{42} & \tau^{-1}\gamma_{43} & \alpha_{44}+\gamma_{44} & \gamma_{45} & \gamma_{46} \\ \tau^{-1}\gamma_{51} & \tau^{-1}\gamma_{52} & \tau^{-1}\gamma_{53} & \gamma_{54} & \alpha_{55}+\gamma_{55} & \gamma_{56} \\ \tau^{-1}\gamma_{61} & \tau^{-1}\gamma_{62} & \tau^{-1}\gamma_{63} & \gamma_{64} & \gamma_{65} & \alpha_{66}+\gamma_{66} \end{pmatrix},$$

and

$$TY^H(B+F)XT^{-1} =$$

$$\begin{pmatrix} \beta_{11}+\eta_{11} & \eta_{12} & \eta_{13} & \tau\eta_{14} & \tau\eta_{15} & \tau\eta_{16} \\ \eta_{21} & \beta_{22}+\eta_{22} & \eta_{23} & \tau\eta_{24} & \tau\eta_{25} & \tau\eta_{26} \\ \eta_{31} & \eta_{32} & \beta_{33}+\eta_{33} & \tau\eta_{34} & \tau\eta_{35} & \tau\eta_{36} \\ \tau^{-1}\eta_{41} & \tau^{-1}\eta_{42} & \tau^{-1}\eta_{43} & \beta_{44}+\eta_{44} & \eta_{45} & \eta_{46} \\ \tau^{-1}\eta_{51} & \tau^{-1}\eta_{52} & \tau^{-1}\eta_{53} & \eta_{54} & \beta_{55}+\eta_{55} & \eta_{56} \\ \tau^{-1}\eta_{61} & \tau^{-1}\eta_{62} & \tau^{-1}\eta_{63} & \eta_{64} & \eta_{65} & \beta_{66}+\eta_{66} \end{pmatrix}.$$

As $\epsilon \to 0$, the first three of the regions $\mathcal{G}_i$ have radii that approach zero. The last three have radii that are bounded by $\sqrt{2}k\epsilon\nu'/\tau$, up to terms of order ϵ^2. Consequently, if we take

$$\tau = \frac{2\sqrt{2}k\nu'\epsilon}{\delta},$$

then for ϵ small enough the first three regions will be disjoint from the last three and hence will contain exactly three eigenvalues. The radius of these disks is bounded by $\sqrt{2}(k-1)\nu\epsilon$ up to terms of order ϵ^2. Since it is easily seen that

$$\chi(\langle\alpha_{ii},\beta_{ii}\rangle,\langle\alpha_{ii}+\gamma_{ii},\beta_{ii}+\eta_{ii}\rangle) \leq \sqrt{2}\nu\epsilon + O(\epsilon^2),$$

we have shown that

there are exactly k eigenvalues $\langle \tilde{\alpha}_i, \tilde{\beta}_i \rangle$ ($i = 1, \ldots, k$) of $(\tilde{A}, \tilde{B})$ satisfying

$$\chi(\langle \alpha_{11}, \beta_{11} \rangle, \langle \tilde{\alpha}_i, \tilde{\beta}_i \rangle + O(\epsilon^2).$$

There are four points to be made about this result. First, the assumption that the pair (A, B) is diagonalizable is not necessary. What is required is that the multiple eigenvalue have k linearly independent eigenvectors. Other multiple eigenvalues corresponding to nontrivial Jordan blocks can be handled as in Section IV.2.

Second, the above development shows that when $k = 1$ there is an eigenvalue of $(\tilde{A}, \tilde{B})$ in the region with center $\langle \alpha + y_1^H E x_1, \beta + y_1^H F x_1 \rangle$ and a radius $O(\epsilon^2)$. Thus, the Gerschgorin theorem gives an independent proof of Theorem 2.2. However, unlike our first proof, this one offers the possibility of computing the bound.

Third, the number ν is a condition number for the multiple eigenvalue in the sense that the bound on the error in the perturbed eigenvalues is proportional to the error times ν. However, the constant of proportionality grows linearly with the multiplicity of the eigenvalue.

Finally, the theory developed here is a worst-case theory depending on the largest of the numbers ν_i. In practice, the perturbed eigenvalues will tend to have condition inversely proportional to of value of ν_i particular to itself. For example the pair

$$\left[\begin{pmatrix} 2 & 0 \\ 0 & 2,000 \end{pmatrix} \begin{pmatrix} 1 & 0 \\ 0 & 1,000 \end{pmatrix} \right]$$

has a double eigenvalue $\langle 2, 1 \rangle$. But one of these eigenvalues is very sensitive to perturbations of order 0.1, whereas the other is not. The question of how to make this observation precise is an open research problem.

2.3. Diagonalizable Pairs

In this subsection we will consider the eigenvalues of diagonalizable pairs, and in particular we will generalize Theorem IV.3.3 — the well-known corollary of the Bauer–Fike theorem.

2. Regular Matrix Pairs

Theorem 2.6. Let (A, B) be a regular pair, and suppose that for some nonsingular X and Y we have

$$(Y^{\mathrm{H}} A X, Y^{\mathrm{H}} B X) = (D_A, D_B),$$

where

$$D_A = \mathrm{diag}(\alpha_1, \ldots, \alpha_n), \qquad D_B = \mathrm{diag}(\beta_1, \ldots, \beta_n).$$

Let $(\tilde{A}, \tilde{B})$ be regular. Then for every eigenvalue $\langle \tilde{\alpha}, \tilde{\beta} \rangle \in \mathcal{L}[(\tilde{A}, \tilde{B})]$ there is an eigenvalue $\langle \alpha, \beta \rangle \in \mathcal{L}[(A, B)]$ that satisfies

$$\chi(\langle \alpha, \beta \rangle, \langle \tilde{\alpha}, \tilde{\beta} \rangle) \leq \kappa_2(X) \rho_{\mathrm{L}}[(A, B), (\tilde{A}, \tilde{B})]. \tag{2.5}$$

Proof. Since both the eigenvalues of (A, B) and the equivalence class $\langle A, B \rangle_{\mathrm{L}}$ are invariant when A and B are premultiplied by a nonsingular matrix, we may assume that $U^{\mathrm{H}} = (A \ B)$ and $\tilde{U}^{\mathrm{H}} = (\tilde{A} \ \tilde{B})$ have orthonormal rows. Moreover, since $(\tilde{A}, \tilde{B})$ is regular, we may assume that $|\tilde{\alpha}|^2 + |\tilde{\beta}|^2 = 1$.

Let $\tilde{x}$ be the right eigenvector corresponding to $\langle \tilde{\alpha}, \tilde{\beta} \rangle$, normalized so that $\|x\|_2 = 1$. Then

$$\tilde{\beta} A x - \tilde{\alpha} B x = \tilde{\beta}(A - U^{\mathrm{H}} \tilde{U} \tilde{A}) x - \tilde{\alpha}(B - U^{\mathrm{H}} \tilde{U} \tilde{B}) x$$

$$= (A - U^{\mathrm{H}} \tilde{U} \tilde{A} \quad B - U^{\mathrm{H}} \tilde{U} \tilde{B}) \begin{pmatrix} \tilde{\beta} x \\ -\tilde{\alpha} x \end{pmatrix}$$

$$= (U^{\mathrm{H}} - U^{\mathrm{H}} \tilde{U} \tilde{U}^{\mathrm{H}}) \begin{pmatrix} \tilde{\beta} x \\ -\tilde{\alpha} x \end{pmatrix}$$

$$= U^{\mathrm{H}}(U U^{\mathrm{H}} - \tilde{U} \tilde{U}^{\mathrm{H}}) \begin{pmatrix} \tilde{\beta} x \\ -\tilde{\alpha} x \end{pmatrix}$$

$$= U^{\mathrm{H}}(P_U - P_{\tilde{U}}) \begin{pmatrix} \tilde{\beta} x \\ -\tilde{\alpha} x \end{pmatrix}.$$

By Theorem I.5.5 the singular values of $P_U - P_{\tilde{U}}$ are the sines of the canonical angles between the column spaces of U and $\tilde{U}$. Hence by Definition 1.23,

$$\|\tilde{\beta} A x - \tilde{\alpha} A x\|_2 \leq \rho_{\mathrm{L}}[(A, B), (\tilde{A}, \tilde{B})]. \tag{2.6}$$

Let $P^{\mathrm{H}} = X^{-1}$ and $Q^{\mathrm{H}} = Y^{-1}$. Then $(A, B) = (QD_A P^{\mathrm{H}}, QD_B P^{\mathrm{H}})$.
Now
$$I = U^{\mathrm{H}} U = Q(D_A P^{\mathrm{H}} P \bar{D}_A + D_B P^{\mathrm{H}} P \bar{D}_B) Q^{\mathrm{H}}.$$
Hence for any w,
$$w^{\mathrm{H}}(Q^{\mathrm{H}} Q)^{-1} w = w^{\mathrm{H}}(D_A P^{\mathrm{H}} P \bar{D}_A + D_B P^{\mathrm{H}} P \bar{D}_B) w$$
$$\leq \|P^{\mathrm{H}} P\|_2 w^{\mathrm{H}}(|D_A|^2 + |D_B|^2) w,$$
and therefore (Exercise 1.5)
$$w^{\mathrm{H}}(Q^{\mathrm{H}} Q) w \geq \|P^{\mathrm{H}} P\|_2^{-1} w^{\mathrm{H}}(|D_A|^2 + |D_B|^2)^{-1} w.$$
Thus
$$\|\tilde{\beta} A x - \tilde{\alpha} A x\|_2 = \|Q(\tilde{\beta} D_A - \tilde{\alpha} D_B) P^{\mathrm{H}} x\|$$
$$= [x^{\mathrm{H}} P(\tilde{\beta} D_A - \tilde{\alpha} D_B)^{\mathrm{H}} Q^{\mathrm{H}} Q(\tilde{\beta} D_A - \tilde{\alpha} D_B) P^{\mathrm{H}} x]^{\frac{1}{2}}$$
$$\geq \|P\|_2^{-1} [x^{\mathrm{H}} P(\tilde{\beta} D_A - \tilde{\alpha} D_B)(|D_A|^2 + |D_B|^2)^{-1} (\tilde{\beta} D_A - \tilde{\alpha} D_B) P^{\mathrm{H}} x]^{\frac{1}{2}}$$
$$\geq \|P\|_2^{-1} (x^{\mathrm{H}} P^{\mathrm{H}} P x)^{\frac{1}{2}} \min_i \frac{|\tilde{\beta} \alpha_i - \tilde{\alpha} \beta_i|}{\sqrt{|\alpha_i|^2 + |\beta_i|^2}}$$
$$\geq \kappa_2(X) \min_i \rho(\langle \alpha_i, \beta_i \rangle, \langle \tilde{\alpha}, \tilde{\beta} \rangle).$$
(2.7)

The theorem now follows on combining (2.6) and (2.7). ∎

Recall that we defined the spectra variation $\mathrm{sv}_A(\tilde{A})$ of $\tilde{A}$ with respect to A as the largest distance of an eigenvalue of $\tilde{A}$ from the nearest eigenvalue of A [see (IV.1.1)]. If we define $\mathrm{sv}_{(A,B)}[(\tilde{A}, \tilde{B})]$ analogously, then the conclusion of Theorem 2.6 can be written

$$\mathrm{sv}_{(A,B)}[(\tilde{A}, \tilde{B})] \leq \kappa_2(X) \rho_{\mathrm{L}}[(A, B), (\tilde{A}, \tilde{B})].$$

Although the bound (2.5) is satisfactory in many ways, it is difficult to use when we only know bounds on the perturbations E and F. One approach is to use Theorem III.4.1 to bound $\rho_{\mathrm{L}}[(A, B), (\tilde{A}, \tilde{B})]$. However, another approach is to adapt the proof of the theorem to give a direct bound.

Theorem 2.7. *In addition to the hypotheses of Theorem 2.6, suppose that the columns of X and Y are normalized so that $|D_A|^2 + |D_B|^2 = I$. Then*

$$\mathrm{sv}_{(A,B)}[(\tilde{A}, \tilde{B})] \leq \|X\|_2 \|Y\|_2 \|(E\ F)\|_2. \tag{2.8}$$

2. REGULAR MATRIX PAIRS

Proof. Consider the equivalent pairs (D_A, D_B) and $(D_A + Y^H E X, D_B + Y^H F X)$. We have

$$\tilde{\beta} D_A x - \tilde{\alpha} D_B x = \tilde{\beta} Y^H E X x - \tilde{\alpha} Y^H F X x$$

$$= Y^H (E \ F) \begin{pmatrix} \tilde{\beta} X x \\ -\tilde{\alpha} X x \end{pmatrix}$$

Hence

$$\|\tilde{\beta} D_A x - \tilde{\alpha} D_B x\|_2 \leq \|X\|_2 \|Y\|_2 \|(E \ F)\|_2.$$

Since D_A and D_B are diagonal and $|D_A|^2 + |D_B|^2 = I$, it is trivial to verify that

$$\|\tilde{\beta} D_A x - \tilde{\alpha} D_B x\|_2 \geq \min_i \chi(\langle \alpha_i, \beta_i \rangle, \langle \tilde{\alpha}, \tilde{\beta} \rangle). \ \blacksquare$$

If in the above theorem we assume that $\|y_i\|_2 = 1$, then $\|x_i\|_2$ is the condition number ν_i. Moreover, $\|Y\|_2 \leq \sqrt{n}$ and $\|X\| \leq \sqrt{n} \max_i \nu_i$. This gives the following corollary.

Corollary 2.8. *Let*

$$\nu_i = \frac{\|x_i\|_2 \|y_i\|_2}{\sqrt{|\alpha_i|^2 + |\beta_i|^2}}.$$

Then

$$\text{sv}_{(A,B)}[(\tilde{A}, \tilde{B})] \leq n \max_i \nu_i \|(E \ F)\|_2.$$

2.4. Eigenspaces

In this section we will treat the perturbation of eigenspaces, which are the natural generalization of invariant subspaces. The theory largely parallels theory of invariant subspaces developed in Chapter V, and the exposition here will be a little terser than usual.

Definition 2.9. *Let (A, B) be a regular matrix pair. The subspace $\mathcal{X}$ is an* EIGENSPACE *if*

$$\dim(A\mathcal{X} + B\mathcal{X}) \leq \dim(\mathcal{X}). \tag{2.9}$$

If $\dim(\mathcal{X}) = l$, then (2.9) implies that both $A\mathcal{X}$ and $B\mathcal{X}$ are contained in a subspace $\mathcal{Y}$ of dimension l. In other words, A and B have essentially the same effect on $\mathcal{X}$.

Eigenspaces have the following characterizations.

Theorem 2.10. *Let (A, B) be a regular matrix pair and let $\mathcal{X}$ be a subspace of dimension l. Then the following statements are equivalent.*

1. *$\mathcal{X}$ is an eigenspace of (A, B).*

2. *There are nonsingular matrices $(X_1 \ U_2)$ and $(V_1 \ Y_2)$ such that $\mathcal{R}(X_1) = \mathcal{X}$ and*

$$\begin{pmatrix} V_1^H \\ Y_2^H \end{pmatrix} A(X_1 \ U_2) = \begin{pmatrix} A_1 & H_A \\ 0 & A_2 \end{pmatrix},$$
$$\begin{pmatrix} V_1^H \\ Y_2^H \end{pmatrix} B(X_1 \ U_2) = \begin{pmatrix} B_1 & H_B \\ 0 & B_2 \end{pmatrix}. \tag{2.10}$$

Moreover the pairs (A_1, B_1) and (A_2, B_2) are regular.

3. *If the columns of X_1 for a basis for $\mathcal{X}$, then there is a regular pair (A_1, B_1) such that*

$$AX_1 B_1 = BX_1 A_1. \tag{2.11}$$

Remark 2.11. The proof will show that we may take $(X_1 \ U_2)$ and $(V_1 \ Y_2)$ to be unitary.

Proof. $1 \Rightarrow 2$: Let $(X_1 \ U_2)$ be a unitary matrix with $\mathcal{R}(X_1) = \mathcal{X}$. Since $\mathcal{X}$ is an eigenspace, both $A\mathcal{X}$ and $B\mathcal{X}$ lie in a subspace $\mathcal{Y}$ of dimension l. If we let $(V_1 \ Y_2)$ be a unitary matrix with $\mathcal{R}(V_1) = \mathcal{Y}$, then $(X_1 \ U_2)$ and $(V_1 \ Y_2)$ are the required nonsingular matrices.

$2 \Rightarrow 3$: From (2.10) we have

$$AX_1 = V_1 A_1 \quad \text{and} \quad BX_1 = V_1 B_1.$$

Since (A_1, B_1) is regular, by Theorem 1.13 there are nonsingular matrices R and S such that $\hat{A}_1 = R^H A_1 S$ and $\hat{B}_1 = R^H B_1 S$ commute. Let

2. Regular Matrix Pairs

$\hat{X}_1 = X_1 S$ and $\hat{V}_1 = V_1 R^{-H}$, so that $\mathcal{R}(\hat{X}_1) = \mathcal{X}$, $A\hat{X}_1 = \hat{V}_1 \hat{A}_1$, and $B\hat{X}_1 = \hat{V}_1 \hat{B}_1$. It follows that $A\hat{X}_1 \hat{B}_1 = B\hat{X}_1 \hat{A}_1$.

If the columns of X_1 form a basis for $\mathcal{X}$ then $X_1 = \hat{X}_1 T$ for some nonsingular matrix T. The conclusion follows on setting $B_1 = T^{-1}\hat{B}_1$ and $A_1 = T^{-1}\hat{A}_1$.

$3 \Rightarrow 1$: By Theorem 1.13 we may assume that $A_1 = \mathrm{diag}(J, I)$ and $B_1 = \mathrm{diag}(I, N)$. Let $P = AX_1$ and $Q = BX_1$. Then with the natural partitioning,

$$(P_1 \; P_2) \begin{pmatrix} I & 0 \\ 0 & N \end{pmatrix} = (Q_1 \; Q_2) \begin{pmatrix} J & 0 \\ 0 & I \end{pmatrix}.$$

It follows that $\mathcal{R}(Q_1) \subset \mathcal{R}(P_1)$ and $\mathcal{R}(P_2) \subset \mathcal{R}(Q_2)$. Hence

$$\dim[\mathcal{R}(P) + \mathcal{R}(Q)] = \dim[\mathcal{R}(P_1) + \mathcal{R}(P_1) + \mathcal{R}(Q_1) + \mathcal{R}(Q_1)]$$
$$= \dim[\mathcal{R}(P_1) + \mathcal{R}(Q_2)] \leq l. \; \blacksquare$$

Equation (2.10) shows that in some sense the pair (A_1, B_1) is a representation of the part of (A, B) associated with $\mathcal{X}$. In particular, if $\langle \alpha, \beta \rangle$ is an eigenvalue (A_1, B_1), then it is an eigenvalue of (A, B). Moreover, if B is nonsingular, then X_1 is an invariant subspace of $B^{-1}A$. See Exercises 2.3 and 2.2.

Equation (2.10) implies that $\mathcal{R}(Y_2)$ is a left eigenspace of (A, B) with representation (A_2, B_2). We will say that $\mathcal{X}$ is a SIMPLE EIGENSPACE if

$$\mathcal{L}[(A_1, B_1)] \cap \mathcal{L}[(A_2, B_2)] = \emptyset.$$

By Theorem 1.11 this is sufficient for the existence of matrices P and Q such that

$$\begin{pmatrix} I & Q \\ 0 & I \end{pmatrix} \begin{pmatrix} A_1 & H_B \\ 0 & A_2 \end{pmatrix} \begin{pmatrix} I & P \\ 0 & I \end{pmatrix} = \begin{pmatrix} A_1 & 0 \\ 0 & A_2 \end{pmatrix}$$

and

$$\begin{pmatrix} I & Q \\ 0 & I \end{pmatrix} \begin{pmatrix} B_1 & H_B \\ 0 & B_2 \end{pmatrix} \begin{pmatrix} I & P \\ 0 & I \end{pmatrix} = \begin{pmatrix} B_1 & 0 \\ 0 & B_2 \end{pmatrix}.$$

If we set
$$X_2 = U_2 + X_1 P$$

and
$$Y_1 = V_1 + Y_2 Q^H,$$
Then we have proved the following spectral resolution theorem.

Theorem 2.12. *Let $\mathcal{X}$ be a simple eigenspace of the regular pair (A, B). Then there are nonsingular matrices $(X_1\ X_2)$ and $(Y_1\ Y_2)$ such that*
$$\begin{pmatrix} Y_1^H \\ Y_2^H \end{pmatrix} A(X_1\ X_2) = \begin{pmatrix} A_1 & 0 \\ 0 & A_2 \end{pmatrix} \qquad (2.12)$$
and
$$\begin{pmatrix} Y_1^H \\ Y_2^H \end{pmatrix} B(X_1\ X_2) = \begin{pmatrix} B_1 & 0 \\ 0 & B_2 \end{pmatrix}. \qquad (2.13)$$

In analogy with the terminology for invariant subspaces, we call $\mathcal{R}(X_2)$ the COMPLEMENTARY EIGENSPACE. The spaces $\mathcal{R}(Y_1)$ and $\mathcal{R}(Y_2)$ are the corresponding left eigenspaces.

Turning now to the perturbation of eigenspaces, we begin with an approximation theorem. Let $(X_1\ U_2)$ and $(V_1\ Y_2)$ be nonsingular and set
$$\begin{pmatrix} V_1^H \\ Y_2^H \end{pmatrix} A(X_1\ U_2) = \begin{pmatrix} A_1 & H_A \\ G_A & A_2 \end{pmatrix},$$
$$\begin{pmatrix} V_1^H \\ Y_2^H \end{pmatrix} B(X_1\ U_2) = \begin{pmatrix} B_1 & H_B \\ G_B & B_2 \end{pmatrix}. \qquad (2.14)$$

If $G_A = G_B = 0$, then $\mathcal{R}(X_1)$ is an eigenspace of (A, B). We now suppose that G_A and G_B are small, and ask how near $\mathcal{R}(X_1)$ is to an eigenspace.

In analogy with the ordinary eigenvalue problem, we introduce perturbations
$$\hat{X}_1 = X_1 + U_2 P \quad \text{and} \quad \hat{Y}_2 = Y_2 + V_1 Q^H \qquad (2.15)$$
and attempt to determine P and Q so that
$$\hat{Y}_2 A \hat{X}_1 = \hat{Y}_2 B \hat{X}_1 = 0.$$
This leads directly to the system of equations
$$\begin{aligned} QA_1 + A_2 P &= -G_A - QH_A P, \\ QB_1 + B_2 P &= -G_B - QH_B P. \end{aligned} \qquad (2.16)$$

2. Regular Matrix Pairs

If we set
$$\mathbf{T} = (P, Q) \mapsto (QA_1 + A_2 P, QB_1 + B_2 P),$$
then (2.16) can be written
$$\mathbf{T}(P, Q) = -(G_A + QH_A P, G_B + QH_B P). \quad (2.17)$$

To establish a perturbation bound we must introduce a norm on the space of pairs (P, Q). Here we will work with the norm $\|\cdot\|_{\mathcal{F}}$ defined by
$$\|(P, Q)\|_{\mathcal{F}} \overset{\text{def}}{=} \max\{\|P\|_{\text{F}}, \|Q\|_{\text{F}}\}.$$
If we define
$$\text{dif}[(A_1, B_1), (A_2, B_2)] \overset{\text{def}}{=} \inf_{\|(P,Q)\|_{\mathcal{F}} = 1} \|\mathbf{T}(P, Q)\|_{\mathcal{F}}, \quad (2.18)$$
then by Theorem 1.11, $\text{dif}[(A_1, B_1), (A_2, B_2)] > 0$ if and only if the spectra of (A_1, B_1) and (A_2, B_2) are disjoint. For later use note that

$$\text{dif}[(A_1 + E_1, B_1 + F_1), (A_2 + E_2, B_2 + F_2)]$$
$$\geq \text{dif}[(A_1, B_1), (A_2, B_2)] - \max\{\|E_1\|_2 + \|E_2\|_2, \|F_1\|_2 + \|F_2\|_2\}. \quad (2.19)$$

With these preliminaries, we may now turn to the approximation theorem.

Theorem 2.13. *Let the regular pair (A, B) be as in (2.14). Set*
$$\gamma = \|(G_A, G_B)\|_{\mathcal{F}}, \quad \eta = \|(H_A, H_B)\|_{\mathcal{F}}.$$
Assume that $\mathcal{L}[(A_1, B_1)] \cap \mathcal{L}[(A_2, B_2)] = \emptyset$ so that
$$\delta = \text{dif}[(A_1, B_1), (A_2, B_2)] > 0.$$
Then if
$$\frac{\eta \gamma}{\delta^2} < \frac{1}{4},$$
there is a unique solution (P, Q) of (2.17) satisfying
$$\|(P, Q)\|_{\mathcal{F}} \leq \frac{2\gamma}{\delta + \sqrt{\delta^2 - 4\gamma\eta}} < 2\frac{\gamma}{\delta}. \quad (2.20)$$

The column spaces of $\hat{X}_1$ and $\hat{Y}_2$ defined by (2.15) are complementary right and left eigenspaces of (A, B) corresponding to the regular pairs $(A_1 + H_A P, B_1 + H_B P)$ and $(A_2 + QH_A, B_2 + QH_B)$, whose spectra are disjoint.

Proof. Let $\varphi[(P,Q)] = (QH_AP, QH_BP)$. Then it is easy to see that the conditions of Theorem V.2.11 are satisfied, which establishes the existence of (P,Q) satisfying (2.17) and (2.20).

To prove the statements about $\hat{X}_1$ and $\hat{Y}_2$, consider the equivalences

$$\begin{pmatrix} I & 0 \\ Q & I \end{pmatrix} \begin{pmatrix} V_1^H \\ Y_2^H \end{pmatrix} A(X_1\ U_2) \begin{pmatrix} I & 0 \\ P & I \end{pmatrix}$$
$$= \begin{pmatrix} V_1^H \\ \hat{Y}_2^H \end{pmatrix} A(\hat{X}_1\ U_2) = \begin{pmatrix} A_1 + H_AP & H_A \\ 0 & A_2 + QH_A \end{pmatrix}$$

and

$$\begin{pmatrix} I & 0 \\ Q & I \end{pmatrix} \begin{pmatrix} V_1^H \\ Y_2^H \end{pmatrix} B(X_1\ U_2) \begin{pmatrix} I & 0 \\ P & I \end{pmatrix}$$
$$= \begin{pmatrix} V_1^H \\ \hat{Y}_2^H \end{pmatrix} B(\hat{X}_1\ U_2) = \begin{pmatrix} B_1 + H_BP & H_B \\ 0 & B_2 + QH_B \end{pmatrix}.$$

This shows that $\hat{X}_1$ and $\hat{Y}_2$ are complementary right and left eigenspaces.

To prove the statement about the spectra, note that by (2.19)

$$\text{dif}[(A_1 + H_AP, B_1 + H_BP), (A_2 + QH_A, B_2 + QH_B)]$$
$$\geq \delta - \max\{\|H_A\|_F(\|P\|_F + \|Q\|_F)\|H_B\|_F(\|P\|_F + \|Q\|_F)\}$$
$$\geq \delta - 2\|(H_A, H_B)\|_{\mathcal{F}}\|(P,Q)\|_{\mathcal{F}}$$
$$> \delta - 4\frac{\eta\gamma}{\delta} > 0. \quad \blacksquare$$

If (X_1, U_2) and (V_1, Y_2) are unitary (see Remark 2.11), then (2.20) bounds the tangents of the canonical angles between $\mathcal{R}(X_1)$ and $\mathcal{R}(\hat{X}_1)$ or $\mathcal{R}(Y_1)$ and $\mathcal{R}(\hat{Y}_1)$, just as in the ordinary eigenvalue problem. Unfortunately, the theorem provides only a single bound for both P and Q. This problem is characteristic of the perturbation theory of matrix pairs.

There is nothing sacred about the norm $\|\cdot\|_{\mathcal{F}}$. Any norm that allows the conditions of Theorem 2.11 to be verified will do.

2. Regular Matrix Pairs 309

Although the function dif is nonzero if and only if the spectra of its arguments are disjoint, its size is not directly related to the distance between the spectra—either in the complex plane or on the Riemann sphere. In fact, multiplying the arguments of dif by a common scalar increases dif by the absolute value of the scalar without changing the spectra of the arguments.

From (2.14), we see that if $(X_1\ U_2)$ and $(V_1\ Y_2)$ are unitary, then η is the $\mathcal{F}$-norm of a perturbation $(E\ F)$ such that $\mathcal{R}(X_1)$ is an eigenspace of $(A+E, B+F)$. Namely, take

$$E = (V_1\ Y_2)\begin{pmatrix} 0 & 0 \\ -G_A & 0 \end{pmatrix}\begin{pmatrix} X_1^{\mathrm{H}} \\ U_2 \end{pmatrix} = -Y_2 G_A X_1^{\mathrm{H}}$$

and

$$F = (V_1\ Y_2)\begin{pmatrix} 0 & 0 \\ -G_B & 0 \end{pmatrix}\begin{pmatrix} X_1^{\mathrm{H}} \\ U_2 \end{pmatrix} = -Y_2 G_B X_1^{\mathrm{H}}.$$

However, this backward perturbation is not necessarily the smallest one with this property.

There is a perturbation theorem corresponding to Theorem 2.13. Its proof is left as an exercise.

Theorem 2.14. *Let $\mathcal{R}(X_1)$ be an eigenspace of the regular pair (A, B), and let the pair have the decomposition (2.10). Given the perturbation (E, F), let*

$$\begin{pmatrix} V_1^{\mathrm{H}} \\ Y_2^{\mathrm{H}} \end{pmatrix} E(X_1\ U_2) = \begin{pmatrix} E_{11} & E_{12} \\ E_{21} & E_{22} \end{pmatrix},$$

$$\begin{pmatrix} V_1^{\mathrm{H}} \\ Y_2^{\mathrm{H}} \end{pmatrix} F(X_1\ U_2) = \begin{pmatrix} F_{11} & F_{12} \\ F_{21} & F_{22} \end{pmatrix}.$$

Set

$$\tilde{\gamma} = \|(E_{21}, F_{21})\|_{\mathcal{F}},$$
$$\tilde{\eta} = \|(H_A + E_{12}, H_B + F_{12})\|_{\mathcal{F}},$$
$$\tilde{\delta} = \mathrm{dif}[(A_1, B_1), (A_2, B_2)]$$
$$\qquad - \max\{\|E_{11}\|_{\mathrm{F}} + \|E_{22}\|_{\mathrm{F}}, \|F_{11}\|_{\mathrm{F}} + \|F_{22}\|_{\mathrm{F}}\}.$$

If $\tilde{\delta} > 0$ and

$$\frac{\tilde{\eta}\tilde{\gamma}}{\tilde{\delta}^2} < \frac{1}{4},$$

then there are matrices P and Q satisfying

$$\|(P,Q)\|_{\mathcal{F}} \leq \frac{2\tilde{\gamma}}{\tilde{\delta} + \sqrt{\tilde{\delta}^2 - 4\tilde{\gamma}\tilde{\eta}}} < 2\frac{\tilde{\gamma}}{\tilde{\delta}}$$

such that the columns of

$$\tilde{X}_1 = X_1 + U_2 P \quad \text{and} \quad \tilde{Y}_2 = Y_2 + V_1 Q^{\mathrm{H}}$$

span left and right complementary eigenspaces of $(A+E, B+F)$ corresponding to the regular pairs

$$[A_1 + E_{11} + (H_A + E_{12})P,\; B_1 + F_{11} + (H_B + F_{12})P]$$

and

$$[A_2 + E_{22} + Q(H_A + E_{12}),\; B_2 + F_{22} + Q(H_B + F_{12})]$$

The spectra of these pairs are disjoint.

If instead of starting with the block triangularization (2.10), we start with a spectral resolution — that is with a block diagonal form — then H_A and H_B vanish and we obtain a sharper bound on the spectra.

Theorem 2.15. *Let $\mathcal{R}(X_1)$ be an eigenspace of the regular pair (A, B), and let the pair have the spectral resolution (2.12) and (2.13). Given the perturbation (E, F), let*

$$\begin{pmatrix} V_1^{\mathrm{H}} \\ Y_2^{\mathrm{H}} \end{pmatrix} E(X_1\; U_2) = \begin{pmatrix} E_{11} & E_{12} \\ E_{21} & E_{22} \end{pmatrix},$$

$$\begin{pmatrix} V_1^{\mathrm{H}} \\ Y_2^{\mathrm{H}} \end{pmatrix} F(X_1\; U_2) = \begin{pmatrix} F_{11} & F_{12} \\ F_{21} & F_{22} \end{pmatrix}.$$

Set

$$\tilde{\gamma} = \|(E_{21}, F_{21})\|_{\mathcal{F}},$$
$$\tilde{\eta} = \|A + E_{12}, F_{12})\|_{\mathcal{F}},$$
$$\tilde{\delta} = \mathrm{dif}[(A_1, B_1), (A_2, B_2)]$$
$$\quad - \max\{\|E_{11}\|_{\mathrm{F}} + \|E_{22}\|_{\mathrm{F}}, \|F_{11}\|_{\mathrm{F}} + \|F_{22}\|_{\mathrm{F}}\}.$$

If $\tilde{\delta} > 0$ and

$$\frac{\tilde{\eta}\tilde{\gamma}}{\tilde{\delta}^2} < \frac{1}{4},$$

2. REGULAR MATRIX PAIRS

then there are matrices P and Q satisfying

$$\|(P,Q)\|_{\mathcal{F}} \leq \frac{2\tilde{\gamma}}{\tilde{\delta} + \sqrt{\tilde{\delta}^2 - 4\tilde{\gamma}\tilde{\eta}}} < 2\frac{\tilde{\gamma}}{\tilde{\delta}}$$

such that the columns of

$$\tilde{X}_1 = X_1 + U_2 P \quad \text{and} \quad \tilde{Y}_2 = Y_2 + V_1 Q^{\mathrm{H}}$$

span left and right complementary eigenspaces of $(A+E, B+F)$ corresponding to the regular pairs

$$(A_1 + E_{11} + E_{12}P, \; B_1 + F_{11} + F_{12}P)$$

and

$$(A_2 + E_{22} + QE_{12}, \; B_2 + F_{22} + QF_{12})$$

The spectra of these pairs are disjoint.

When X_1 has only one column—i.e., when it is an eigenvector—Theorem 2.15 shows that the approximation $(\tilde{\alpha}_1, \tilde{\beta}_1) \cong (\alpha_1 + y_1^{\mathrm{H}} E x_1, \beta_1 + y_1^{\mathrm{H}} F x_1)$ is accurate up to terms of second order in the error. Thus the theorem gives another proof of Theorem 2.2.

Notes and References

The first order perturbation analysis is new, as is the systematic use of the condition number ν defined by (2.2).

The Gerschgorin theory and its application to multiple eigenvalues is from a paper by Stewart [204, 1975]. The simplified bounds are due to Sun [228, 1985].

The generalization of the Bauer–Fike theorem is due to Elsner and Sun [67, 1982]. This paper also contains generalizations of Henrici's theorem and of the Hoffman–Wielandt theorem for "normal" pairs—pairs for which $B^{-1}A$ is normal (in the case where B is nonsingular).

The perturbation theory for eigenspaces is taken from a paper by Stewart [201, 1972], where eigenspaces were called deflating subspaces. Although this theory is asymptotically sharp, it is complicated by the fact that the function dif is not easy to interpret. When the concern is with eigenvectors, it is possible to write out explicit perturbation expansions (Exercise 2.6).

Exercises

1. Let (A, B) be a regular matrix pair. Then there is a permutation matrix P such that no diagonal of (AP, BP) is $(0,0)$. [Hint: Use Theorem II.3.14.]

2. Let $\mathcal{X}$ be an eigenspace of the regular pair (A, B) and let $AX_1B_1 = BX_1A_1$ as in (2.11). Show that if z is an eigenvector of (A_1, B_1) then X_1B_1z (or X_1A_1z if $B_1z = 0$) is an eigenvector of (A, B). Conversely if $x \in \mathcal{X}$ is an eigenvector of (A, B), then there is an eigenvector z of (A_1, B_1) such that $x = X_1B_1z$ (or $x = X_1A_1z$ if $B_1z = 0$).

3. Let $\mathcal{X}$ be an eigenspace of the regular pair (A, B). Show that if B is nonsingular then $\mathcal{X}$ is an invariant subspace of $B^{-1}A$.

4. Show that
$$\text{dif}[(A_1 + E_1, B_1 + F_1), (A_2 + E_2, B_2 + F_2)]$$
$$\geq \text{dif}[(A_1, B_1), (A_2, B_2)] - \max\{\|E_1\|_2 + \|E_2\|_2, \|F_1\|_2 + \|F_2\|_2\}$$

5. Show that
$$\text{dif}[(A_1, I), (A_2, I)] \leq \text{sep}_\text{F}(A_1, A_1).$$
Moreover, if $\|A_1\|_2, \|A_1\|_2 \leq 1$, then
$$\text{dif}[(A_1, I), (A_2, I)] \geq \frac{1}{2}\text{sep}_\text{F}(A_1, A_1).$$

6. Under the hypotheses of Theorem 2.13, show that when $A_1 = \alpha_1$ is a scalar,
$$\hat{x}_1 = x_1 - U_2(\beta_1 A_2 - \alpha_1 B_2)^{-1}(\beta_1 g_A - \alpha_1 g_B) + O(\|(g_A, g_B)\|_\text{F}^2).$$

3. Definite Matrix Pairs

In this the concluding section we will treat the perturbation of eigenvalues and eigenspaces of definite matrix pairs. We begin with the analogue of Corollary IV.4.6, which gives a uniform bound for all the eigenvalues of a definite pair. We then look at the specialization to definite pairs of our general theory of eigenspaces developed in the last section. Finally we consider some direct bounds for eigenspaces.

> Throughout this section (A, B) will denote a definite matrix pair of order n.

3. Definite Matrix Pairs

3.1. Eigenvalues of Definite Pairs

Let us begin with some general observations on the condition of eigenvalues of a definite matrix pair. If x is an eigenvector of (A, B) corresponding to the eigenvalue $\langle \alpha, \beta \rangle = \langle x^H A x, x^H B x \rangle$, then the number

$$\nu = \frac{\|x\|_2^2}{\sqrt{(x^H A x)^2 + (x^H B x)^2}}$$

is a condition number for $\langle \alpha, \beta \rangle$ in the chordal metric. This fact has two consequences.

First, the eigenvalues of a definite pair, unlike the eigenvalues of a Hermitian matrix, are not automatically well conditioned. As in the Hermitian case, small eigenvalues can be ill conditioned in a relative sense; but eigenvalues of ordinary size can be ill conditioned in an absolute sense. For example, the eigenvalue $\langle 1 \rangle$ of the pair

$$\left[\begin{pmatrix} 1 & 0 \\ 0 & 0.002 \end{pmatrix}, \begin{pmatrix} 1 & 0 \\ 0 & 0.001 \end{pmatrix} \right]$$

is insensitive to perturbations of magnitude 10^{-4}, but the eigenvalue $\langle 2 \rangle$ is quite sensitive.

Second, we defined (A, B) to be definite if the number

$$\gamma(A, B) = \min_{\|x\|_2 = 1} \sqrt{(x^H A x)^2 + (x^H B x)^2}$$

is nonzero. We now see that $\gamma^{-1}(A, B)$ is an upper bound on the condition of the eigenvalues. Thus although the eigenvalues of a definite pair can be ill conditioned, the degree of ill conditioning is bounded.

The motor that drives the perturbation theory of Hermitian matrices is the natural ordering of the real line, which defines an association between the eigenvalues of a Hermitian matrix and its perturbation. Eigenvalues of definite pairs also have an ordering, although it is not as natural. To define it, let (A, B) and $(\tilde{A}, \tilde{B})$ be definite pairs. By Theorem 1.18 the field of values $\mathcal{F}(A + iB)$ lies in a half plane that does not contain the origin and $\mathcal{F}(\tilde{A} + i\tilde{B})$ lie in another such half plane. Therefore, there is a ray $\mathcal{O}$, emanating from the origin that lies in neither half plane. Given any real pair $(\alpha, \beta) \neq (0, 0)$, define $\theta(\alpha, \beta)$ to be the angle the line from the origin to (α, β) makes with $\mathcal{O}$, measured clockwise.

This construction allows us to associate angles with the eigenvalues of a definite pair and a perturbation of the pair. Specifically, we will suppose the pair (A, B) has eigenvalues $\langle \alpha_i, \beta_i \rangle$ ($i = 1, \ldots, n$) and set

$$\theta_i = \theta(\alpha_i, \beta_i).$$

The numbers θ_i are called the EIGENANGLES of (A, B). We will assume that the eigenangles are ordered so that

$$0 \le \theta_1 \le \cdots \le \theta_n < \pi. \qquad (3.1)$$

The eigenangles of the pair $(\tilde{A}, \tilde{B})$ are defined similarly.

Eigenangles have a variational characterization.

Lemma 3.1. *With the ordering (3.1), the eigenangles of the definite pair (A, B) satisfy*

$$\theta_i = \min_{\dim(\mathcal{X})=i} \max_{\substack{x \in \mathcal{X} \\ x \ne 0}} \theta(x^H A x, x^H B x) \qquad (3.2)$$

and

$$\theta_i = \max_{\dim(\mathcal{X})=n-i+1} \min_{\substack{x \in \mathcal{X} \\ x \ne 0}} \theta(x^H A x, x^H B x). \qquad (3.3)$$

Proof. By Theorem 1.18, we may assume that B is positive definite. Then for some fixed angle θ_0,

$$\theta(x^H A x, x^H B x) = \theta_0 + \cot^{-1}\left(\frac{x^H A x}{x^H B x}\right).$$

The lemma now follows from Fischer's min-max characterization (Corollary 1.16). ∎

The main theorem of this section bounds the chordal metric of the perturbation of the eigenvalues in terms of the metric ρ_D introduced in Section 1.

Theorem 3.2. *Let (A, B) be a definite pair and let $(\tilde{A}, \tilde{B}) = (A + E, B + F)$. If*

$$\zeta \equiv \max_{\|x\|_2=1} \sqrt{\frac{(x^H E x)^2 + (x^H F x)^2}{(x^H A x)^2 + (x^H B x)^2}} < 1,$$

3. Definite Matrix Pairs

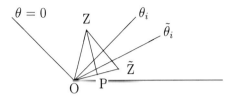

Figure 3.1: Eigenangles and Their Bounds

then $(\tilde{A}, \tilde{B})$ is definite. Moreover, if the eigenvalues $\langle \alpha_i, \beta_i \rangle$ are ordered so that their eigenangles θ_i are nondecreasing and the eigenvalues $\langle \tilde{\alpha}_i, \tilde{\beta}_i \rangle$ $(i = 1, \ldots, n)$ are ordered similarly, then $|\theta_i - \tilde{\theta}_i| < \frac{\pi}{2}$ and

$$\chi(\langle \alpha_i, \beta_i \rangle, \langle \tilde{\alpha}_i, \tilde{\beta}_i \rangle) \leq \rho_{\mathrm{D}}[(A, B), (\tilde{A}, \tilde{B})], \qquad i = 1, \ldots, n. \tag{3.4}$$

Proof. Recalling that $(\tilde{A}, \tilde{B})$ is definite if and only if $\gamma(\tilde{A}, \tilde{B}) > 0$, we have

$$\gamma(\tilde{A}, \tilde{B}) \geq \min_{\|x\|_2 = 1} \left\{ \sqrt{(x^{\mathrm{H}} A x)^2 + (x^{\mathrm{H}} B x)^2} - \sqrt{(x^{\mathrm{H}} E x)^2 + (x^{\mathrm{H}} F x)^2} \right\}$$
$$\geq (1 - \zeta)\gamma(A, B) > 0.$$

Hence $(\tilde{A}, \tilde{B})$ is definite.

Now suppose that $\tilde{\theta}_i \geq \theta_i$. Let $\mathcal{X}$ be a subspace for which the minimum is attained in (3.2). Then

$$\tilde{\theta}_i \leq \max_{\substack{x \in \mathcal{X} \\ x \neq 0}} \theta(x^{\mathrm{H}} \tilde{A} x, x^{\mathrm{H}} \tilde{B} x).$$

Let x be a vector for which the above maximum is attained. Then

$$\theta(x^{\mathrm{H}} A x, x^{\mathrm{H}} B x) \leq \theta_i \leq \tilde{\theta}_i \leq \theta(x^{\mathrm{H}} \tilde{A} x, x^{\mathrm{H}} \tilde{B} x).$$

These inequalities are pictured in Figure 3.1, in which $(x^{\mathrm{H}} A x, x^{\mathrm{H}} B x)$ is denoted by Z and $(x^{\mathrm{H}} \tilde{A} x, x^{\mathrm{H}} \tilde{B} x)$ by $\tilde{\mathrm{Z}}$.

Since $\zeta < 1$, we have $|\mathrm{Z}\tilde{\mathrm{Z}}| < |\mathrm{OZ}|$. Hence the angle $\mathrm{Z}\mathrm{O}\tilde{\mathrm{Z}}$ is less than $\frac{\pi}{2}$, which implies that $\tilde{\theta}_i - \theta_i < \frac{\pi}{2}$. Moreover, if we let ZP be the line from Z perpendicular to OZ̃, then elementary geometry gives

$$\tilde{\theta}_i - \theta_i \leq \text{ZOZ̃} = \sin^{-1}\sqrt{1 - \left(\frac{|\text{OP}|}{|\text{OZ}|}\right)^2}$$

$$= \sin^{-1}\sqrt{1 - \frac{(x^{\text{H}}Axx^{\text{H}}\tilde{A}x + x^{\text{H}}Axx^{\text{H}}\tilde{A}x)^2}{[(x^{\text{H}}Ax)^2 + (x^{\text{H}}Bx)^2][(x^{\text{H}}\tilde{A}x)^2 + (x^{\text{H}}\tilde{B}x)^2]}}$$

$$= \sin^{-1}\chi[(x^{\text{H}}Ax, x^{\text{H}}Bx),(x^{\text{H}}\tilde{A}x, x^{\text{H}}\tilde{B}x)]$$

$$\leq \sin^{-1}\rho_{\text{D}}[(A,B),(\tilde{A},\tilde{B})]. \tag{3.5}$$

But is is easily verified that $\chi[(\alpha_i,\beta_i),(\tilde{\alpha}_i,\tilde{\beta}_i)] = \sin(\tilde{\theta}_i - \theta_i)$. Hence (3.4) holds when $\theta_i \leq \tilde{\theta}_i$. The case $\theta_i \geq \tilde{\theta}_i$ is established in a similar manner, beginning with the characterization (3.3). ∎

We may obtain a bound in terms of $\|E\|$ and $\|F\|$ by observing that

$$\rho_{\text{D}}[(A,B),(\tilde{A},\tilde{B})] \leq \zeta \leq \frac{\sqrt{\|E\|^2 + \|F\|^2}}{\gamma(A,B)}.$$

Thus, we have the following corollary.

Corollary 3.3. *If*

$$\frac{\sqrt{\|E\|^2 + \|F\|^2}}{\gamma(A,B)} < 1,$$

the pair $(\tilde{A},\tilde{B})$ is definite and

$$\rho(\langle\alpha_i,\beta_i\rangle,\langle\tilde{\alpha}_i,\tilde{\beta}_i\rangle) \leq \frac{\sqrt{\|E\|^2 + \|F\|^2}}{\gamma(A,B)}, \qquad i = 1,\ldots,n. \tag{3.6}$$

Theorem 3.2 and its Corollary 3.3 have pretty forms, but their content is less than satisfactory. The bound (3.6), for example, depends on $\gamma(A,B)^{-1}$, which is greater than the largest individual condition number of the eigenvalues. Of course it is to be expected that a bound for all the eigenvalues would depend on $\max_i \nu_i$, since it must take into account the worst case. The trouble is that $\gamma(A,B)^{-1}$ can be arbitrarily larger than $\max_i \nu_i$, as the following example shows.

3. DEFINITE MATRIX PAIRS

Example 3.4. Let
$$A = \begin{pmatrix} 1 & 0 \\ 0 & -1 \end{pmatrix},$$
and for $\eta > 0$ let
$$B = \frac{1}{2}\begin{pmatrix} 1+\eta & \eta-1 \\ \eta-1 & 1+\eta \end{pmatrix}.$$
The eigenvalues of B are 1 and η, the latter corresponding to the eigenvector $\mathbf{1}$. Since $\mathbf{1}^H A \mathbf{1} = 0$,
$$\gamma(A, B) = \eta^{-1}.$$
Since $\operatorname{trace}(A^{-1}B) = 0$ and $\det(A^{-1}B) = \eta$, the eigenvalues of the pair (A, B) are $\langle \pm 1/\sqrt{\eta} \rangle$.

Now the corresponding eigenvectors are of the form $x = (1\ \xi)^T$, where
$$\xi = \frac{1-\eta}{1 \pm 2\sqrt{\eta} - \eta} \cong 1 \mp 2\sqrt{\eta}.$$
It follows that
$$\max_i \nu_i \lesssim \frac{x^T x}{|x^T A x|} \lesssim \frac{1}{2\sqrt{\eta}}.$$
Thus $\gamma(A, B) = O(\eta^{-1})$ while $\max_i \nu_i = O(\eta^{-\frac{1}{2}})$.

Even this example would not be damning if $\gamma(A, B)$ really reflected the effects of perturbations when ζ is near one. However, even in this case the condition numbers will give a more realistic estimate. The reason is that for the second inequality in (3.5) to be realistic, the values of $(x^H A x)^2 + (x^H B x)^2$ and $(x^H \tilde{A} x)^2 + (x^H \tilde{B} x)^2$ must be nearly minimal, which is unlikely.

3.2. Eigenspaces

The theory of eigenspaces for definite pairs, like its counterpart for Hermitian matrices, is both simpler and more complex than the general theory. On the one hand, the assumption of definiteness simplifies the general theory; on the other hand the same assumption gives us more structure to exploit in extending the theory.

The basic fact of eigenspaces of definite pairs is that a right eigenspace is also a left eigenspace.

Theorem 3.5. Let (A, B) be definite. Let the columns of X_1 span an eigenspace of (A, B). Then there is a matrix X_2 such that $(X_1 \ X_2)$ is nonsingular and the pair (A, B) has the spectral resolution

$$\begin{pmatrix} X_1^H \\ X_2^H \end{pmatrix} A (X_1 \ X_2) = \begin{pmatrix} A_1 & 0 \\ 0 & A_2 \end{pmatrix} \quad (3.7)$$

and

$$\begin{pmatrix} X_1^H \\ X_2^H \end{pmatrix} B (X_1 \ X_2) = \begin{pmatrix} B_1 & 0 \\ 0 & B_2 \end{pmatrix} \quad (3.8)$$

Moreover, X_1 and X_2 may be chosen so that A_1, A_2, B_1, B_2 are diagonal (i.e., the columns of $(X_1 \ X_2)$ are eigenvectors.)

Proof. It is easily verified that $\mathcal{R}(X_1)$ is an eigenspace of the pair $(A\cos\phi - B\sin\phi, A\sin\phi + B\cos\phi)$. Hence by Theorem 1.18 we may assume that B is positive definite. It follow that X_1 and BX_1 are acute (Definition III.3.2). Hence if the columns of X_2 form a basis for $\mathcal{R}(X_1)_\perp$, the matrix $(X_1 \ X_2)$ is nonsingular (Exercise III.3.2) and $X_2^H B X_1 = 0$. Since $\mathcal{R}(AX_1) \subset \mathcal{R}(BX_1)$ it follows that $X_2^H A X_1 = 0$, which establishes (3.7) and (3.8).

Since the pairs (A_i, B_i) $(i = 1, 2)$ are definite, by Corollary 1.19 there are nonsingular matrices U_i such that $(U_i^H A_i U_i, U_i^H B_i U_i)$ are diagonal. If we make the substitutions $X_i \leftarrow X_i U_i$ $(i = 1, 2)$, then $(X_1 \ X_2)$ diagonalizes (A, B). ∎

The second part of the theorem allows to assume that the matrices A_1, A_2, B_1, and B_2 of the spectral resolution (3.7)–(3.8) have the form

$$A_1 = \mathrm{diag}(\alpha_1, \ldots, \alpha_k), \quad A_2 = \mathrm{diag}(\alpha_{k+1}, \ldots, \alpha_n),$$
$$B_1 = \mathrm{diag}(\beta_1, \ldots, \beta_k), \quad B_2 = \mathrm{diag}(\beta_{k+1}, \ldots, \beta_n), \quad (3.9)$$
$$\alpha_i^2 + \beta_i^2 = 1, \quad i = 1, \ldots, n.$$

In this case we will say that the spectral resolution (3.7)–(3.8) is NORMALIZED. Among other things, normalization implies that the columns of $X = (X_1 \ X_2)$ satisfy

$$\|x_i\|_2^2 = \nu_i, \quad i = 1, \ldots, n, \quad (3.10)$$

3. DEFINITE MATRIX PAIRS

and
$$\max_i \|x_i\|_2^2 \leq \gamma^{-1}(A, B).$$

The normalization of the resolution allows us to give an explicit bound for the function dif in terms of the eigenvalues of the pair (A, B).

Theorem 3.6 *Let A_1, A_2, B_1, and B_2 satisfy (3.9), and let*
$$\delta = \min_{\substack{1 \leq i \leq k \\ k+1 \leq j \leq n}} \chi[\langle \alpha_i, \beta_i \rangle, \langle \alpha_j, \beta_j \rangle].$$

Then
$$\frac{\delta}{\sqrt{2}} \leq \operatorname{dif}[(A_1, B_1), (A_2, B_2)] \leq \delta.$$

Proof. To establish the lower bound we must show that for all R and S the solution of the system
$$\begin{aligned} QA_1 + A_2 P &= R \\ QB_1 + B_2 P &= S \end{aligned} \tag{3.11}$$

satisfies
$$\|(P, Q)\|_{\mathcal{F}} \leq \frac{\sqrt{2}\|(R, S)\|_{\mathcal{F}}}{\delta}. \tag{3.12}$$

If we postmultiply the first of the equations (3.11) by B_1 and the second by A_1 and subtract, we get (remember that since A_1 and B_1 are diagonal, they commute)
$$A_2 P B_1 - B_2 P A_1 = R B_1 - S A_2.$$

Hence the (i, j)-element of P is given by
$$\pi_{ij} = \frac{\rho_{ij} \beta_j - \sigma_{ij} \alpha_j}{\alpha_{i+k} \beta_j - \beta_{i+k} \alpha_j}. \tag{3.13}$$

Since the diagonal pairs (A_1, B_1) and (A_2, B_2) are normalized,
$$|\pi_{ij}|^2 \leq \frac{|\rho_{ij}|^2 + |\sigma_{ij}|^2}{\delta^2}.$$

Hence
$$\|P\|_{\mathrm{F}} \leq \frac{\sqrt{\|R\|_{\mathrm{F}}^2 + \|S\|_{\mathrm{F}}^2}}{\delta} \leq \frac{\sqrt{2}\|(R, S)\|_{\mathcal{F}}}{\delta}.$$

By a similar argument,

$$\|Q\|_{\mathrm{F}} \leq \frac{\sqrt{2}\|(R,S)\|_{\mathcal{F}}}{\delta},$$

and (3.12) follows.

To establish the upper bound, we must show that there are matrices R and S such that the solution of the system (3.11) satisfies

$$\|(P,Q)\|_{\mathcal{F}} \geq \frac{\|(R,S)\|_{\mathcal{F}}}{\delta}.$$

Let the minimum in the definition of δ occur for the pairs $\langle \alpha_{k+i}, \beta_{k+i}\rangle$ and $\langle \alpha_j, \beta_j\rangle$. Let $R = \text{sign}(\beta_j)\mathbf{1}_i\mathbf{1}_j^{\mathrm{T}}$ and $R = \text{sign}(\alpha_j)\mathbf{1}_i\mathbf{1}_j^{\mathrm{T}}$, so that $\|(R,S)\|_{\mathcal{F}} = 1$. Then from (3.13),

$$\|(P,Q)\|_{\mathcal{F}} \geq |\pi_{ij}| = \frac{|\alpha_j| + |\beta_j|}{\delta} \geq \frac{1}{\delta} = \frac{\|(R,S)\|_{\mathcal{F}}}{\delta}. \blacksquare$$

We may now combine all these facts into a perturbation theorem which is essentially a corollary of Theorem 2.14.

Theorem 3.7. *Let the definite pair* (A, B) *have the spectral resolution (3.7)–(3.8) satisfying (3.9). Let* ν_i $(i = 1, \ldots, n)$ *be the condition numbers of the eigenvalues of* (A, B), *and set*

$$\nu = \max_i \nu_i.$$

Given the Hermitian perturbations $\tilde{A} = A + E$ *and* $\tilde{B} = B + F$ *set*

$$\begin{pmatrix} X_1^{\mathrm{H}} \\ X_2^{\mathrm{H}} \end{pmatrix} E (X_1\ X_2) = \begin{pmatrix} E_{11} & E_{21}^{\mathrm{H}} \\ E_{21} & E_{22} \end{pmatrix},$$

$$\begin{pmatrix} X_1^{\mathrm{H}} \\ X_2^{\mathrm{H}} \end{pmatrix} F (X_1\ X_2) = \begin{pmatrix} F_{11} & F_{21}^{\mathrm{H}} \\ F_{21} & F_{22} \end{pmatrix}.$$

Let

$$\tilde{\gamma} = \|(E_{21}, F_{21})\|_{\mathcal{F}}$$

3. Definite Matrix Pairs

and

$$\tilde{\delta} = \frac{1}{\sqrt{2}} \min_{\substack{1 \leq i \leq k \\ k+1 \leq j \leq n}} \rho[\langle \alpha_i, \beta_i \rangle, \langle \alpha_j, \beta_j \rangle] - n\nu \max\{\|E\|_2, \|F\|_2\}.$$

If
$$\frac{\tilde{\gamma}}{\tilde{\delta}} < \frac{1}{2},$$

then there are matrices P and Q satisfying

$$\|(P,Q)\|_{\mathcal{F}} \leq \frac{2\tilde{\gamma}}{\tilde{\delta} + \sqrt{\tilde{\delta}^2 - 4\tilde{\gamma}^2}} < 2\frac{\tilde{\gamma}}{\tilde{\delta}} \qquad (3.14)$$

such that the columns of

$$\tilde{X}_1 = X_1 + X_2 P \quad \text{and} \quad \tilde{X}_2 = X_2 + X_1 Q^{\mathrm{H}}$$

span complementary eigenspaces of $(\tilde{A}, \tilde{B})$ corresponding to the pairs

$$(A_1 + E_{11} + E_{21}^{\mathrm{H}} P + P^{\mathrm{H}} E_{21} + P^{\mathrm{H}} E_{22} P, B_1 + F_{11} + F_{21}^{\mathrm{H}} P + P^{\mathrm{H}} F_{21} + P^{\mathrm{H}} F_{22} P) \qquad (3.15)$$

and

$$(A_2 + E_{22} + E_{21} Q^{\mathrm{H}} + Q E_{12}^{\mathrm{H}} + Q E_{11} Q^{\mathrm{H}}, B_2 + F_{22} + E_{21} Q^{\mathrm{H}} + Q F_{12}^{\mathrm{H}} + Q F_{11} Q^{\mathrm{H}}). \qquad (3.16)$$

Proof. We have

$$\|E_{11}\|_{\mathrm{F}} \leq \|X_1\|_{\mathrm{F}}^2 \|E\|_2 \leq k\nu \|E\|_2,$$

the last inequality following from (3.10). Similarly $\|E_{22}\|_{\mathrm{F}} \leq (n-k)\nu \|E\|_2$, $\|F_{11}\|_{\mathrm{F}} \leq k\nu \|F\|_2$, and $\|F_{22}\|_{\mathrm{F}} \leq (n-k)\nu \|F\|_2$, so that $\tilde{\delta}$ is a lower bound on on dif$[(A_1 + E_{11}, B_1 + F_{11}), (A_2 + E_{22}, B_2 + F_{22})]$. The inequality (3.14) now follows from Theorem 2.14. The pairs (3.15) and (3.15) are obtained by considering the diagonalizing congruences

$$\begin{pmatrix} I & P^{\mathrm{H}} \\ Q & I \end{pmatrix} \begin{pmatrix} A_{11} + E_{11} & E_{21}^{\mathrm{H}} \\ E_{21} & A_{22} + E_{22} \end{pmatrix} \begin{pmatrix} I & Q^{\mathrm{H}} \\ P & I \end{pmatrix}$$

and

$$\begin{pmatrix} I & P^{\mathrm{H}} \\ Q & I \end{pmatrix} \begin{pmatrix} B_{11} + F_{11} & F_{21}^{\mathrm{H}} \\ F_{21} & B_{22} + F_{22} \end{pmatrix} \begin{pmatrix} I & Q^{\mathrm{H}} \\ P & I \end{pmatrix}. \blacksquare$$

When E and F are sufficiently small, the bound (3.14) assumes the asymptotic form
$$\|(P,Q)\|_{\mathcal{F}} \lesssim \frac{\|(E_{21}, F_{21})\|_{\mathcal{F}}}{\delta},$$
where
$$\delta = \frac{1}{\sqrt{2}} \min_{\substack{1 \leq i \leq k \\ k+1 \leq j \leq n}} \rho[\langle \alpha_i, \beta_i \rangle, \langle \alpha_j, \beta_j \rangle].$$
Since $\|E_{21}\|_{\mathrm{F}} \leq \|X_1\|_{\mathrm{F}} \|X_2\|_{\mathrm{F}} \|E\|_2$ and similarly for $\|F_{21}\|_{\mathrm{F}}$, we have
$$\|P\|_{\mathrm{F}} \lesssim \frac{\|X_1\|_{\mathrm{F}} \|X_2\|_{\mathrm{F}} \max\{\|E\|_2, \|F\|_2\}}{\delta}.$$
But $\tilde{X}_1 - X_1 = X_2 P$. Hence
$$\begin{aligned}\frac{\|\tilde{X}_1 - X_1\|_{\mathrm{F}}}{\|X_1\|_{\mathrm{F}}} &\lesssim \frac{\|X_2\|_{\mathrm{F}}^2 \max\{\|E\|_2, \|F\|_2\}}{\delta} \\ &\lesssim \frac{(n-k)\nu \max\{\|E\|_2, \|F\|_2\}}{\delta}.\end{aligned} \quad (3.17)$$

Thus the ratio of the overall condition of the eigenvalues to their separation is a condition number for the problem.

3.3. Direct Bounds

We conclude this chapter with three direct bounds for eigenspaces, which we state without proof.

The first bound is for an eigenvector.

Theorem 3.8. *Let x_1 be a eigenvector of the definite pair (A, B) with eigenvalue $\langle \lambda_1 \rangle$. Suppose that the definite pair $(\tilde{A}, \tilde{B})$ has $n-1$ eigenvalues $\langle \tilde{\lambda}_i \rangle$ $(i = 1, \ldots, n)$ such that*
$$\delta \equiv \min_{i>1} \chi(\langle \lambda_1 \rangle, \langle \tilde{\lambda}_i \rangle) > 0.$$
Let
$$\epsilon = \sqrt{\|E\|_2^2 + \|F\|_2^2}.$$
If $\epsilon/\delta < \gamma(\tilde{A}, \tilde{B})$, then there is a vector p_1 satisfying
$$\frac{\|p_1\|_2}{\|x_1\|_2} \leq \frac{\epsilon}{\delta \gamma(\tilde{A}, \tilde{B})} < 1 \quad (3.18)$$

3. Definite Matrix Pairs 323

such that $x_1 + q_1$ is an eigenvector of $(\tilde{A}, \tilde{B})$ corresponding to the remaining eigenvalue $\langle \tilde{\lambda}_1 \rangle$.

Note that this bound is closely related to the asymptotic bound (3.17). The factor $(n-k)\nu/\delta$, which multiplies the error in (3.17), corresponds to the factor $\gamma(\tilde{A}, \tilde{B})^{-1}/\delta$ in (3.18).

We turn now to theorems that bound the sin of the canonical angles between eigenspaces. As usual they come in two varieties: one in the Frobenius norm that requires no restrictions on the situation of the eigenvalues and one in any unitarily invariant norm that requires the eigenvalues to be suitably clustered.

Theorem 3.9. *Let the definite pair (A, B) be decomposed as in (3.7) and (3.8), where X_1 and X_2 have orthonormal columns. Let the analogous decomposition be given for the pair $(\tilde{A}, \tilde{B}) = (A + E, B + F)$. If*

$$\delta \equiv \min\{\chi(\langle\lambda\rangle, \langle\tilde{\lambda}\rangle) : \lambda \in \mathcal{L}[(A_1, B_1)], \tilde{\lambda} \in \mathcal{L}[(\tilde{A}_1, \tilde{B}_1)]\} > 0$$

then

$$\|\sin\Theta[\mathcal{R}(X_1), \mathcal{R}(\tilde{X}_1)]\|_F \leq \frac{\sqrt{\|A^2 + B^2\|_2}}{\gamma(A,B)\gamma(\tilde{A},\tilde{B})} \frac{\sqrt{\|EX_1\|_F^2 + \|FX_1\|_F^2}}{\delta}.$$

Finally, we state a $\sin\Theta$ theorem that is valid for all unitarily invariant norms.

Theorem 3.10. *Let the definite pair (A, B) be decomposed as in (3.7) and (3.8), where X_1 and X_2 have orthonormal columns. Let the analogous decomposition be given for the pair $(\tilde{A}, \tilde{B}) = (A + E, B + F)$. Suppose that there are numbers $\alpha \geq 0$ and $\delta > 0$ with $\alpha + \delta \leq 1$ such that for some real number γ*

$$\mathcal{L}[(A_1, B_1)] \subset \{\langle\lambda\rangle : \chi(\langle\lambda\rangle, \langle\gamma\rangle) \leq \alpha\}$$

and

$$\mathcal{L}[(\tilde{A}_2, \tilde{B}_2)] \subset \{\langle\lambda\rangle : \chi(\langle\lambda\rangle, \langle\gamma\rangle) \geq \alpha + \delta\}.$$

Then

$$\|\sin\Theta[\mathcal{R}(X_1), \mathcal{R}(\tilde{X}_1)]\|_F$$

$$\leq \frac{\pi(\alpha, \delta; \gamma)\sqrt{\|A^2 + B^2\|_2}}{\gamma(A,B)\gamma(\tilde{A},\tilde{B})} \frac{\sqrt{\|EX_1\|_F^2 + \|FX_1\|_F^2}}{\delta},$$

where

$$\pi(\alpha, \delta; \gamma) = \begin{cases} \sqrt{2} \dfrac{(\alpha+\delta)\sqrt{1-\alpha^2} + \alpha\sqrt{1-(\alpha+\delta)^2}}{2\alpha+\delta} & \text{if } \gamma \neq 0, \\ \dfrac{(\alpha+\delta)\sqrt{1-\alpha^2} + \alpha\sqrt{1-(\alpha+\delta)^2}}{2\alpha+\delta} & \text{if } \gamma = 0. \end{cases}$$

Notes and References

The connection of the number $\gamma(A, B)$ with perturbation theory for definite pairs was first noted by Crawford [48, 1976], who used it to derive bounds on the spectral variation of matrix pencils. Stewart [207, 1979] introduced the angles associated with the eigenvalues and used the induced ordering to bound the matching distance. The sharper form of the theorem given here is due to Sun [222, 1982].

Theorem 3.8 is a special case of a theorem of Stewart [207, 1979]. The $\sin\Theta$ theorems are due to Sun [223, 225, 1983]. The second paper also contains $\sin 2\Theta$ theorems. A troublesome feature of the $\sin\Theta$ theorems is the appearance of two infima γ in the denominator of the bounds. Whether both of them should be there is an open question.

Just as the singular value decomposition is related to an associated semidefinite matrix, the generalized singular value decomposition (Exercise I.5.8) is related to an associated definite pair. Sun [223, 224, 1983] gives perturbation bounds for the generalized singular value decomposition. Paige [173, 1984] gives a different derivation and bounds on the CS decomposition.

Exercises

1. Show that the ill-conditioning of one eigenvalue of a definite pair can infect the others by considering the matrices $A = \mathrm{diag}(1, 10^{-8})$, $B = \mathrm{diag}(1, 2 \cdot 10^{-8})$,

$$\tilde{A} = \begin{pmatrix} 1 & \sqrt{2} \cdot 10^{-8} \\ \sqrt{2} \cdot 10^{-8} & 2 \cdot 10^{-8} \end{pmatrix},$$

and $\tilde{B} = B$.

2. Show that the eigenangles $\theta(\alpha, \beta)$ and $\theta(\tilde{\alpha}, \tilde{\beta})$ satisfy

$$\sin|\theta(\tilde{\alpha}, \tilde{\beta}) - \theta(\alpha, \beta)| = \chi(\langle \alpha, \beta \rangle, \langle \tilde{\alpha}, \tilde{\beta} \rangle).$$

References

[1] N. N. Abdelmalek (1974). "On the Solution of Least Squares Problems and Pseudo-Inverses." *Computing* **13**, 215–228.

[2] S. N. Afriat (1956). "On the Latent Vectors and Characteristic Values of Products of Pairs of Symmetric Idempotents." *Quarterly Journal of Mathematics* **7**, 76–78.

[3] S. N. Afriat (1957). "Orthogonal and Oblique Projectors and the Characteristics of Pairs of Vector Spaces." *Proceedings of the Cambridge Philosophical Society* **53**, 800–816.

[4] A. R. Amir-Moéz (1956). "Extremal Properties of Eigenvalues of a Hermitian Transformation and the Singular Values of the Sum and Product of Linear Transformations." *Duke Mathematical Journal* **23**, 463–476.

[5] T. Ando (1989). "Majorization, Doubly Stochastic Matrices, and Comparison of Eigenvalues." *Linear Algebra and Its Applications* **118**, 163–248.

[6] M. Arioli, I. S. Duff, and P. P. M. van Rijk (1989). "On the Augmented System Approach to Sparse Least-Squares Problems." *Numerische Mathematik* **55**, 667–685.

[7] L. Autonne (1902). "Sur les groupes linéaires, réels et orthogonaux." *Bulletin de la Société Mathématique de France* **30**, 121–134.

[8] L. Autonne (1913). "Sur les matrices hypohermitiennes et les unitairs." *Comptes Rendus de l'Academie des Sciences, Paris* **156**, 858–860.

[9] S. Banach (1922). "Sur les opérations dans les ensembles abstraits et leur application aux équations integrales." *Fundementa Mathematicae* **3**, 133–181.

[10] S. Banach (1929). "Sur les functionnelles linéaires II." *Studia Mathematica* **1**, 223–239.

[11] R. H. Bartels and G. W. Stewart (1972). "Algorithm 432: The Solution of the Matrix Equation $AX - BX = C$." *Communications of the ACM* **8**, 820–826.

[12] H. Bateman (1908). "A Formula for the Solving Function of a Certain Integral Equation of the Second Kind." *Cambridge Philosophical Transactions* **20**, 179–187.

[13] F. L. Bauer (1963). "Optimally Scaled Matrices." *Numerische Mathematik* **5**, 73–87.

[14] F. L. Bauer (1966). "Genauigkeitsfragen bei der Lösung linear Gleichungssysteme." *Zeitschrift für angewandte Mathematik und Mechanik* **46**, 409–421.

[15] F. L. Bauer and C. T. Fike (1960). "Norms and Exclusion Theorems." *Numerische Mathematik* **2**, 137–141.

[16] F. L. Bauer and A. S. Householder (1960). "Moments and Characteristic Roots." *Numerische Mathematik* **2**, 42–43.

[17] H. Baumgärtel (1972). *Endlichdimensionale Analytische Störungstheorie*. Akademie-Verlag, Berlin. Cited in [28].

[18] C. Bavely and G. W. Stewart (1979). "An Algorithm for Computing Reducing Subspaces by Block Diagonalization." *SIAM Journal on Numerical Analysis* **16**, 359–367.

[19] A. E. Beaton, D. B. Rubin, and J. L. Barone (1976). "The Acceptability of Regression Solutions: Another Look at Computational Accuracy." *Journal of the American Statistical Association* **71**, 158–168.

[20] E. F. Beckenbach and R. Bellman (1971). *Inequalities*. Springer, New York.

[21] R. Bellman (1970). *Introduction to Matrix Analysis*. Mc Graw-Hill, New York.

[22] D. A. Belsley, A. E. Kuh, and R. E. Welsch (1980). *Regression Diagnostics: Identifying Influential Data and Sources of Collinearity*. John Wiley and Sons, New York.

[23] E. Beltrami (1873). "Sulle Funzioni Bilineari." *Giornale di Matematiche ud uso Degli Studenti Delle Universita*, **11**, 98–106.

[24] A. Ben-Israel (1966). "On Error Bounds for Generalized Inverses." *SIAM Journal on Numerical Analysis* **3**, 585–592.

[25] I. Bendixson (1902). "Sur les racines d'une equation fondemental." *Acta Mathematica* **3**, 359–366.

[26] P. G. Bergman, R. Penfield, R. Schiller, and H. Zatkis (1950). "The Hamiltonian of the General Theory of Relativity with Electromagnetic Field." *Physical Review* **52**, 1950.

[27] E. Berkson (1963). "Some Metrics on the Subspaces of a Banach Space." *Pacific Journal of Mathematics* **13**, 7–22.

[28] R. Bhatia (1987). *Perturbation Bounds for Matrix Eigenvalues.* Pitman Research Notes in Mathematics. Longmann Scientific & Technical, Harlow, Essex. Published in the USA by John Wiley.

[29] R. Bhatia and C. Davis (1984). "A Bound for the Spectral Variation of a Unitary Operator." *Linear and Multilinear Algebra* **15**, 71–76.

[30] R. Bhatia, C. Davis, and P. Koosis (1987). "An Extremal Problem in Fourier Analysis with Applications to Operator Theorem." Preprint cited in [28].

[31] R. Bhatia, C. Davis, and A. McIntosh (1983). "Perturbation of Spectral Subspaces and Solution of Linear Operator Equations." *Linear Algebra and Its Applications* **52-53**, 45–67.

[32] R. Bhatia and J. A. R. Holbrook (1985). "Short Normal Paths and Spectral Variation." *Proceedings of the American Mathematical Society* **94**, 377–382.

[33] G. D. Birkhoff (1946). "Tres Observaciones Sobre el Algebra Lineal." *Universidad Nacional de Tucuman Revista, Serie A* **5**, 147–151.

[34] A. Bjerhammer (1951). "Rectangular Reciprocal Matrices with Special Reference to Geodetic Calculations." *Bulletin Géodésique* **52**, 118–220.

[35] Å. Björck (1967). "Solving Linear Least Squares Problems by Gram–Schmidt Orthogonalization." *BIT* **7**, 1–21.

[36] Å. Björck (1989). "Componentwise Backward Errors and Condition Estimates for Linear Least Square Problems." Technical Report LiTH-MATH-R-1989-13, Department of Mathematics, Linköping University.

[37] Å. Björck and G. H. Golub (1973). "Numerical Methods for Computing Angles between Linear Subspaces." *Mathematics of Computation* **27**, 579–594.

[38] Åke Björck (1987). "Least Squares Methods." Working paper, Department of Mathematic, Linköping University. To appear in *Handbook of Numerical Analysis, V.1: Solution of Equations in R^n*, P. G. Ciarlet and J. L. Lions editors, Elsevier, North Holland.

[39] C. W. Borchart (1857). "Bemerkung über die beiden vorstehenden Aufsätze." *Journal für die reine und angewandte Mathematik* **53**, 281–283.

[40] J. R. Bunch, J. W. Demmel, and C. F. Van Loan (1989). "The Strong Stability of Algorithms for Solving Symmetric Linear Systems." *SIAM Journal on Matrix Analysis and Applications* **10**, 494–499.

[41] A. L. Cauchy (1821). "Cours d'analyse de l'École Royale Polytechnique." In *Oeuvres Complétes (IIe Série)*, volume 3.

[42] A. L. Cauchy (1829). "Sur l'équation á l'aide de laquelle on détermine les inégalités séculaires des mouvements des planètes." In *Oeuvres Complétes (IIe Série)*, volume 9.

[43] F. Chatelin (1983). *Spectral Approximation of Linear Operators*. Academic Press, New York.

[44] P. L. Chebyshev (1859). "Sur l'interpolation par la methode des moindres carrés." *Mémoires de l'Academie Impériale des sciences de St.-Petersbourg, VIIe série* **15**, 1–24.

[45] A. K. Cline, C. B. Moler, G. W. Stewart, and J. H. Wilkinson (1979). "An Estimate for the Condition Number of a Matrix." *SIAM Journal on Numerical Analysis* **16**, 368–375.

[46] R. Courant (1920). "Ueber die Eigenwert bei den Differentialgleichungen der Mathematischen Physik." *Mathematische Zeitschrift* **7**, 1–57.

[47] P. J. Courtois and P. Semal (1984). "Error Bounds for the Analysis by Decomposition of Non-Negative Matrices." In G. Iazeolla, P. J. Courtois, and A. Hordijk, editors, *Mathematical Computer Performance and Reliability*, pages 287–302. Elsevier, North Holland.

[48] C. R. Crawford (1976). "A Stable Generalized Eigenvalue Problem." *SIAM Journal on Numerical Analysis* **13**, 854–860.

[49] R. B. Davies and B. Hutton (1975). "The Effects of Errors in the Independent Variables in Linear Regression." *Biometrika* **62**, 383–391.

[50] C. Davis (1963). "The Rotation of Eigenvectors by a Perturbation." *Journal of Mathematical Analysis and Applications* **6**, 159–173.

[51] C. Davis (1965). "The Rotation of Eigenvectors by a Perturbation. II." *Journal of Mathematical Analysis and Applications* **11**, 20–27.

[52] C. Davis, W. Kahan, and H. Weinberger (1982). "Norm-Preserving Dilations and Their Applications to Optimal Error Bounds." *SIAM Journal on Numerical Analysis* **19**, 445–469.

[53] C. Davis and W. M. Kahan (1970). "The Rotation of Eigenvectors by a Perturbation. III." *SIAM Journal on Numerical Analysis* **7**, 1–46.

[54] H. P. Decell (1972). "On the Derivative of the Generalized Inverse of a Matrix." *Linear and Multilinear Algebra* **1**, 357–359.

[55] J. W. Demmel (1987). "On Condition Numbers and the Distance to the Nearest Ill-Posed Problem." *Numerische Mathematik* **51**, 251–290.

[56] J. E. Dennis and J. J. Moré (1977). "Quasi-Newton Methods, Motivations and Theory." *SIAM Review* **19**, 46–89.

[57] J. E. Dennis and R. B. Schnabel (1979). "Least Change Secant Updates for Quasi-Newton Methods." *SIAM Review* **21**, 443–459.

[58] J. Desplanques (1887). "Théorème d'algèbra." *J. de Math. Spec.* **9**, 12–13. Cited in [152].

[59] J. J. Dongarra, J. R. Bunch, C. B. Moler, and G. W. Stewart (1979). *LINPACK User's Guide*. SIAM, Philadelphia.

[60] P. Van Dooren (1979). "The Computation of Kronecker's Canonical Form." *Linear Algebra and Its Applications* **1979**, 103–140.

[61] M. P. Drazin (1958). "Pseudo-Inverses in Associative Rings and Semigroups." *American Mathematical Monthly* **65**, 506–514.

[62] L. Dulmage and I. Halperin (1955). "On a Theorem of Frobenius–König and J. von Neuman's Game of Hide and Seek." *Transactions of the Royal Society of Canada, Section 3, Third Series* **49**, 23–25.

[63] C. Eckart and G. Young (1936). "The Approximation of One Matrix by Another of Lower Rank." *Psychometrika* **1**, 211–218.

[64] L. Eldń (1983). "A Weighted Pseudoinverse, Generalized Singular Values, and Constrained Least Squares Problems." *BIT* **22**, 487–502.

[65] L. Elsner (1982). "On the Variation of the Spectra of Matrices." *Linear Algebra and Its Applications* **47**, 127–138.

[66] L. Elsner (1985). "An Optimal Bound for the Spectral Variation of Two Matrices." *Linear Algebra and Its Applications* **71**, 77–80.

[67] L. Elsner and J.-G. Sun (1982). "Perturbation Theorems for the Generalized Eigenvalue Problem." *Linear Algebra and Its Applications* **48**, 341–357.

[68] V. N. Faddeeva (1959). *Computational Methods of Linear Algebra.* Dover, New York. Translated from the Russian by C. D. Benster.

[69] S. Falk (1965). "Einschliessungssätze für die Eigenvektoren normaler Matrizenpaare." *Zeitschrift für Angewandte Mathematik und Mechanik* **45**, 47–56.

[70] K. Fan (1951). "Maximum Properties and Inequalities for the Eigenvalues of Completely Continuous Operators." *Proceedings of the National Academy of Sciences* **37**, 760–766.

[71] K. Fan and A. J. Hoffman (1955). "Some Metric Inequalities in the Space of Matrices." *Proceedings of the American Mathematical Society* **6**, 111–116.

[72] D. B. Feingold and R. S. Varga (1962). "Block Diagonally Dominant Matrices and Generalizations of the Gerschgoring Circle Theorem." *Pacific Journal of Mathematics* **12**, 1241–1250.

[73] W. Feller and G. E. Forsythe (1951). "New Matrix Transformations for Obtaining Characteristic Vectors." *Quarterly of Applied Mathematics* **8**, 325–331.

[74] E. Fischer (1905). "Über quadratische Formen mit reelen Koffizienten." *Monatshefte für Mathematik und Physik* **16**, 234–249.

[75] J. G. F. Francis (1961, 1962). "The QR Transformation, Parts I and II." *Computer Journal* **4**, 265–271, 332–345.

[76] F. G. Frobenius (1911). "Über den von L. Bieberbach gefundenen Beweis eines Satzes von C. Jordan." *Sitzungsberichte der Königlich Preubischen Akademie der Wisenschaften zu Berlin*, 492–501. In [79, v. 3, pp. 492–501].

[77] F. G. Frobenius (1911). "Über die unzerlegbaren diskreten Beweguugsgruppen." *Sitzungsberichte der Königlich Preusischen Akademie der Wissenschaften zu Berlin*, 507–518. In [79, v. 3, pp. 507–518].

[78] F. G. Frobenius (1912). "Über Matrizen aus nicht negativen Elementen." *Sitzungsberichte der Königlich Preusischen Akademie der Wissenschaften zu Berlin*, 456–477. In [79, v. 3, pp. 546–567].

REFERENCES

[79] F. G. Frobenius (1968). *Ferdinand Georg Frobenius. Gesammelte Abhandlungen* (J.-P. Serre editor). Springer Verlag, Berlin.

[80] W. A. Fuller (1987). *Measurement Error Models*. John Wiley, New York.

[81] F. R. Gantmacher (1959). *The Theory of Matrices, Vols. I, II*. Chelsea Publishing Company, New York.

[82] C. F. Gauss (1809). *Theoria Motus Corporum Coelestium in Sectionibus Conicis Solem Ambientium*. Perthes and Besser, Hamburg.

[83] C. F. Gauss (1809). *Theory of the Motion of the Heavenly Bodies Moving about the Sun in Conic Sections*. Dover, New York (1963). C. H. Davis, Trans.

[84] C. F. Gauss (1821). "Theoria Combinations Observationum Erroribus Minimis Obnoxiae, Pars Prior." In *Werke, IV*, pages 1–26. Königlichen Gessellshaft der Wissenschaften zu Göttinging (1880).

[85] S. A. Gerschgorin (1931). "Über die Abgrenzung der Eigenwerte einer Matrix." *Izv. Akad. Nauk SSSR, Ser. Fiz.-Mat.* **6**, 749–754.

[86] I. Gohberg, P. Lancaster, and L. Rodman (1982). *Matrix Polynomials*. Academic Press, New York.

[87] I. Gohberg, P. Lancaster, and L. Rodman (1986). *Invariant Subspaces of Matrices with Applications*. John Wiley, New York.

[88] G. H. Golub (1965). "Numerical Methods for Solving Least Squares Problems." *Numerische Mathematik* **7**, 206–216.

[89] G. H. Golub, S. Nash, and C. Van Loan (1979). "Hessenberg-Schur Method for the Problem $AX + XB = C$." *IEEE Transactions on Automatic Control* **AC-24**, 909–913.

[90] G. H. Golub and V. Pereyra (1973). "The Differentiation of Pseudoinverses and Nonlinear Least Squares Problems Whose Variables Separate." *SIAM Journal on Numerical Analysis* **10**, 413–432.

[91] G. H. Golub and V. Pereyra (1976). "Differentiation of Pseudoinverses, Separable Nonlinear Least Squares Problems and Other Tales." In M. Z. Nashed, editor, *Generalized Inverses and Applications*, pages 303–324. Academic Press, New York.

[92] G. H. Golub and C. F. Van Loan (1980). "An Analysis of the Total Least Squares Problem." *SIAM Journal on Numerical Analysis* **17**, 883–893.

[93] G. H. Golub and C. F. Van Loan (1983). *Matrix Computations*. Johns Hopkins University Press, Baltimore, Maryland.

[94] G. H. Golub and J. H. Wilkinson (1966). "Note on the Iterative Refinement of Least Squares Solution." *Numerische Mathematik* **9**, 139–148.

[95] G. H. Golub and J. H. Wilkinson (1976). "Ill-Conditioned Eigensystems and the Computation of the Jordan Canonical Form." *SIAM Review* **18**, 578–619.

[96] W. B. Gragg and G. W. Stewart (1976). "A Stable Variant of the Secant Method for Solving Nonlinear Equations." *SIAM Journal on Numerical Analysis* **14**, 880–903.

[97] J. P. Gram (1883). "Über die Entwickelung reeler Functionen in Reihen mittelst der Methode der kleinsten Quadrate." *Journal für die reine und angewandte Mathematik* **94**, 41–73.

[98] W. H. Greub (1967). *Linear Algebra*. Springer-Verlag, New York.

[99] H. Hahn (1927). "Über lineare Gleichungssysteme in linearen Räumen." *Journal für die reine und angewandte Mathematik* **157**, 214–229.

[100] P. Hall (1935). "On Representation of Subsets." *Journal of the London Mathematical Society* **10**, 26–30.

[101] P. R. Halmos (1950). "Normal Dilations and Extensions of Operators." *Summa Brasiliensis Math.* **2**, 125–134. Cited in [237].

[102] R. J. Hanson and C. L. Lawson (1969). "Extensions and Applications of the Householder Algorithm for Solving Linear Least Squares Problems." *Mathematics of Computation* **23**, 787–812.

[103] G. H. Hardy, J. E. Littlewood, and G. Pólya (1934). *Inequalities*. Cambridge University Press, Cambridge, England.

[104] F. Hausdorff (1914). *Grundzüge der Mengenlehre*. Chelsea, New York. Reprinted by Chelsea, 1949.

[105] F. Hausdorff (1919). "Das Wertvorrat einer Bilinearform." *Mathematische Zeitschrift* **3**, 314–316.

[106] M. Haviv and L. van der Heyden (1984). "Perturbation Bounds for the Stationary Probabilities of a Finite Markov Chain." *Advances in Applied Probability* **16**, 804–818.

[107] J. Z. Hearon and J. W. Evans (1968). "Differentiable Generalized Inverses." *Journal of Research of the National Bureau of Standards, Series B* **72**, 109–113.

[108] H. V. Henderson and S. R. Searle (1981). "On Deriving the Inverse of a Sum of Matrices." *SIAM Review* **23**, 53–60.

[109] P. Henrici (1962). "Bounds for Iterates, Inverses, Spectral Variation and Fields of Values of Nonnormal Matrices." *Numerische Mathematik* **4**, 24–39.

[110] C. Hermite (1857). "Extrait d'une lettre de M. C. Hermite à M. Borchardt sur l'invariabilité des carrés positifs et des carrés négatifs dans la transformation des polynomes homogènes du second degré." *Journal für die reine und angewandte Mathematik* **53**, 271–274.

[111] N. J. Higham (1987). "A Survey of Condition Number Estimation for Triangular Matrices." *SIAM Review* **29**, 575–596.

[112] N. J. Higham (1989). "Computing Error Bounds for Regression Problems." Manuscript to appear in the Proceedings of the AMS Conference on Measurement Error Models, Humboldt, CA.

[113] N. J. Higham (1989). "How Accurate is Gaussian Elimination?" Technical Report TR 89-1024, Department of Computer Science, Cornell University. To appear in the proceedings of the 13th Dundee Biennial Conference on Numerical Analysis.

[114] N. J. Higham and G. W. Stewart (1987). "Numerical Linear Algebra in Statistical Computing." In A. Iserles and M. J. D. Powell, editors, *The State of the Art in Numerical Analysis*, pages 41–57. Clarendon Press, Oxford.

[115] A. Hirsch (1902). "Sur les racines d'une equation fondementale (Extrait d'une lettre de M. A. Hirsch à M. I. Bendixson)." *Acta Mathematica* **25**, 367–370.

[116] S. D. Hodges and P. G. Moore (1972). "Data Uncertainties and Least Squares Regression." *Applied Statistics* **21**, 185–195.

[117] A. J. Hoffman and H. W. Wielandt (1953). "The Variation of the Spectrum of a Normal Matrix." *Duke Mathematical Journal* **20**, 37–39.

[118] O. Hölder (1899). "Über einen Mittelwertsatz." *Götting Nachr.*, pages 38–47. Cited in [20].

[119] H. Hotelling (1933). "Analysis of a Complex of Statistical Variables into Principal Components." *Journal of Educational Psychology* **24**, 417–441 and 498–520.

[120] A. S. Householder (1958). "Unitary Triangularization of a Nonsymmetric Matrix." *Journal of the Association for Computing Machinery* **5**, 339–342.

[121] A. S. Householder (1964). *The Theory of Matrices in Numerical Analysis*. Dover Publishing, New York. Originally published by Ginn Blaisdell.

[122] Vasile I. Istrăţescu (1981). *Introduction to Linear Operator Theory*. Marcel Decker, New York.

[123] C. G. J. Jacobi (1857, posthumous). "Über eine elementare Transformation eines in Buzug jedes von zwei Variablen-Systemen linearen und homogenen Ausdrucks." *Journal für die reine und angewandte Mathematik* **53**, 265–270.

[124] C. Jordan (1870). *Traité des Substitutions et des Équations Algébriques*. Paris. Cited in [150].

[125] C. Jordan (1874). "Mémoire sur les formes bilinéaires." *Journal de Mathématiques Pures et Appliquées, Deuxiéme Série* **19**, 35–54.

[126] C. Jordan (1875). "Essai sur la géométrie à n dimensions." *Bulletin de la Société Mathématique* **3**, 103–174.

[127] P. Jordan and J. von Neumann (1935). "On Inner Products in Linear Metric Spaces." *Annals of Mathematics* **36**, 719–723.

[128] B. Kågström and A. Ruhe (1980). "An Algorithm for Numerical Computation of the Jordan Normal Form of a Complex Matrix." *Transactions on Mathematical Software* **6**, 398–419.

[129] W. Kahan (1966). "Numerical Linear Algebra." *Canadian Mathematical Bulletin* **9**, 757–801.

[130] W. Kahan (1967). "Inclusion Theorems for Clusters of Eigenvalues of Hermitian Matrices." Technical report, Computer Science Department, University of Toronto.

[131] W. Kahan (1972). "Conserving Confluence Curbs Ill-Conditioning." Technical Report 6, Computer Science Department, University of California, Berkeley.

[132] W. Kahan (1973). "Every $n \times n$ Matrix Z with Real Spectra Satisfies $\|Z - Z^*\| \leq \|Z + Z^*\|(\log n + 0.038)$." *Proceedings of the American Mathematical Society* **39**, 235–241.

[133] W. Kahan (1975). "Spectra of Nearly Hermitian Matrices." *Proceedings of the American Mathematical Society* **48**, 11–17.

[134] W. Kahan, B. N. Parlett, and E. Jiang (1982). "Residual Bounds on Approximate Eigensystems of Nonnormal Matrices." *SIAM Journal on Numerical Analysis* **19**, 470–484.

[135] T. Kato (1966). *Perturbation Theory for Linear Operators*. Springer Verlag, New York.

[136] D. König (1916). "Über Graphen und ihre Anwendung auf Determinantentheory und Mengenlehre." *Mathematische Annalen* **77**, 453–465.

[137] M. G. Krein and M. A. Krasnoselski (1947). "Fundamental Theorems Concerning the Extension of Herminian Operators and Some of Their Applications to the Theory of Orthogonal Polynomials and the Moment Problem." *Uspekhi Mat. Nauk.* **2**. In Russian. Cited in [27].

[138] M. G. Krein, M. A. Krasnoselski, and D. P. Milman (1948). "Concerning the Deficiency Numbers of Linear Operators in Banach Space and Some Geometric Questions." *Sbornik Trudov Inst. A. N. Ukr. S. S. R.* **11**. In Russian. Cited in [27].

[139] L. Kronecker (1890). "Algebraische Reduction der Schaaren bilinearer Formen." *Sitzungberichte der Königlich Preußischen Akademie der Wissenschaften zu Berlin*, pages 1225–1237.

[140] Peter Lancaster and Miron Tismenetski (1985). *The Theory of Matrices*. Academic Press, New York.

[141] P. S. Laplace (1820). *Théoria analytique des probabilities (3rd ed.) premier supplement: Sur l'application du calcul des probabilités a la philosophie naturelle. Oeuvres, v. 7.* Gauthier-Villars. Supplement published before 1820.

[142] C. L. Lawson and R. J. Hanson (1974). *Solving Least Squares Problems*. Prentice Hall, Englewood Cliffs, New Jersey.

[143] A. M. Legendre (1805). *Nouvelle méthodes pour la détermination des orbites des comètes*. Courcier, Paris. Cited in [219].

[144] N. J. Lehmann (1963). "Optimale Eigenwerteinschiessungen." *Numerische Mathematik* **5**, 246–272.

[145] N. J. Lehmann (1966). "Zur Verwendung optimaler Eigenwerteingrenzungen bei der Lösung symmetrischer Matrizenaufgaben." *Numerische Mathematik* **8**, 42–55.

[146] L. Lévy (1881). "Sur la possibilité du l'équilibre électrique." *Comptes Rendus de l'Académie des Sciences, Paris* **93**, 706–708.

[147] V. B. Lidskii (1950). "The Proper Values of the Sum and Product of Symmetric Matrices." *Doklady Akademii Nauk SSSR* **75**, 769–772. In Russian. Translation by C. Benster available from the National Translation Center of the Library of Congress.

[148] A. Loewy (1898). "Sur les formes quadratique défines à indétermininées conjugées de M Hermite." *C. R. Acad. Sci. Paris* **123**, 168–171. Cited in [150, p. 79].

[149] Qi-keng Lu (1963). "The Elliptic Geometry of Extended Space." *Chinese Mathematics* **4**, 54–69. Translation of an article appearing in *Acta mathematica Sinica*, 13 (1963).

[150] C. C. Mac Duffee (1946). *The Theory of Matrices*. Chelsea, New York.

[151] M. Marcus (1960). *Basic Theorems in Matrix Theory*. Applied Mathematics Series #57. National Bureau of Standards, Washington, D.C.

[152] M. Marcus and H. Minc (1964). *A Survey of Matrix Theory and Matrix Inequalities*. Allyn and Bacon, Boston.

[153] J. L. Massera and J. J. Schäffer (1958). "Linear differential equations and functional analysis I." *Annals of Math.* **67**, 517–573. Cited in [27].

[154] H. I. Medley and R. S. Varga (1968). "On Smallest Isolated Gerschgorin Disks for Eigenvalues. III." *Numerische Mathematik* **11**, 361–369.

[155] C. Meyer and G. W. Stewart (1988). "Derivatives and Perturbations of Eigenvectors." *SIAM Journal on Numerical Analysis* **25**, 679–691.

[156] H. Minkowski (1896). *Geometrie der Zahlen. I.* B. G. Teubner, Leipzig. Cited in [20].

[157] H. Minkowski (1911, posthumous). "Theorie der Konvexen Körper, inbesondere Begründung ihres Oberflächenbegriffs." In David Hilbert, editor, *Minkowski Abhandlung*. Teubner Verlag.

[158] L. Mirsky (1960). "Symmetric Gage Functions and Unitarily Invariant Norms." *Quarterly Journal of Mathematics* **11**, 50–59.

[159] L. Mirsky (1963). "Results and Problems in the Theory of Doubly Stochastic Matrices." *Zeitschrift für Wahrscheinlichkeitstheorie und verwandte Gebiete* **1**, 319–334.

[160] D. S. Mitrinović (1970). *Analytic Inequalities*. Springer, New York.

[161] C. Moler and G. W. Stewart (1973). "An Algorithm for Generalized Matrix Eigenvalue Problems." *SIAM Journal on Numerical Analysis* **10**, 241–256.

[162] E. H. Moore (1920). "On the Reciprocal of the General Algebraic Matrix." *Bulletin of the American Mathematical Society* **26**, 394–395. Abstract.

[163] M. Z. Nashed and L. B. Rall (1976). "Annotated Bibliography on Generalized Inverses and Applications." In M. Z. Nashed, editor, *Generalized Inverses and Applications*, pages 771–1041. Academic Press, New York.

[164] W. Oettli and W. Prager (1964). "Compatibility of Approximate Solution of Linear Equations with Given Error Bounds for Coefficients and Right-Hand Sides." *Numerische Mathematik* **6**, 405–409.

[165] D. P. O'Leary (1989). "On Bounds For Scaled Projections and Pseudo-Inverses." To appear in *Linear Algebra and Its Applications*.

[166] J. M. Ortega and W. C. Rheinboldt (1970). *Iterative Solution of Nonlinear Equations in Several Variables*. Academic Press, New York.

[167] A. Ostrowski (1951). "Ueber das Nichtverschwinden einer Klasse von Determinanten und die Lokalisierung der charakteristischen Wurzel von Matrizen." *Compositio Mathematica* **9**, 209–226.

[168] A. Ostrowski (1952). "Sur quelques applications des fonctions convexes et concaves au sens de I. Schur." *Journal de Mathématiques Pures et Appliquées* **117**, 253–292.

[169] A. Ostrowski (1957). "Über die Stetigkeit von charakteristischen Wurzeln in Abhängigkeit von den Matrizenelementen." *Jahresberichte der Deutsche Mathematische Verein* **60**, 40–42.

[170] D. V. Ouellette (1981). "Schur Complement and Statistics." *Linear Algebra and its Applications* **36**, 187–295.

[171] C. C. Paige (1979). "Computer Solution and Perturbation Analysis of Generalized Linear Least Squares Problems." *Mathematics of Computation* **33**, 171–184.

[172] C. C. Paige (1979). "Fast Numerically Stable Computations for Generalized Least Squares Problems." *SIAM J. on Numerical Analysis* **16**, 165–171.

[173] C. C. Paige (1984). "A Note on a Result of Sun Ji-guang: Sensitivity of the CS and GSV Decomposition." *SIAM Journal on Numerical Analysis* **21**, 186–191.

[174] C. C. Paige and M. A. Saunders (1981). "Toward a Generalized Singular Value Decomposition." *SIAM Journal on Numerical Analysis* **18**, 398–405.

[175] B. N. Parlett (1980). *The Symmetric Eigenvalue Problem.* Prentice-Hall, Englewood Cliffs, New Jersey.

[176] M. Pavel-Parvu and A. Korganoff (1969). "Iteration Functions for Solving Polynomial Equations." In B. Dejon and P. Henrici, editors, *Constructive Aspects of the Fundemental Theorem of Algebra.* John Wiley, New York.

[177] G. Peano (1888). "Intégration par séries des équations différentielles linéaires." *Mathematische Annallen* **32**, 450–456.

[178] R. Penrose (1955). "A Generalized Inverse for Matrices." *Proceedings of the Cambridge Philosophical Society* **51**, 406–413.

[179] R. Penrose (1956). "On Best Approximate Solutions of Linear Matrix Equations." *Proceedings of the Cambridge Philosophical Society* **52**, 17–19.

[180] V. Pereyra (1969). "Stability of General Systems of Linear Equations." *Aequationes Mathematicae* **2**, 194–206.

[181] É. Picard (1910). "Sur un théorèm général relatif aux équations intégrales de premièr espèce et sur quelques problèmes de physique mathématique." *Rendicondi del Circolo Matematico di Palermo* **25**, 79–97.

[182] L. Qi (1984). "Some Simple Estimates for Singular Values of a Matrix." *Linear Algebra and Its Applications* **56**, 105–119.

[183] Lord Rayleigh (J. W. Strutt) (1899). "On the Calculation of the Frequency of Vibration of a System in its Gravest Mode, with an Example from Hydrodynamics." *The Philosophical Magazine* **47**, 556–572. Cited in [175].

[184] F. Riesz and B. Sz.-Nagy (1955). *Functional Analysis.* Ungar, New York. L. F. Boron, Translator.

[185] J. L. Rigal and J. Gaches (1967). "On the Compatibility of a Given Solution with the Data of a Linear System." *Journal of the Association for Computing Machinery* **14**, 543–548.

[186] W. Ritz (1909). "Über eine neue Method zur Lösung gewisser Variationsprobleme der mathematischen Physik." *Journal für die reine und angewandte Mathematik* **135**, 1–61.

[187] H. Rohrbach (1931). "Bemerkungen zu einem Determinantensatz von Minkowski." *Jahresbericht der Deutschen Mathematiker Vereinigung* **40**, 49–53.

References

[188] M. Rosenblum (1956). "On the Operator Equation $BX - XA = Q$." *Duke Mathematical Journal* **23**, 263–269.

[189] A. Ruhe (1970). "An Algorithm for Numerical Determination of the Structure of a General Matrix." *BIT* **10**, 196–216.

[190] A. Ruhe (1970). "Perturbation Bounds for Means of Eigenvalues and Invariant Subspaces." *BIT* **10**, 343–354.

[191] H. Rutishauser (1955). "Une méthode pour la determination des valeurs propres d'une matrice." *Comptes Rendus de l'Académie des Sciences, Paris* **240**, 34–36.

[192] E. Schmidt (1907). "Zur Theorie der linearen und nichtlinearen Integralgleichungen. I Tiel. Entwicklung willkürlichen Funktionen nach System vorgeschriebener." *Mathematische Annalen* **63**, 433–476.

[193] I. Schur (1909). "Über die charakteristischen Würzeln einer linearen Substitution mit einer Anwendung auf die Theorie der Integralgleichungen." *Mathematische Annalen* **66**, 448–510.

[194] P. J. Schweitzer (1968). "Perturbation Theory and Finite Markov Chains." *Journal of Applied Probability* **5**, 401–413.

[195] C. J. Scriba (1973). "Carl Gustav Jacob Jacobi." In C. C. Gillispe, editor, *Dictionary of Scientific Biography, VII.* Charles Scribner's Sons, New York.

[196] G. A. F. Seber (1977). *Linear Regression Analysis.* John Wiley, New York.

[197] R. D. Skeel (1979). "Scaling for Numerical Stability in Gaussian Elimination." *Journal of the Association for Computing Machinery* **26**, 494–526.

[198] F. Smithes (1937). "The Eigen-values and Singular Values of Integral Equations." *Proceedings of the London Mathematical Society* **43**, 255–279.

[199] G. W. Stewart (1969). "On the Continuity of the Generalized Inverse." *SIAM Journal on Applied Mathematics* **17**, 33–45.

[200] G. W. Stewart (1971). "Error Bounds for Approximate Invariant Subspaces of Closed Linear Operators." *SIAM Journal on Numerical Analysis* **8**, 796–808.

[201] G. W. Stewart (1972). "On the Sensitivity of the Eigenvalue Problem $Ax = \lambda Bx$." *SIAM Journal on Numerical Analysis* **4**, 669–686.

[202] G. W. Stewart (1973). "Error and Perturbation Bounds for Subspaces Associated with Certain Eigenvalue Problems." *SIAM Review* **15**, 727–764.

[203] G. W. Stewart (1974). *Introduction to Matrix Computations*. Academic Press, New York.

[204] G. W. Stewart (1975). "Gerschgorin Theory for the Generalized Eigenvalue Problem $Ax = \lambda Bx$." *Mathematics of Computation* **29**, 600–606.

[205] G. W. Stewart (1977). "On the Perturbation of Pseudo-Inverses, Projections, and Linear Least Squares Problems." *SIAM Review* **19**, 634–662.

[206] G. W. Stewart (1977). "Research Development and LINPACK." In J. R. Rice, editor, *Mathematical Software III*, pages 1–14. Academic Press, New York.

[207] G. W. Stewart (1979). "The Effects of Rounding Error on an Algorithm for Downdating a Cholesky Factorization." *Journal of the Institute for Mathematics and Applications* **23**, 203–213.

[208] G. W. Stewart (1979). "A Note on the Perturbation of Singular Values." *Linear Algebra and Its Applications* **28**, 213–216.

[209] G. W. Stewart (1982). "Computing the CS Decomposition of a Partitioned Orthogonal Matrix." *Numerische Mathematik* **40**, 297–306.

[210] G. W. Stewart (1984). "On the Invariance of Perturbed Null Vectors under Column Scaling." *Numerische Mathematik* **44**, 61–65.

[211] G. W. Stewart (1984). "Rank Degeneracy." *SIAM Journal on Scientific and Statistical Computing* **5**, 403–413.

[212] G. W. Stewart (1984). "A Second Order Perturbation Expansion for Small Singular Values." *Linear Algebra and Its Applications* **56**, 231–235.

[213] G. W. Stewart (1987). "Collinearity and Least Squares Regression." *Statistical Science* **2**, 68–100.

[214] G. W. Stewart (1987). "Invariant Subspaces and Capital Punishment." Technical Report TR-1923, Department of Computer Science, University of Maryland.

[215] G. W. Stewart (1988). "Stochastic Perturbation Theory." Technical Report CS-TR2129, Department of Computer Science, University of Maryland. To appear in *SIAM Review*.

[216] G. W. Stewart (1989). "On Scaled Projections and Pseudo-Inverses." *Linear Algebra and Its Applications* **112**, 189–194.

[217] G. W. Stewart (1989). "Perturbation Theory and Least Squares with Errors in the Variables." Technical Report UMIACS-TR-89-97, CS-TR 2326, Department of Computer Science, University of Maryland. To appear in the Proceedings of the AMS Conference on Measurement Error Models, Humboldt, California.

[218] G. W. Stewart (1989). "Two Simple Residual Bounds for the Eigenvalues of Hermitian Matrices." Technical Report CS-TR 2364, Department of Computer Science, University of Maryland.

[219] S. M. Stigler (1986). *The History of Statistics*. Harvard University Press, Cambridge, Massachusetts.

[220] J.-G. Sun (1979). "A Theorem on the Perturbation of Generalized Eigenvalues." Report on the Conference Numerical Mathematics, Guangzhou, China.

[221] J.-G. Sun (1980). "Invariant Subspaces and Generalized Invariant Subspaces (II)." *Math. Numer. Sinica* **2**, 113–123. Cited in [67].

[222] J.-G. Sun (1982). "A Note on Stewart's Theorem for Definite Matrix Pairs." *Linear Algebra and Its Applciations* **48**, 331–339.

[223] J.-G. Sun (1983). "Perturbation Analysis for the Generalized Eigenvalue Problem and the Generalized Singular Value Problem." In B. Kågström and A. Ruhe, editors, *Matrix Pencils*, pages 221–244. Springer Verlag, New York.

[224] J.-G. Sun (1983). "Perturbation Analysis for the Generalized Singular Value Decomposition." *SIAM Journal on Numerical Analysis* **20**, 611–625.

[225] J.-G. Sun (1983). "Perturbation Bounds for Eigenspaces of a Definite Matrix Pair." *Numerische Mathematik* **41**, 321–343.

[226] J.-G. Sun (1984). "Estimation of the Separation of Two Matrices." *Journal of Computational Mathematics* **2**, 189–200.

[227] J.-G. Sun (1984). "On the Perturbation of the Eigenvalues of a Normal Matrix." *Math. Numer. Sinca* **6**, 334–336.

[228] J.-G. Sun (1985). "Gerschgorin Type Theorem and the Perturbation of the Eigenvalues of a Singular Pencil." *Math. Numer. Sinica* **7**, 253–264. In Chinese. Translation in *Chinese Journal of Numerical Mathematics and Applications* **10** (1988) 1–13.

[229] J.-G. Sun (1987). *Matrix Perturbation Analysis.* Academic Press, Beijing. In Chinese.

[230] J.-G. Sun (1988). "A Note on Simple Non-Zero Singular Values." *Journal of Computational Mathematics* **6**, 258–266.

[231] J.-G. Sun (1989). "A Note on Local Behavior of Multiple Eigenvalues." *SIAM Journal on Matrix Analysis and Applications* **10**, 533–541.

[232] C. A. Swanson (1961). "An Inequality for Linear Transformations with Eigenvalues." *Bulletin of the American Mathematical Society* **67**, 607–608.

[233] J. J. Sylvester (1852). "A Demonstration of the Theorem that Every Homogeneous Quadratic Polynomial is Reducible by Real Orthogonal Substitutions to the Form of a Sum of Positive and Negative Squares." *Philosopical Magazine* **2**, 138–142.

[234] J. J. Sylvester (1853). "On a Theory of the Syzygetic Relations etc." *Phil. Trans.*, pages 481–84. Cited in [39].

[235] J. J. Sylvester (1889). "Sur la réduction biorthogonale d'une forme linéo-linéaire à sa forme cannonique." *Comptes Rendus de l'Academie des Sciences, Paris* **108**, 651–653.

[236] J. J. Sylvester (1890). "On the Reduction of a Bilinear Quantic of the n^{TH} Order to the Form of a Sum of n Products by a Double Orthogonal Substitution." *Messenger of Mathematics* **19**, 42–46.

[237] Bela Sz.-Nagy (1960). *Extensions of Linear Transformations in Hilbert Space which Extend beyond This Space.* Ungar, New York. An appendix to Riesz and Sz.-Nagy [184] issued as a separate pamphlet.

[238] O. Taussky (1948). "Bounds for Characteristic Roots of Matrices." *Duke Mathematical Journal* **15**, 1043–1044.

[239] O. Taussky (1949). "A Recurring Theorem on Determinants." *American Mathematical Monthly* **46**, 672–675.

[240] J. Todd (1950). "The Condition of a Certain Matrix." *Proceedings of the Cambridge Philosophical Society* **46**, 116–118.

[241] O. Toeplitz (1918). "Das algebraische Analogon zu einem Satze von Fejér." *Mathematische Zeitschrift* **2**, 187–197.

[242] A. M. Turing (1948). "Rounding-off Errors in Matrix Processes." *The Quarterly Journal of Mechanics and Applied Mathematics*, **1**, 287–308.

REFERENCES

[243] F. Uhlig (1979). "A Recurring Theorem about Pairs of Quadratic Forms and Extensions: A Survey." *Linear Algebra and Its Applications* **25**, 219–237.

[244] A. van der Sluis (1969). "Condition Numbers and Equilibration of Matrices." *Numerische Mathematik* **14**, 14–23.

[245] A. van der Sluis (1975). "Stability of the Solutions of Linear Least Squares Problems." *Numerische Mathematik* **23**, 241–254.

[246] B. L. van der Warden (1927). "Ein Satz über Klasseneinteilungen von Endlicher Mengen." *Abhandlungen aus dem Mathematischen Seminar der Hamburgischen Universität* **5**, 185–188.

[247] C. F. Van Loan (1975). "A General Matrix Eigenvalue Algorithm." *SIAM Journal on Numerical Analysis* **12**, 819–834.

[248] C. F. Van Loan (1976). "Generalizing the Singular Value Decomposition." *SIAM Journal on Numerical Analysis* **13**, 76–83.

[249] C. F. Van Loan (1985). "On the Method of Weighting for Equality Constrained Least Squares." *SIAM Journal on Numerical Analysis* **22**, 851–864.

[250] J. M. Varah (1967). "The Computation of Bounds for the Invariant Subspaces of a General Matrix Operator." Technical Report CS 66, Computer Science Department, Stanford University.

[251] J. M. Varah (1970). "Computing Invariant Subspaces of a General Matrix When the Eigensystem Is Poorly Determined." *Mathematics of Computation* **24**, 137–149.

[252] R. S. Varga (1962). *Matrix Iterative Analysis*. Prentice-Hall, Englewood Cliffs, New Jersey.

[253] R. S. Varga (1964). "On Smallest Isolated Gerschgorin Disks for Eigenvalues." *Numerische Mathematik* **6**, 366–376.

[254] J. von Neuman (1937). "Some Matrix-Inequalities and Metrization of Matrix-Space." *Tomsk. Univ. Rev.* **1**, 286–300. In [255, v.4, pp.205–219].

[255] J. von Neuman (1962). *Collected Works* (A. H. Taub, editor). Pergamon, New York.

[256] J. von Neuman and H. H. Goldstine (1947). "Numerical Inverting of Matrices of High Order." *Bulletin of the American Mathematical Society* **53**, 1021–1099.

[257] J. H. M. Wedderburn (1934). *Lectures on Matrices.* American Mathematical Society Colloquium Publications, V. XVII. American Mathematical Society, New York.

[258] P.-Å. Wedin (1969). "On Pseudoinverses of Perturbed Matrices." Technical report, Department of Computer Science, Lund University.

[259] P.-Å. Wedin (1972). "Perturbation Bounds in Connection with Singular Value Decomposition." *BIT* **12**, 99–111.

[260] P.-Å. Wedin (1973). "Pertubation Theory for Pseudo-Inverses." *BIT* **13**, 217–232.

[261] P.-Å. Wedin (1983). "On Angles Between Subspaces." In B. Kagstrom and A. Ruhe, editors, *Matrix Pencils*, pages 263–285. Springer, New York.

[262] P.-Å. Wedin (1987). "Perturbation Theory and Condition Numbers for Generalized and Constrained Least Squares Problems." Technical Report S-901-87, Institute of Information Processing, University of Umeå.

[263] K. Weierstrass (1867). "Zur Theorie der bilinearen und quadratischen Formen." *Monatsh. Akad. Wiss. Berlin*, pages 310–38. Cited in [81].

[264] H. F. Weinberger (1974). *Variational Methods for Eigenvalue Approximation.* Society for Industrial and Applied Mathematics, Philadelphia. Cited in [175].

[265] H. Weyl (1912). "Das asymptotische Verteilungsgestez der Eigenwert linearer partieller Differentialgleichungen (mit einer Anwendung auf der Theorie der Hohlraumstrahlung)." *Mathematische Annalen* **71**, 441–479.

[266] H. Weyl (1949). "Inequalities between the Two Kinds of Eigenvalues of a Linear Transformation." *Proceedings of the National Academy of Sciences* **35**, 408–411.

[267] H. W. Wielandt (1955). "An Extremum Property of Sums of Eigenvalues." *Proceedings of the American Mathematical Society* **6**, 106–110.

[268] N. Wiener (1922). "Limit in Terms of Continuous Transformations." *Bulletin de le Société Mathématique de France* **50**, 119–134.

[269] J. H. Wilkinson (1965). *The Algebraic Eigenvalue Problem.* Clarendon Press, Oxford, England.

[270] J. H. Wilkinson (1971). "Modern Error Analysis." *SIAM Review* **13**, 548–568.

[271] J. H. Wilkinson (1972). "Note on Matrices with a Very Ill-Coditioned Eigenproblem." *Numerische Mathematik* **19**, 176–178.

[272] J. H. Wilkinson (1979). "Kronecker's Canonical Form and the QZ Algorithm." *Linear Algebra and Its Applications* **28**, 285–303.

[273] J. H. Wilkinson and G. H. Golub (1976). "Ill-Conditioned Eigensystems and the Computation of the Jordan Canonical Form." *SIAM Review* **18**, 578–619.

[274] H. Wittmeyer (1936). "Einfluß der Änderung einer Matrix auf der Lösung des zugehörigen Gleichungssystems, sowie auf die charakteristischen Zahlen und die Eigenvektoren." *Zietschrift für angewandte Mathematik und Mechanik* **16**, 287–300.

[275] T. Yamamoto (1980). "Error Bounds for Computed Eigenvalues and Eigenvectors." *Numerische Mathematik* **34**, 189–199.

[276] D. M. Young (1971). *Iterative Solution of Large Linear Systems*. Academic Press, New York.

[277] Z. Zhang (1986). "On the Perturbation of the Eigenvalues of a Non-Defective Matrix." *Math. Numer. Sinica* **6**, 106–108. In Chinese.

Notation

$\mathbf{R}$	The set of real numbers	1
$\mathbf{R}^n$	The set of real n-dimensional vectors	1
$\mathbf{R}^{m \times n}$	The set of real $m \times n$ matrices	1
$\mathbf{C}$	The set of complex numbers	1
$\mathbf{C}^n$	The set of complex n-dimensional vectors . . .	1
$\mathbf{C}^{m \times n}$	The set of complex $m \times n$ matrices	1
I (I_n)	The identity matrix (of order n)	2
$\mathbf{1}$	The vector $(1, \ldots, 1)^\mathrm{T}$	2
$\mathbf{1}_i$	The ith unit vector	2
A^T	The transpose of A	2
A^H	The conjugate transpose of A	2
A^{-1}	The inverse of A	2
$A^{-\mathrm{T}}, A^{-\mathrm{H}}$	The inverse transpose and conjugate transpose of A	2
$\mathcal{X} + \mathcal{Y}$	The set $\{x + y : x \in \mathcal{X}, y \in \mathcal{Y}\}$. Other operations on sets are defined similarly	2
$\mathcal{R}(A)$	The column space of A	2
$\mathcal{N}(A)$	The null space of A	2
$\mathrm{rank}(A)$	The rank of A	2
$\dim(\mathcal{X})$	The dimension of the subspace $\mathcal{X}$	2
$\det(A)$	The determinant of A	2
$\mathrm{trace}(A)$	The trace of A	2
$\|x\|_2$	The 2-norm of x	3

$|A|$	The matrix whose elements are the absolute values of the elements of A	3
$A > B$	A is component-wise greater than B. Other relations are defined similarly.	3
$\stackrel{\text{def}}{=}$	Formal definition	3
$\equiv$	Implicit definition	3
$\mathrm{diag}(\delta_1,\ldots,\delta_n)$	A diagonal matrix	4
$\mathcal{R}(A)^\perp, \mathcal{X}_\perp$	The orthogonal complement of $\mathcal{R}(A), \mathcal{X}$	8
$P_A, P_\mathcal{X}$	The orthogonal projection onto $\mathcal{R}(A), \mathcal{X}$	10
$P_A^\perp, P_\mathcal{X}^\perp$	The complementary projectors $I - P_A, I - P_\mathcal{X}$	10
$\mathcal{L}(A)$	The set of eigenvalues of A	14
$\phi_A(\lambda)$	The characteristic polynomial of A	15
$A^{\frac{1}{2}}, A^{-\frac{1}{2}}$	The square root of A and its inverse	20
$J_k(\lambda)$	A Jordan block of order k	20
$\mathcal{F}(\mathcal{A})$	The field of values of A	23
$\mathcal{H}(\mathcal{L})$	The convex hull of $\mathcal{L}$	24
$A \otimes B$	The Kronecker or tensor product of A and B	30
$\mathcal{S}(A)$	The set of singular values of A	31
$\|A\|_2$	The spectral norm of A	33
$\inf_2(A)$	The smallest singular value of A	33
$\Theta(\mathcal{X}, \mathcal{Y})$	The matrix of canonical angles between $\mathcal{X}$ and $\mathcal{Y}$	43
$\angle(x, y)$	The angle between x and y	45
$\|x\|_1$	The 1-norm of x	51
$\|x\|_\infty$	The ∞-norm norm of x	51
$\|x\|_p$	The Hölder p-norm of x	51
$\|A\|_F$	The Frobenius norm of A	65
$\rho(A)$	The spectral radius of A	66
$\|A\|_p$	The Hölder p-norm of A	69
$\|A\|_1$	The 1-norm or row-sum norm of A	69
$\|A\|_\infty$	The ∞-norm or the column-sum norm of A	69
$\|\cdot\|_\Phi$	The unitarily invariant norm defined by the symmetric gauge function Φ	76
$x \succ y$	x majorizes y	81
Φ_k	Fan's symmetric gauge functions	87
$\mathbf{C}_l^n\ (\mathbf{R}_l^n)$	The set of l-dimensional subspaces of $\mathbf{C}^n\ (\mathbf{R}^n)$	90
$\rho_{\mathrm{g},\nu}(\mathcal{X}, \mathcal{Y})$	The ν-gap between $\mathcal{X}$ and $\mathcal{Y}$	91

NOTATION

$A^{(i,j,k)}$	A generalized inverse satisfying Penrose's conditions i, j, and k	102
$A^\dagger$	The pseudo-inverse of A	102
$\mathrm{ae}(\alpha,\tilde\alpha)$, $\mathrm{re}(\alpha,\tilde\alpha)$	The absolute and relative errors errors in $\tilde\alpha$	115
$\mathrm{ae}(A,\tilde A)$, $\mathrm{re}(A,\tilde A)$	The absolute and relative errors errors in $\tilde A$	116
$\kappa(A)$	The condition number of A: $\|A\|\|A^\dagger\|$	119
$\kappa_{\mathrm{BS}}(A)$	The Bauer–Skeel condition number of A	128
$\mathrm{sv}_A(\tilde A)$	The spectra variation of $\tilde A$ with respect to A	167
$\mathrm{hd}(A,\tilde A)$	The Hausdorff distance between the eigenvalues of A and $\tilde A$	167
$\mathrm{md}(A,\tilde A)$	The matching distance between the eigenvalues of A and $\tilde A$	167
$\delta_\nu(A)$	The ν-departure of A from normality	171
$\mathrm{md}_2(A,\tilde A)$	The 2-norm matching distance between the eigenvalues of A and $\tilde A$	189
$\mathrm{inertia}(A)$	The inertia of A	196
$\mathrm{sep}(L,M)$	The separation of the spectra of L and M	231
$\langle\alpha,\beta\rangle$, $\langle\lambda\rangle$	Generalized eigenvalues	273
$\gamma(A,B)$	Crawford's number for the definite pair (A,B)	282
$\chi(\langle\alpha,\beta\rangle,\langle\gamma,\delta\rangle)$	The chordal distance between $\langle\alpha,\beta\rangle$ and $\langle\gamma,\delta\rangle$	283
ρ_D	The chordal metric for definite pairs.	288
ρ_L, ρ_R	The left and right equivalence metrics for regular pairs	288
$\|(P,Q)\|_\mathcal{F}$	The combined Frobenius norm	307
$\mathrm{dif}[(A,B),(C,D)]$	The difference between the spectra of (A,B) and (C,D)	307

Index

2-norm 51, 59, 71, 72
  as largest singular value 33,
    69
  consistency with Frobenius
    norm 66
  matrix 2-norm 69
  properties 36, 51, 70
  relation to the 1-norm and
    ∞-norm 55
  symmetric gauge function 79
  unitary invariance 52, 60, 72,
    74
  vector 2-norm 3

Abdelmalek, N. N. 163
absolute and relative error
  in individual elements 128
  matrix 116, 117, 119, 134
    limitations 117
    properties 116
  scalar 115
    properties 115
absolute value 49
acute perturbation 136–140, 151,
  152
  continutiy of pseudo-inverse
    140
  definition 139

  in reduced form 139
acute subspaces 151, 152, 255
Afriat, S. N. 45
Amir-Moéz, A. R. 209
approximation by matrix of lower
  rank (see
  Schmidt–Mirsky
  theorem) 208
Ariolo, M. 163
arithmetic-geometric mean
  inequality 61
artificial ill-conditioning 125, 193
Autonne, L. 35, 36

backward perturbation (see under
  linear system, least
  squares, etc. 128
Banach space 60, 98
Banach, S. 60
Bateman, H. 35
Bauer–Fike theorem 171, 181,
  192, 294, 300
Bauer–Skeel theorem 127
Bauer, F. L. 133, 177, 194, 195
Baumgärtel, H. 176
Beckenbach, E. F. 60
Bellman, R. 4, 60
Beltrami, E. 34

Ben-Israel, A. 152
Bendixson–Hirsch–Toeplitz
    theorem 30
Bendixson, I. 30
Bergman, P. G. 108
Berkson, E. 98
Bhatia, R. 176, 194, 227, 228
Birkhoff's theorem 50, 85, 87,
    190, 193, 211
Birkhoff, G. D. 88
Bjerhammer, A. 108, 109
Björck, Å. 163
Borchart, C. W. 209
Bunch, J. R. 136
Bunyakovski 60

canonical angle 43, 45, 94, 98, 99,
    226, 232, 240, 250, 255,
    260, 323
    basis metric 95
    computation 43, 45
    gap function 92
    pairs of projections 43
    variational characterization
        45
canonical bases 40
Cauchy's interlacing theorem
    196–198, 209
Cauchy inequality 5, 60
    generalized 67
Cauchy sequence 63, 64, 99
Cauchy, A. L. 60, 209
Cayley–Hamilton theorem 27
characteristic equation 15
characteristic polynomial 15
Chatelin, F. 4
Chebyshev, P. L. 11
Cholesky factor 13
chordal metric 283, 284, 290, 314
column space 2, 4

column sum norm (see norm,
    matrix 1-norm) 70
companion matrix 28
complete space 63, 64, 99
condition estimation 133
condition number (see under
    matrix inverse,
    eigenvalue, etc.) 118
congruence transformation 196
consistency
    between norms (see norm,
        consistency) 66
    foolish 26
contraction matrix 46
Courant–Fischer theorem (see
    Fischer's theorem) 209
Courant, R. 209
Crawford, C. R. 290, 324
CS decomposition 37–40, 45
    computation 45
    existence 37
    generalized singular value
        decomposition 47
    perturbation theory 324

Davis, C. 45, 46, 151, 194, 227,
    228, 244, 258, 259
Decell, H. P. 152
defective matrix 16
    Jordan form 21
definite matrix pair (see matrix
    pair (definite)) 281
Demmel, J. 133, 136, 177
Dennis, J. E. 135
departure from normality 171,
    172, 177, 178
Desplanques, J. 186
determinant as poor measure of
    condition 122
diagonalizable matrix 21, 28

perturbation of eigenvalues
  192, 215–217
diagonally dominant matrix
  186–188
diagonal matrix 3, 5
  block 4
dif 307, 309, 311, 312
  bounds for definite matrix
    pair 319
dilation 209, 211
direct rotation 46
doubly stochastic matrix 83, 85,
  88, 195
  Birkhoff's theorem 85
  definition 81
Drazin, M. P. 108, 113
Duff, I. S. 163
Dulmage, L. 88

Eckart–Young theorem (see
  Schmidt–Mirsky
  theorem) 210
Eckart, C. 35, 210
eigenpair 14
eigenproblem 26
eigenspace (see under matrix
  pair) 303
eigensystem 26
eigenvalue 14
  algebraic multiplicity 15, 16
  backward perturbation 175
    optimal 176
  Cauchy's interlacing theorem
    (q.v.) 196
  complex 15
  condition number 186, 188,
    192, 226, 240
  continuity 166, 167, 176, 178,
    244
  defective 16, 176

Fischer's theorem (q.v.) 27
geometric multiplicity 15, 16
Gerschgorin's theorem (q.v.)
  181
Gerschgorin disks (q.v.) 181
Hausdorff distance 167–169,
  177, 178
inclusion region 195, 210
matching distance 167–169,
  174, 177, 178, 217
md$_2$ 189
matrix pair (q.v.) 271
multiple 15, 26, 27
nomenclature 14, 26
of matrix functions 29
perturbation theory 165, 176,
  203, 241
  Bauer–Fike theorem (q.v.)
    171
  diagonalizable matrix 192
  Elsner's theorem (q.v.) 168
  generalized Rayleigh
    quotient 249
  Henrici's theorem (q.v.)
    172
  Hermitian matrix 258, 263
  Hoffman–Wielandt theorem
    (q.v.) 189
  matrices similar to a
    Hermitian matrix 215,
    216
  matrices similar to a
    normal matrix 216
  Mirsky's theorem (q.v.)
    205
  non-Hermitian
    perturbations of
    Hermitian matrices
    212–214, 217
  normal matrix 192, 195

orthogonal matrix 195
Ostrowski–Elsner theorem
  (q.v.) 170
simple eigenvalue 183
Weyl's theorem (q.v.) 203
Rayleigh quotient (q.v.) 185
residual bound 174, 176, 191,
  205–207, 209, 211, 235,
  248, 254–257
set of eigenvalues 26
simple 15, 29, 183
differentiability 185
spectral variation 167–169,
  177
eigenvector 14
backward pertubation 179
condition 219
condition number 241
left 15
nonuniqueness 89, 219, 220
perturbation theory 240, 241,
  244
  Hermitian matrix 258, 263
  stochastic matrix 241, 244,
    246
residual bound 211
right 15
simple eigenvalue 29
elliptic norm 109
Elsner's theorem 167, 168, 170,
  181
Elsner, L. 177, 178, 311
error (see absolute and relative
  error) 115
Euclidean norm (also see 2-norm)
  3, 53
Evans, J. W. 152
exponential of a matrix 73

Faddeeva, V. N. 71

Fan's theorem 50, 81, 86, 254
Fan, K. 88
Feingold, D. B. 188
Feller, W. 10
field of values 23, 24, 27
  convex hull of eigenvalues 24
  convexity of 23
  definition 23
  of a Hermitian matrix 28
  of a normal matrix 24
Fike, C. T. 177, 194
first order approximation (also
  see under least squares,
  eigenvalue, etc.) 131,
  134, 292
Fischer's theorem 27, 196, 198,
  201, 209, 281, 289
Fischer, E. 209, 289
Forsythe, G. E. 10
Francis, J. G. F. 11
Frobenius norm 65, 71, 110, 131,
  135, 172, 177, 180, 247,
  258
  consistency 65, 69, 71
  consistency with 2-norm 66
  relation to eigenvalues 72
  symmetric gauge function 79
  unitary invariance 71, 72, 74
Frobenius, F. G. 71, 88
full rank factorization 12, 32, 105

Gaches, J. 134
Gantmacher, F. R. 4
gap function (see under metrics
  for subspaces) 90
Gastinel, N. 71, 133
Gatlinburg Conferences 71
Gaussian elimination 132
Gauss, C. F. 108–110, 134

generalized eigenvalue problem
    (see matrix pair) 271
generalized inverse 102
    (i,j,k)-inverse 102
    discontinuity 108
    Drazin inverse 108, 113, 241
    from singular value
        decomposition 103, 104,
        110
    group inverse 241
    limitiations 108
    projections 110
generalized singular value
        decomposition 46
    perturbation theory 324
Gerschgorin's theorem 177, 181,
        186, 187, 203
    block variant 188
    compared with Elsner's
        theorem 181
    generalized (see under matrix
        pair (regular)) 291
Gerschgorin disks 181, 186, 187
    irreducible matrix 188
    isolated 187
    reduction by diagonal
        similarity 182–187
Gerschgorin, S. A. 177, 186
Gohberg, I. 227
Golub, G. H. 4, 151, 152, 155, 163
Gragg, W. B. 133
Gram–Schmidt algorithm 11, 12
    modified 12
Gram, J. P. 11

Hadamard's inequality 8, 14, 168,
        177
Hahn–Banach theorem 57, 63
Hahn, H. 60

Hall's theorem 84, 88, 89, 170,
        178
Hall, P. 88, 89
Halmos, P. R. 46
Halperin, I. 88
Hanson, R. J. 152, 163
Hardy–Littlewood Pólya theorem
        81, 87
Hardy, G. H. 88
Hausdorff distance (see under
        eigenvalue) 167
Hausdorff, F. 27, 177
Hearon, J. Z. 152
Henrici's theorem 172–174, 177
Henrici, P. 177, 178
Hermite, C. 209
Hermitian matrix 3, 5
    Cauchy's interlacing theorem
        (q.v.) 196
    eigenvalues 19
        bounds 30
        of sums 25
    field of values 28
    Fischer's theorem (q.v.) 27
    perturbation of eigenvales
        Mirsky's theorem (q.v.)
        204
    perturbation of eigenvalues
        203, 258, 263
        generalized Rayleigh
            Quotient 249
        matrices similar to a
            Hermitian matrix 215,
            216
        non-Hermitian
            perturbations 212–214,
            217
        positive definite
            perturbation 203
        Weyl's theorem (q.v.) 203

perturbation of eigenvectors
    258, 263
perturbation of invariant
    subspaces 244
residual bound for
    eigenvalues 205–207
residual bounds for
    eigenvalues 248, 254–257
residual bounds for invariant
    subspaces 249–254
    first sin Θ theorem 250, 258
    nonorthonormal baisis 251
    second sin Θ theorem 251,
        255, 258
    tan Θ theorem 253, 258,
        259
skew Hermitian matrix 5
spectral decomposition 19,
    226
sums of 27
Hessian matrix 134
Higham, N. J. 133, 164
Hilbert space 64, 98
Hirsh, H. 30
Hoffman–Wielandt theorem 189,
    193, 205, 213, 218, 257
    generalizations 193, 194
    limitations 191
Hoffman, A. J. 88, 193
Holbrook, J. R. A. 194
Hölder's inequality 61
Hölder norms 192
Hölder, O. 61
Hotelling, H. 35
Householder transformation 5, 6,
    10
Householder, A. S. 4, 6, 10, 11,
    71, 176, 195

idempotent matrix 28

inertia of a matrix 196
$\inf_2$
    in terms of the inverse matrix
        36
    in terms of the pseudo-inverse
        110
inner product 53, 62
invariant subspace 21, 22
    approximate (see invariant
        subspace, residual
        bound) 230
    backward perturbation 175,
        178–180
    optimal 176
    characterization 220, 227
    complementary 222, 225
    complex conjugate
        eigenvalues 29
    definition 22
    left 221, 225
    normalization 239, 244
    perturbation theory 229,
        236–240, 244, 254
        canonical angles 237
        first sin Θ theorem 250, 258
        Hermitian matrix 244
        second sin Θ theorem 251,
            255, 258
        tan Θ theorem 253, 258,
            259
    reduced form of matrix 221
    representation of matrix 22,
        220, 231, 235, 237
        condition number 240
    residual bound 174, 206,
        229–236, 246, 249–254
        canonical angles 232
        nonorthonormal basis 251
    sensitivity 21, 90, 219
    simple 220, 221, 226, 227,

231, 235, 238
  existence of complementary
    invariant subspace 225
  reduction to block diagonal
    form 224
  spectral projection 114, 225,
    240
  canonical angles 226
  spectral resolution 223–228,
    237, 244
  Sylvester's equation (q.v.)
    222
inverse
  matrix
    condition number
inverse matrix 102
  asymptotic forms and
    derivatives 130–132
  condition 17
  condition number 119, 127,
    133
  artificial ill-conditioning
    122–124
  distance from singularity
    120, 121, 133
  in the 2-norm 121
  optimal 133, 135, 193
  relation to determinant 134
  relation to eigenvalues 122,
    134, 135
  significant digits in inverse
    120
  left and right inverses 134
  perturbation theory 117–124
    linear system 124
    random perturbations 131
    well-conditioned 120
irreducible matrix 186, 188

Jacobian matrix 134

Jacobi, C. G. J. 209
Jiang, E. 178, 180
Jordan–Wielandt matrix 32, 34,
  35, 259
Jordan block 20, 28, 174, 180,
  186, 300
  function of 73
  powers of 28
Jordan canonical form 20, 21, 26,
  174, 227, 280
  associated invariant
    subspaces 21
  computation 27
  Drazin generalized inverse
    113
  limitations 21, 227
  principal vector 21
Jordan, C. 26, 34, 35, 45, 289
Jordan, P. 63

Kahan, W. 27, 45, 46, 133, 151,
  178, 180, 209, 211, 217,
  218, 244, 258, 259
Kato, T. 4, 176, 178, 244
König, D. 88
Korganoff, A. 152
Krasnoselski, M. A. 98
Krein, M. G. 98
Kronecker product 30, 228, 258
Kronecker, L. 289

Lancaster, P. 227
Laplace, P. S. 11
Lawson, C. L. 152, 163
least squares 10, 11, 101
  asymptotic forms and
    derivatives 162
  backward perturbation
    160–163
  Björck's theorem 162

condition
  condition number 156, 163
    reflected by solution
      156–158, 163
    square of κ 158, 163
  constrained 109, 112
  cross product matrix
    backward perturbation 164
  elliptic norm 109
  errors in the variables 163
    bias 267
    expanded equations 107, 161, 163
    Gauss-Markov theorem 110
    Gauss, C. F. 108
    measurement error models 163
    normal equations 107
    numerical methods 152, 163
    perturbation theory 156–160
      Björck's theorem 158
      errors in a column 163
      perturbation in A 157
      perturbation in b 156
      perturbation of the residual vector 160
      structured perturbation 158, 163
    priority dispute between Gauss and Legendre 109
    reduced form 155
    regression diagnostics 163
    residual vector 107
    solution by pseudo-inverse 107
    statisticians' notation 109
Legendre, A. M. 109
Lévy, L. 186
Lidskii, V. B. 209, 210
linear functional 56

linear system 101, 114
  artificial ill-conditioning 125
  asymptotic forms and derivatives 130–132
  backward perturbation
    128–130, 133, 135, 136
    Oettli–Prager theorem 130
    Rigal–Gaches theorem 128
    structured 129
  condition number 125, 127
    artificial ill-conditioning 128
    Bauer–Skeel 128, 133
    reflected by solution 126
    perturbation theory 124–128, 132
  Bauer–Skeel theorem 127
  component-wise bounds 125, 126
  from inverse matrix 124
  perturbation in matrix 124
  perturbation in the right-hand side 126, 127
  structured perturbation 127, 128
  residual bound 128–130
  structured perturbation 133
LINPACK 133
Littlewood, J. E. 88
Loewy, A. 27
LR algorithm 11
Lu, Q.-k. 99

Mac Duffee, C. C. 4
majorization 81, 88, 89
Marcus, M. 4
matching distance (see under eigenvalue) 167
matrix function 20
  exponential 73

INDEX

Jordan block 73
Neumann series 73
power series 73
rational 29
spectral resolution 229
matrix norm (see under norm) 64
matrix pair
  characteristic equation 274
  eigenvalue 273
  $\langle \alpha, \beta \rangle$ notation 273
  eigenvector 273
  equivalence 276
  generalized Schur form 277
  Hermitian 281
  infinite eigenvalues 271, 273
  linear transformations of
    matrix pairs 275
  metrics 284–288
    limitations 288
  nonsingular B 274
  numerical methods 274
  reduction to ordinary
    eigenvalue problem 274
  scaling problems 272
  singular 273, 274, 289
  systems of differential
    equations 289
matrix pair (definite) 281–283, 289
  bounds for dif 319
  definition 282
  diagonalizability 283
  eigenangle 314
    perturbation theory 324
    variational characterization 314
  eigenangles 324
  eigenvector 318
  failure of definition in real
    case 290

matching distance 324
perturbation of eigenspaces 317–322
  condition number 322
perturbation of eigenvalues 314–317
  condition number 313, 316, 324
  limitations of the theory 316
perturbation of eigenvectors 322
positive definite B 281, 282, 289
  Fischer's theorem 281
projective metric 285–287, 290, 314
right and left eigenspaces 317
sin Θ theorems 323, 324
spectral resolution 318
  normalized 318
matrix pair (regular) 273, 274
  approximate eigenspace 306–309
  continuity of eigenvalues 291
  deflating subspace 311
  diagonalizable pair 280, 281, 297
  eigenspace 312
    backward perturbation 309
    characterizations 304
    complementary 306
    definition 303
    eigenvectors 312
    simple 305, 306
  eigenvalue
    simple 277
  generalized Bauer–Fike
    theorem 294, 301, 311
  generalized Henrici Theorem

311
generalized
  Hoffman–Wielandt
  theorem 311
generalized Schur form 276,
  289
real pair 200
Gerschgorin theory 294–300
  diagonal similarity 297
  generalized Gerschgorin
    theorem 295
  simplification of bounds
    296
  infinite eigenvalues 274
  left projective metric 288, 290
  normal pair 311
  number of eigenvalues 275
  perturbation of eigenspaces
    309–311
  perturbation of eigenvalues
    311
    condition number 294, 300,
      303
    diagonalizable pair 300–303
    first order approximation
      292
    first order error bounds 293
    generalized Bauer–Fike
      theorem 301
    multiple eigenvalue 297–300
    spectral variation 302
  perturbation of eigenvectors
    312
  recovery of bounds for
    ordinary eigenvalue
    problem 294, 297
  right projective metric 288,
    290
  spectral resolution 306, 310
  Weierstrass canonical form

280, 290
matrix pencil (also see matrix
  pair) 209, 271
  rectangular 272
  singular 271
McIntosh, A. 194, 227, 228
metric 53
  from a norm 62
  pseudo-metric 62
metrics for subspaces 50, 90
  θ-metric 96, 98, 99
  basis metric 95, 98, 99
  completeness 99
  distance from a vector to a
    subspace 90, 99
  gap function 90–94, 98, 99
    canonical angles 92
    definition 91
    equivalence of gap
      functions 91
    projections 93
  gap topology 91, 93, 94
  projection metrics 93, 94
  Schäffer's metric 99
  unitarily invariant metrics
    94–98
    failure to generate the gap
      topology 94
    unitary invariance 99
Meyer, C. 244
Milman, D. P. 98
Minc, H. 4
Minkowski's inequality 61
Minkowski, H. 59, 61
Mirsky's theorem 194, 204, 206,
  208, 209
Mirsky, L. 71, 72, 194, 209, 210
Moler, C. B. 290
Moore-Penrose generalized inverse
  (see pseudo-inverse) 102

INDEX   361

Moore, E. H. 108, 109
Moré, G. 135

Nashed, M. Z. 108
nearly singular matrix 120
Neumann series 73
nilpotent matrix 29
nondefective matrix (also see
    diagonalizable matrix)
    21, 28
norm
    2-norm (q.v.) 3
    absolute norm 52, 72, 76
    and spectral radius 73
    combining norms 61
    consistency 65, 67, 71
    definition 66
    failure 65
    family of consistent norms
        69
    unitarily invariant norms
        80
    vector norm consistent with
        a matrix norm 66
    convexity and norms 59, 63
    dual 57, 59, 67
    dual of dual 58
    elementary properties 51
    elliptic 53, 62, 111
    equivalence of norms 54, 55,
        59, 65, 72
    failure 64
    Frobenius norm (q.v.) 65
    generated by an inner
        product 62
    generated by linear
        transformation 53
    generated by positive definite
        matrix 53, 62
    Hilbert–Schmidt norm 210

    Hölder norms 51, 57, 60, 61,
        64, 69, 121, 134
    infinite dimensional spaces 71
    limits and norms 55, 65
    matrix ∞-norm 69, 71, 72
    matrix 1-norm 69, 71, 72
    matrix norm 49, 64
    of a linear functional 56
    operator norm 67, 68, 72
    consistency 67, 68
    polar 59
    spectral radius and norms 67,
        72, 73
    subordinate norm (see norm,
        operator norm) 68
    unitarily invariant (q.v.) 50
    vector ∞-norm 51, 55, 69
    vector 1-norm 51, 55
    vector norm 49, 50
normalizable matrix (see
    diagonalizable matrix)
    189
normal matrix 3, 5, 171, 191, 194
    condition number 134
    departure from normality
        (q.v.) 171
    eigenvectors 19
    field of values 24
    perturbation of eigenvalues
        192, 195
    Hoffman–Wielandt theorem
        (q.v.) 189
    matrices similar to a
        normal matrix 216
    residual bound 191
    Schur decomposition 18
null space 2

O'Leary, D. P. 112
Oettli–Prager theorem 130, 161

Oettli, W. 133
orthogonal matrix 3, 194
   perturbation of eigenvalues 195
orthonormal basis 8
Ostrowski–Elsner theorem 170
Ostrowski, A. 88, 177, 187

Paige, C. C. 45, 46, 99, 111, 324
Parlett, B. N. 4, 178, 180, 209
Pavel-Parvu, M. 152
Peano, G. 71
Penrose's conditions 102, 110
Penrose, R. 108, 110, 151
Pereyra, V. 152, 155, 163
permutation matrix 3, 83–85
permutation vector 83–85
Picard, É 35
polar decomposition 36
Pólya, G. 88
positive definite matrix 3–5, 27, 73, 74
   condition number 122
   norm generated by 53
positive semi-definite matrix 3, 5
   square root 20
powers of a matrix 73
Prager, W. 133
principal vector 21
projection (oblique) 11, 14, 152
   generalized inverse 110
   spectral projection 114
   with respect to an inner product 111
projection (orthogonal) 9
   acute perturbation 137–140, 153
   as Hermitian idempotent 10
   asymptotic forms and derivatives 154

canonical angles 43
complementary 10
condition number 154
continuity 153
generalized inverse 110
least squares 10
perturbation of products 141
perturbation theory 153, 154
pseudo-inverse 106
reduced form 153
projection (with respect to a norm) 91
pseudo-inverse 101, 102
   application to least squares 107
   asymptotic forms and derivatives 150–152
   Bjerhammer's characterization 109
   condition number 146, 149, 163
   distance from matrix of lower rank 152
   continuity 136, 140, 146, 151
   counterexamples 105
   elementary properties 104
   elliptic 109, 111
   existence and uniqueness 104, 110
   expressions for perturbed pseudo-inverse 142
   full rank case 108
   Gauss, C. F. 108
   minimality 110
   Moore's characterization 109
   nonacute perturbations 140
   orthogonal projections 106
   perturbation theory 140–151
   acute perturbations 146–150

general results 140–146
  Wedin's bounds 142–146
QR decomposition 110
  reduced form 137
  scaled 111
  weighted 111
Pythagorean equality 10

Qi, L. 187
QR algorithm 11, 18
QR decomposition 6–8, 11, 30
  existence 7
  pseudo-inverse 110
  uniqueness 13
  with pivoting 11
QR factorization 8, 13
  generalized singular value
    decomposition 47
  partitioned 13
quasi-Newton method 134

Rall, L. B. 108
random perturbation 131, 134, 163
Rayleigh–Ritz approximation 207, 209
Rayleigh quotient 185, 241
  generalized 240, 244, 248, 249
Rayleigh, Lord (J. W. Strutt) 210
residual bounds (see under linear system, eigenvalue, etc.) 128
Riesz, F. 60
Rigal–Gaches theorem 128
Rigal, J. L. 134
right inverse 110
Ritz vectors 210
Ritz, W. 210
Rodman, L. 227
Rohrback, H. 186

Rosenblum, M. 227, 228
Rouché's theorem 167, 176
rounding-error analysis 132, 133
rounding error 274
row sum norm (see norm, matrix ∞-norm) 70
Ruhe, A. 244

Saunders, M. A. 45, 46
Schäffer, J. J. 99
Schmidt–Mirsky theorem 208, 210
Schmidt, E. 11, 35, 209, 210
Schur complement 13
Schur decomposition 17–20, 26, 28, 171, 222
  existence 17
  of a real matrix 26, 29
  of normal matrix 18
  uniqueness 26
Schur, I. 13, 26, 71
Schwarz 60
sep 244
  continuity 234, 236
  definition 231
  Hermitian matrices 247, 258
  properties 245
  relation to separation of eigenvalues 233, 247, 258
set operations 2
Sherman–Morrison–Woodbury formula 5
similarity transformation 16
  ill conditioned 17, 21
  unitary 17
singular subspace 259
  residual bound 260–262, 266, 267
  Wedin's Φ–Θ theorems 260, 262, 267
singular value 31

behavior of small singular
values 263, 266
determinant 36
expression for condition
number 121
from cross-product matrix 31,
259, 263
Gerschgorin-like theorem 187
largest as 2-norm 33
nondifferentiability 259
partitioned matrix 33
perturbation expansion
263–266
perturbation theory 209
Mirsky's theorem (q.v.)
204
product of two matrices 34,
36
relation to eigenvalues 36, 87,
88, 252
set of singular values 26, 31
smallest as $\inf_2$ 33
von Neumann's lemma 76
singular value decomposition
30–33
basis for column space 32
basis for null space 32
existence 30
full rank factorization 32
generalized inverse 103, 110
history 34, 35
principal components 35
projections 32
pseudo-inverse 104
Schmidt–Mirsky theorem
(q.v.) 208
singular value factorization 32,
105
singular vector 31
perturbation expansion
263–266
perturbation theory 267
uniqueness 31
Skeel, R. D. 133
skew-Hermitian matrix 27
Smithes, F. 35
spectral norm (see 2-norm) 70
spectral projection (see under
invariant subspace) 225
spectral radius 66, 72, 73
spectral variation (see under
eigenvalue) 167
spectrum (see eigenvalue) 26
stable matrix 74
Stewart, G. W. 4, 45, 112, 133,
151, 152, 155, 163, 164,
244, 258, 267, 290, 311,
324
stochastic matrix 81, 244, 246
perturbation of eigenvector
241
structured perturbation bounds
117
Sun, J.-G. 210, 217, 267, 290,
311, 324
Swanson, C. A. 244
Sylvester's equation 222, 223,
227, 251
generalized 278, 289
integral representation 227,
228
Kronecker product 228
numerical solution 227
Sylvester–Jacobi theorem 196
Sylvester, J. J. 35, 209
symmetric gauge function 75, 76,
78
definition 75
generating the 2-norm 79
generating the Frobenius

norm 79
symmetric matrix 3
Sz.-Nagy, B. 60

Taussky, O. 186, 188
Tensor product 30
tilde convention 16, 114
Toeplitz–Hausdorff theorem 23
Toeplitz, O. 27, 30
topology 54, 65
triangle inequality 5, 49, 59
triangular matrix 5
   block 4
      eigenvalues 15
   eigenvalues 15
   lower 3
   strictly upper 5
   upper 3
Turing, A. M. 133

unitarily invariant norm 50, 137, 175, 177, 192, 206, 209
   2-norm 52, 60, 72, 74
   minimality 80
   analogy with absolute value 88
   characterization by dual gauge function 78
   consistency 80
   definition 74
   family of norms 79
   Fan's theorem 86
   Frobenius norm 71, 72, 74
   normalized 74
   partitioned matrix 80, 88
   rectangular matrices 79
   symmetric gauge function (q.v.) 75
   von Neumann's theorem 50, 78

unitary matrix 3, 27, 105, 194
   CS decomposition 37
   eigenvalues 19
   with specified first column 6

van der Sluis, A. 133, 135, 163
van der Waerden, B. L. 88
Van Loan, C. F. 4, 46, 136
van Rijk, P. P. M. 163
Varah, J. M. 244
Varga, R. S. 4, 188
vector norm (see under norm) 50
von Neumann's theorem 205
von Neumann, J. 63, 71, 87

Wedderburn, J. H. M. 4
Wedin's Φ–Θ theorems (see under singular subspace) 260
Wedin, P.-Å. 45, 113, 151–153, 163, 266
Weierstrass canonical form 280, 289
   computational methods 289
Weierstrass, K. 289
Weyl's theorem 203, 209
Weyl, H. 88, 209, 210
Wielandt's theorem 196, 199, 209
Wielandt, H. W. 35, 193, 209
Wiener, N. 60
Wilkinson, J. H. 4, 27, 71, 133, 151, 163, 176–178, 187, 188, 193, 217
Wittmeyer, H. 132, 194

Yamamoto, T. 244, 246
Young, G. 35, 210

Zhang, Z. 217